P9-BZD-476

APPLIED

FOODSERVICE SANITATION

FOURTH EDITION

THE NATIONAL FOOD SAFETY CERTIFICATION PROGRAM

A Foundation Textbook

John Wiley & Sons, Inc.
in cooperation with
The Educational Foundation of the National Restaurant Association

Recognizing the importance of preserving what has been written, it is a policy of John Wiley & Sons, Inc., to have books of enduring value published in the United States printed on acid-free paper, and we exert our best efforts to that end.

Cover photo by Matt Dinerstein

Published by John Wiley & Sons, Inc.
Published simultaneously in Canada.

Library of Congress Cataloging-in-Publication Data

Applied foodservice sanitation : a foundation textbook.—4th ed.
p. cm.
"Developed in collaboration with the National Sanitation Foundation."
Includes index.
ISBN 0-471-54221-0 (alk. paper)
1. Foodservice—Sanitation. I. Educational Foundation (National Restaurant Association) II. National Sanitation Foundation (U.S.)
TX911.3.S3A68 1992
363.72'96—dc20 91-23570

Printed in the United States of America

10 9 8 7 6 5 4 3 2 1

Printed and bound by R.R.Donnelley & Sons, Inc.

Contents

On behalf of the
people in the foodservice industry
who will benefit, The Educational Foundation
is pleased to thank

AMERICAN EGG BOARD

BEEF INDUSTRY COUNCIL OF THE
NATIONAL LIVE STOCK & MEAT BOARD

HEINZ U.S.A.

NATIONAL FISHERIES INSTITUTE

PRISM
A SUBSIDIARY OF JOHNSON WAX

THE PROCTER & GAMBLE COMPANY

SPARTA BRUSH COMPANY, INC.

WORLD DRYER CORPORATION
A DIVISION OF SPECIALTY EQUIPMENT COMPANIES, INC.

for the financial support
which has made possible
the development of this book.

Foreword

The Educational Foundation of the National Restaurant Association is continuing to provide leadership in foodservice sanitation and foodborne disease prevention through the updating of this text. This leadership was highlighted by the development of SERVSAFE®, a training and certification program for foodservice personnel, and the production of the coursebook *Applied Foodservice Sanitation.*

The text covers the principles of food microbiology, important foodborne diseases, standards that are enforced by regulatory agencies, and applied measures for the prevention of foodborne diseases and other microbiological problems. Additionally, this fourth edition describes the Hazard Analysis Critical Control Point (HACCP) system. The HACCP system is the most effective approach to food safety yet known, which when properly devised and implemented, provides high assurance that foodborne disease outbreaks and food spoilage will not occur. This concept is weaved into chapters that deal with successive operations from receipt of foods until service to customers. Hazards, standards, and corrective actions are presented for important operations (e.g., cooking, cooling, hot holding, and reheating) that are critical control points for food safety. This is a major advancement.

Foodservice industry professionals, regulatory agencies, and instructors in academic institutions who either apply or teach others to apply the principles and practices presented in *Applied Foodservice Sanitation, Fourth Edition,* will contribute to the reduction of foodborne diseases and to the improvement of food quality. This text, therefore, continues to serve as a comprehensive force to help these groups work together for these two important common goals.

FRANK L. BRYAN, M.P.H., PH.D

Director, Food Safety
Consultation and Training,
Lithonia, Georgia
(Formerly, Scientist
Director, Centers for Disease
Control, U.S. Public Health
Service, Recipient of four
Norbert Sherman Awards
for Applied Foodservice
Research)

A Message from The Educational Foundation

The Educational Foundation of the National Restaurant Association is proud to present *Applied Foodservice Sanitation, Fourth Edition,* a book for foodservice managers, supervisors, and employees aspiring to management jobs in the foodservice industry. Earlier editions have firmly established it as the nation's leading text for foodservice manager training programs by emphasizing key principles and practices of food safety. The fourth edition incorporates new information, concepts, and procedures vital to food safety.

Today's consumers demand more and more from the restaurants they patronize. They want top-rate, efficient service; varied, well-prepared menu offerings; and an attractive, clean, and comfortable dining atmosphere. An effective food safety program will complement and reinforce operators' efforts to provide all of these benefits to their guests. *Applied Foodservice Sanitation* provides information and methods to help the foodservice manager apply sanitation procedures to foodhandling functions from food purchasing and storage, to preparation and service to patrons. Only by implementing a complete food safety program can the operator be assured that food is safe right up to the time it is served.

Applied Foodservice Sanitation, Fourth Edition, introduces new material on implementing a foodservice sanitation management system in your operation. For the first time in any foodservice text, this system, known as Hazard Analysis Critical Control Point (HACCP), is detailed in a simple, straightforward manner so that managers may easily incorporate it into their operating procedures. The book itself has been organized according to the HACCP principle of following the flow of food through the operation. Students first learn to identify the foods at risk within an operation, then learn about the types of hazards to safe food, and finally learn how to protect food from those hazards during receiving, storing, preparing, holding, and serving.

This edition also incorporates changes in FDA standards, new information on emerging pathogens, updated information on equipment and cleaning and sanitizing operations, and a new chapter on crisis management. Like the third edition, the fourth edition focuses on *working through people* to maintain a sanitary operation, emphasizing such factors as training, influencing employee attitudes, working with vendors, and dealing with local health officers.

In some states and several county and municipal jurisdictions, training and certification for managers in foodservice sanitation is required by law. Nationally, however, the restaurant industry is exercising leadership in ensuring that food is handled safely. Recently, the National Restaurant Association strongly urged all of its members to voluntarily undertake training and testing for all management personnel to document that necessary food protection knowledge has been secured.

Applied Foodservice Sanitation is a critical component of SERVSAFE®, The Educational Foundation of the National Restaurant Association's comprehensive crisis management program that concentrates on three key areas of potential risk—food safety, responsible alcohol service, and employee and customer safety. SERVSAFE courses focus on the manager's role in assessing risks, establishing policies, and training employees. To help managers convey proper procedures to their employees, SERVSAFE programs include a host of employee training materials for each subject, such as employee study guides, videos, and a manager's kit that includes a leader's guide and other teaching aids.

As the manager training component in the SERVSAFE Food Safety program, *Applied Foodservice Sanitation* may be adapted for administration in a variety of ways. Foodservice employees who purchase the Certification Coursebook and register with The Foundation may take the course in seminar format, traditional classroom situations, or a combination of home study and group learning, which minimizes the time that managers have to spend in a classroom. Courses may be administered for in-house group training programs sponsored by foodservice companies, trade associations, and other industry organizations. The text and related course materials are available to college and community college students of hotel, restaurant, and institutional management as part of the regular academic curriculum.

All who complete the course will be eligible for an Educational Foundation SERVSAFE certificate upon satisfactory completion of a certification examination administered under the auspices of The Foundation.

The development of this book was greatly aided by members of the industry, educators, and regulatory representatives who served on the *Applied Foodservice Sanitation* Advisory Committee. They include Dr. Frank L. Bryan, M.P.H., Director, Food Safety Consultation and Training (formerly Scientist Director, Centers for Disease Control, U.S. Health Service); Gary P. DuBois, National Manager, Field and Regulatory Quality Assurance, Taco Bell Corporation; Larry M. Eils, R.S., Director of Public Health and Safety, National Automatic Merchandisers Association; Dr. Robert L. Flentge, D.V.M., M.S., Chief, Division of Food, Drugs, and Dairies, Illinois Department of Public Health; Glen A. Graf, Training/Human Resources, Bakers Square Restaurants; Robert E. Harrington, Assistant Director, Technical Services, Public Health and Safety, National Restaurant Association; Edward LaClair, Quality Assurance Manager, Taco Bell Corporation; Dr. David McSwane, H.S.D., Assistant Professor of

Public and Environmental Affairs, Indiana University; Dr. Peter B. Manning, Associate Department Head and Professor, Department of Hotel, Restaurant, and Travel, University of Massachusetts at Amherst; Marsha Robbins, R.S.; Joel Simpson, Director of Quality Assurance, Dobbs International Services, Inc.; and Edward Sherwin, Chairman, Hotel/Motel & Restaurant-Club Management Department, Essex Community College. Special recognition is due The Foundation's "sanitation team," Paul F. Martin, Director of Educational Programs; Marianne Gajewski, Manager, Educational Projects; and particularly to Susan Marts Myers, Project Editor; and Elizabeth Teeter, Assistant Editor, whose expertise resulted in the smooth integration of new material and reviewer comments in this edition.

We also appreciate the technical review and information received from our SERVSAFE sponsor companies: American Egg Board; Beef Industry Council of the National Live Stock & Meat Board; Heinz U.S.A.; National Fisheries Institute; PRISM (A subsidiary of Johnson Wax); Procter and Gamble Company; Sparta Brush Company, Inc.; and World Dryer Corporation (A Division of Specialty Equipment Companies, Inc.).

The Educational Foundation of the National Restaurant Association is dedicated to the advancement of professionalism in the foodservice industry through education and training. Our objective is to provide the resources managers need to reach the highest possible level of achievement in a very competitive environment.

Education contributes to both the growth of the industry and to the professional and personal growth of the individual who enjoys the rewards. *Applied Foodservice Sanitation, Fourth Edition,* is one way we can contribute to your professional growth.

DANIEL A. GESCHEIDLE

President
The Educational Foundation
of the National Restaurant
Association

PART I

The Sanitation Challenge

1

Providing Safe Food

Eating out is fun. People love to get together over food. Friends meet for lunch, for cocktails after work, or go to dinner after an evening at the theater. Meeting over food is often conducive to negotiating and planning. Conferences are often arranged around a schedule that includes at least one, and often several, meals.

It is also true that many of us simply eat away from home at one time or another for convenience or by necessity. Two-income families and singles often opt to eat out after a long day at work. Passengers on transportation systems, guests in hotels, patients in hospitals, residents of nursing homes or other institutions, students in schools and universities, and the men and women serving in the military all need to have food provided for them.

WHAT PEOPLE EXPECT

Diners walking into a commercial facility for the first time bring with them a number of expectations. They expect good, safe food, clean surroundings, and pleasant service. Together these elements make up a pleasant dining experience.

It is a challenge to managers of commercial and noncommercial establishments to direct a number of activities at once, including employee training and management; and the purchasing, preparation, and service of food. Foodservice managers generally expect to meet the diners' expectations. Managers *assume* that they are going to provide good, safe food in clean surroundings with friendly service. This assumption, especially regarding safe food and clean surroundings, should be based not only on a foundation of goodwill and good intentions, but on a sound understanding of sanitary policies and procedures.

THE CHALLENGE TO FOOD SAFETY

Food safety has always been a concern of foodservice operators. By and large the industry has done an excellent job of providing safe food to the American public; however, the number of foodborne illnesses attributed to the foodservice industry is still significant. The foodservice industry is serving a public that is increasingly intolerant of any food safety risk.

A shortage of foodservice employees trained and motivated to follow safe food practices makes the job of food safety management even more difficult. Foodservice operators cannot assume workers learned proper personal and food hygiene practices in their homes or that government health inspections will ensure safe operations.

A food safety management program is an organized system developed for all levels of the foodservice operation. The system controls factors that can compromise food safety and contribute to incidents of foodborne illness. A key component of this system is to provide adequate training for all foodservice personnel in safe foodhandling practices.

The National Restaurant Association has reaffirmed its position that the responsibility for providing safe and wholesome foods rests with the foodservice industry. It has called for the establishment of and adherence to effective standards of safe foodhandling practices. The Association strongly urges its membership to train and test its managers to a standard of food safety knowledge and commits itself to offering comprehensive training and testing to the industry.

In this chapter we will:

- Learn the reasons for managing a sanitary foodservice operation.
- Define the terms *foodborne illness* and *outbreak.*
- Define *sanitation.*
- Introduce the Hazard Analysis Critical Control Point/Sanitary Assessment of Food Environment (HACCP/S.A.F.E.) system of ensuring food safety.
- Discuss how to market sanitation to your employees and guests.

FOODBORNE ILLNESS

A *foodborne illness* is a disease that is carried or transmitted to human beings by food. Most victims of foodborne illnesses do not readily identify the source of their symptoms, but the public is becoming more aware that certain types of illnesses may be food-related.

All foodservice operations have the potential to cause foodborne illness through errors in purchasing, receiving, storing, preparing, and serving food. None of the many types of operations

Exhibit 1.1 Classification of foodservice operations

Group I Commercial Foodservice

Eating Places

Restaurants, lunchrooms
Limited-menu restaurants
Commercial cafeterias
Social caterers
Ice cream, frozen custard and yogurt stands
Bars and taverns

Food Contractors

Manufacturing and industrial plants
Commercial and office buildings
Hospitals and nursing homes
Colleges and universities
Primary and secondary schools
In-transit feeding (airlines, trains, cruise ships)
Recreation and sports centers

Lodging Places

Hotel restaurants
Motor-hotel restaurants
Motel restaurants

Other Commercial

Retail host restaurants
Recreation and sports
—Includes drive-in movies, bowling lanes, recreation and sports centers
Mobile caterers
Vending and non-store retailers
—Includes sales of hot foods, sandwiches, pastries, coffee, and other hot beverages

Source: Adapted from "The Foodservice Industry—Food and Drink Purchases Projected to 1991." NRA Forecast 1991. *Restaurants USA,* December 1990. Reprinted by permission.

(see Exhibit 1.1) is exempt from causing foodborne illness in the absence of adequate precautions.

Consider the following:

- A once popular restaurant in the Midwest is forced into bankruptcy after a botulism outbreak caused by contaminated onions kills a 73-year-old woman. Lawsuits filed against the restaurant reach well into millions of dollars.
- A major food chain in the Southwest is sued by parents of children who were served food that was prepared too far in advance, not adequately reheated, and contaminated with toxin from *Bacillus cereus.*
- At a hospital in the East, 34 patients, including two in traction with their jaws wired shut, are stricken with the foodborne illness salmonellosis. The cause is traced to eggnog prepared in the hospital kitchen.
- At a school in the East, more than 400 children suddenly become ill from staphylococcal foodborne illness. The bacteria are traced to egg salad sandwiches.

Television and newspapers report outbreaks connected to institutions and other public gatherings. Mostly, though, it is up to the various governmental agencies and the state and local equivalents of the U.S. Public Health Service to obtain and release to the public information on foodborne illnesses.

Since 1938, the U.S. Public Health Service Centers for Disease Control (CDC) have published annual summaries on outbreaks of foodborne illness. An *outbreak* is defined as an incidence of foodborne illness that involves two or more people who eat a common food, which is confirmed through laboratory analysis as the source of the illness. There are two exceptions: (1) a single incidence of botulism; or (2) a chemical-caused illness. Each one of these qualifies as an outbreak. CDC reports are based on summaries of foodborne illnesses reported by individual states.

The U.S. Public Health Service has three objectives in gathering information on foodborne illnesses: (1) disease prevention and control, including correction of faulty food preparation practices in foodservice establishments; (2) knowledge of the disease and the causing element; and (3) administrative guides.

CAUSES OF FOODBORNE ILLNESS

Any kind of food can be the vehicle for foodborne illness. Some of the foods implicated in foodborne illnesses are poisonous by nature, for example, certain types of mushrooms. However, it is generally high-protein foods that we eat regularly that are responsible for most foodborne illnesses. These high-protein foods are classified as *potentially hazardous* by the U.S. Public Health Service and include *any food that consists in whole or in part of milk or milk products, shell eggs, meats, poultry, fish, shellfish, edible crustacea (shrimp, lobster, crab, etc.), baked or boiled potatoes, tofu and other soy-protein foods, plant foods that have been heat-treated, raw seed sprouts, or synthetic ingredients.* Potentially hazardous foods are capable of supporting rapid and progressive growth of infectious or disease-causing micro-organisms. The term does not include foods that have a pH (acidity) of 4.6 or below, or an A_w (water activity) level of 0.85 or less, under standard conditions. (*Water activity* is the amount of moisture available for bacterial growth and will be explained in more detail in Chapter 2.)

Critical Offenses

In addition to paying special attention to potentially hazardous foods, the foodservice manager should focus on errors in foodhandling that are most often implicated in a foodborne illness outbreak. The following is a list of the *most frequently cited factors.*

1. Failure to properly cool food.
2. Failure to thoroughly heat or cook food.
3. Infected employees who practice poor personal hygiene at home and at the workplace.
4. Foods prepared a day or more before they are served.
5. Raw, contaminated ingredients incorporated into foods that receive no further cooking.
6. Foods allowed to remain at bacteria-incubation temperatures.
7. Failure to reheat cooked foods to temperatures that kill bacteria.
8. Cross-contamination of cooked foods with raw foods, or by employees who mishandle foods, or through improperly cleaned equipment.

WHAT IS SANITATION?

Sanitation is the creation and maintenance of healthful, or hygienic, conditions. Sanitation comes from the Latin word *sanitas* meaning health. In a foodservice situation, the word *sanitation* means wholesome food, handled and prepared in a way that the food is not contaminated with disease-causing agents. In other words, sanitation is what helps food stay safe.

But does sanitary simply mean clean? Not necessarily. That which appears to be clean may not always be sanitary.

Clean means free of visible soil. *Sanitary* means free of harmful levels of disease-causing micro-organisms and other harmful contaminants. Clean refers to aesthetics and concerns outward appearance—a face without a smudge, a glass that sparkles, a shelf wiped clear of dust. However, although clean on the surface, objects can harbor invisible disease-causing agents or harmful chemicals. Baby bottles boiled in water for ten minutes may be splotched and water-marked. They may not look clean on the surface, but they are free of disease-causing agents and can accurately be referred to as sanitary.

Let's look at the factors involved in food safety, as well as the inherent risks. They generally fall into three categories:

- Food—its safe condition initially, and its protection in preparation and service
- People—those involved in handling food both as employees and as customers
- Facilities—the sanitary condition of the physical plant, and the equipment used in a foodservice operation.

Food

Not all food is safe when it arrives in the operation. Any animal products, such as fresh poultry, may already be contaminated by the time the items are received. The foodservice manager must work with reputable suppliers and implement tight receiving procedures to help ensure safe food. Once the food arrives, it must be stored, prepared, and served using methods that maintain its safety.

Chapters 2 and 3 will cover what can cause *contamination*—the unintended presence of harmful substances or micro-organisms in food—the effects of contaminated food, and cross-contamination.

People

People pose the major risk to safe food (see Exhibit 1.2). The success of a foodservice manager in dealing with the foodborne-illness problem depends on how the human factor is handled. A manager must carefully train, monitor, and reinforce by example the principles presented during training.

Not only does a manager have to be concerned about the employees, a manager also has to overcome the problems of the unsanitary customer. Well-trained employees and well-designed facility are positive preventive measures against the customer who returns to the salad bar and wants to use the same plate and the customer who has an uncontrolled and unshielded cough.

Well-trained employees are essential when a crisis, accident, injury, or an illness happens in the establishment. For example, employees need to be informed and know what is expected of them in a crisis situation, such as a foodborne illness outbreak.

To help control this human factor in contamination and the challenge to safety, Chapters 4 and 16 cover in detail the elements of personal hygiene, and employee training and motivation respectively. Chapter 13 details accident prevention and emergency action, and Chapter 14 discusses crisis management.

Facilities and Equipment

Eliminating hard-to-clean work areas, faulty or overloaded refrigerators or other equipment, dirty surroundings, poor housekeeping, and conditions attractive to pest infestation make up the third focus of our analysis—the facilities and equipment.

Chapter 9 treats the subject of built-in sanitation in terms of materials, design, construction, and installation and layout of equipment. Adequacy of utilities and services, as well as other environmental factors in safeguarding people and food products, are also covered.

Chapters 10 and 11 present principles and procedures for the cleaning and sanitizing of equipment, utensils, and food preparation areas, and the development of a master cleaning schedule. Chapter 12 gives some practical information on pest control.

HAZARD CONTROL FOOD SAFETY PROGRAMS

Food safety programs need to incorporate a new system based on the actual

Exhibit 1.2 Transmission of a foodborne illness from infected human beings to food and back to other human beings

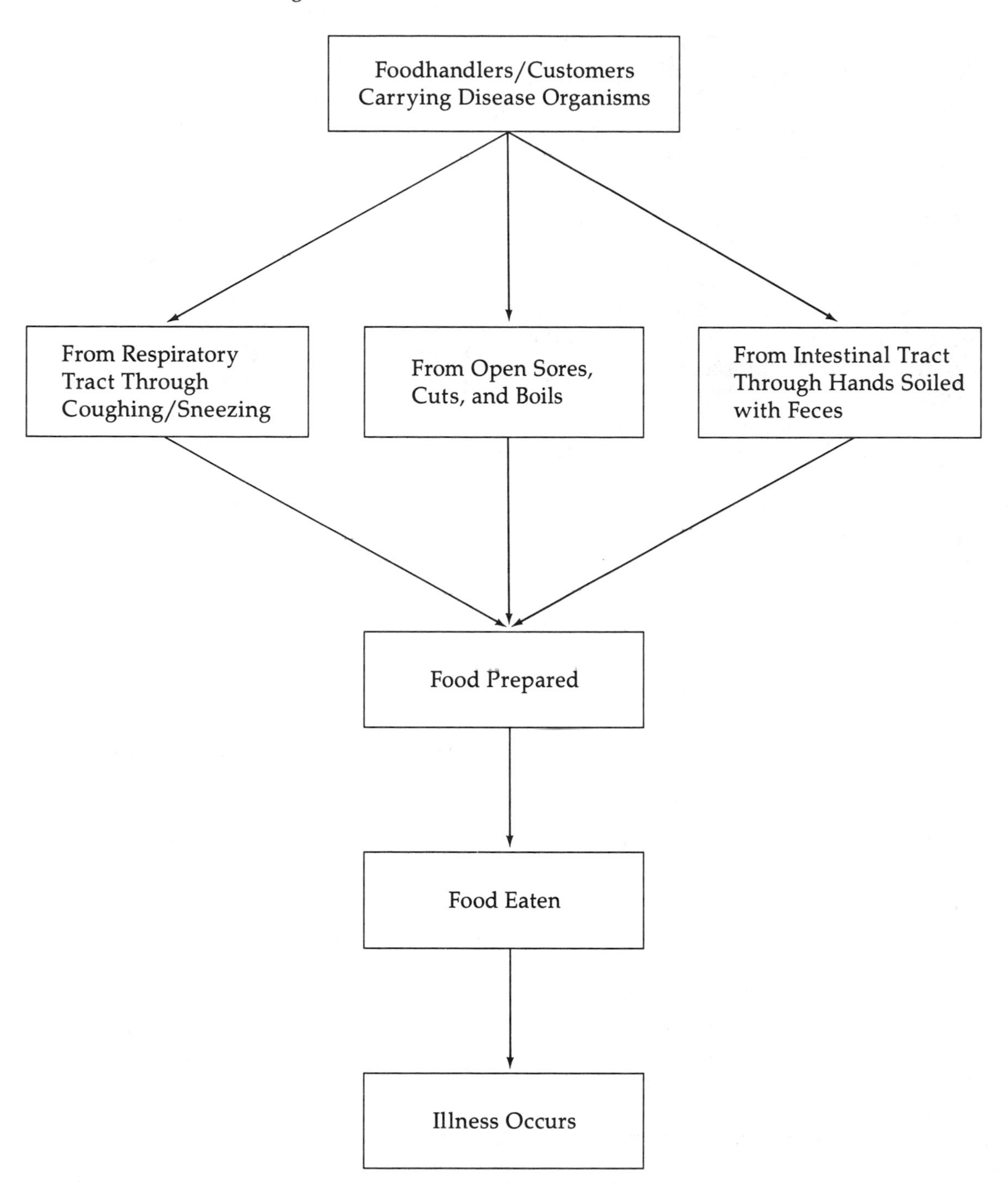

risk of acquiring a foodborne disease. This system, called *Hazard Analysis Critical Control Point* (HACCP), is being adapted from the food-processing industry. The National Restaurant Association has developed a HACCP-consistent system called *Sanitary Assessment of the Food Environment* (S.A.F.E.), which allows foodservice operators to monitor their operations and reduce the opportunity for foodborne illness.

The key to both the HACCP and S.A.F.E. systems is the emphasis on the food itself and how it flows through the operation from purchasing through serving. The HACCP/S.A.F.E. method identifies the points in an operation where contamination or growth of micro-organisms can occur and then implements a control procedure based on that hazard.

This system will be explained in greater detail in Chapter 5. The format of this text follows the focus of HACCP/S.A.F.E. by ordering chapters according to the way food flows through the operation.

FOOD QUALITY

Food safety and *food quality* are critical to the bottom line. Food that has off flavors, is dry, or appears stale is not likely to impress customers. Preserving food quality is another objective of sanitation.

Sanitary handling is a major factor in both food safety and quality. Food that is stored, prepared, and served properly is more likely to retain its quality. The standards of a food's quality include its safety, appearance, chemical properties, texture, consistency, nutritional values, and flavor. Any one of these can be destroyed by unsanitary procedures from purchase to service. Proper foodhandling with food safety in mind is important to preserve food quality.

THE ROLE OF THE FOODSERVICE MANAGER IN SANITATION

The foodservice manager's basic role is to take responsibility for both serving safe food to customers and patrons and for training employees on a continual basis. Adding the responsibility for implementing food safety procedures to an already multifaceted job is necessary. Your own attitude toward the subject is of the utmost importance. This attitude must be based on a knowledge of the industry with an appreciation of the impact of the foodborne illness problem, and your professional and legal position in relation to it.

The following are the major objectives of the foodservice manager:

1. Establishing a HACCP-based sanitation program.
2. Training, motivating, and supervising employees to maintain a sanitary facility serving safe food.
3. Regularly inspecting the facility and employees to make sure that sanitation standards are being met, and revising the system if necessary.
4. Cooperating with local public health officials during inspections.

Ways to meet each of these objectives are discussed throughout this book.

DOLLARS AND SENSE

Why Foodservice Managers Should Protect Customers

Why should the foodservice manager be interested in protecting patrons and customers from foodborne illness? Apart from the obvious moral responsibility to keep the public healthy, protecting customers makes good sense. Establishing safe foodhandling practices increases profitability, helps maintain patronage, reduces legal liability, enhances the operation's competitive position, and promotes goodwill. When all these factors are taken into account, protecting customers helps save dollars. The costs of foodborne illness can be high (see Exhibit 1.3).

Profitability

There is a definite monetary reward in protecting customers. The legal fees, medical claims, lost wages, and loss of business associated with foodborne illness can be staggering. The National Restaurant Association estimates that an average outbreak can cost an implicated

Exhibit 1.3 Cost of foodborne illness to the foodservice industry

operation in excess of $75,000. Additional losses can include top-to-bottom cleaning and food items that must be thrown out after an outbreak. The losses of profit and business can be much higher; some operations can be forced to close.

In addition, consider the bad publicity that can follow an outbreak of foodborne illness. Even if only one person gets ill, he or she may warn others to stay away from an operation.

Immediate returns can be gained through good food-protection practices. Food costs are directly lowered when waste is reduced and when portion control is improved. Quality control is improved when procedures and practices are standardized. Providing high quality food is an important strategy in an increasingly competitive industry. Increased productivity is achieved when employees are trained on a continual basis and are held accountable for their actions. Thus, improved food protection results in improved operating efficiency, which ultimately results in improved profitability.

Sanitary procedures will prevent outbreaks of foodborne illness, maintain goodwill, and keep the financial bottom line from bottoming out.

Regulation

One powerful motive for protecting customers and patrons is the law. Simply put, if you serve food to the public, it is illegal to be unsanitary. State laws, county or municipal health codes, and municipal ordinances with provision for fines or closure for violations, prohibit unsafe foodhandling practices for food services. These laws are aimed at protecting the public. Some local health services may also serve as consultants on safe foodhandling practices as well as provide employee training and advice to operators on meeting requirements for sanitary facilities. Federal regulations that cover food produced and processed for interstate commerce may or may not directly affect foodservice operations.

Liability

The law provides yet another incentive to serve safe food. Today, consumers are increasingly willing to use the law to seek compensation for products that have caused them harm. This is true for items from cars to baby cribs, and it is certainly true for unsafe food products. The Uniform Commercial Code (UCC) provides an option to people who want compensation for illness or injury from unsafe food products. People suing must prove that the food was unfit, that it caused them harm, and that in serving them unfit food the operator violated the warranty of sale.

If an operation is sued, two types of damages can be awarded to the plaintiff. *Compensatory damages* are awarded for the lost work, lost wages, and medical bills that the plaintiff might have experienced. *Punitive damages* are awarded in excess of normal compensation; they are awarded to punish the defendant for wanton and willful neglect.

An establishment that has developed standards to ensure safe food and that follows procedures to prevent illness can present a *reasonable care defense* in the event that it is sued for a foodborne-illness outbreak. This defense requires that an operation establish that it did everything that can be reasonably expected to ensure the food was safe and to prevent illness.

Operators of noncommercial facilities share similar concerns regarding the law, and in some cases liability. However, particular establishments have an additional reason for wanting to protect patrons. For example, most people go to hospitals because they are already ill. The last thing hospital patients need is a foodborne illness. The results could be serious, even deadly.

The very young and the very old are especially susceptible to serious problems from foodborne illness; the former because they have not built up adequate immune systems to cope with such diseases and the latter simply because they have less immunity and resistance as they age. Other foodservice patrons with compromised immune systems include AIDS patients, people on antibiotics, pregnant women, and cancer patients.

Customers and Goodwill

One of the best forms of advertising is the type that generates *goodwill.* Goodwill is measured by the patron's recognition that the foodservice operator means well and wants them to return.

Without regular patrons, foodservice operations would close their doors. Without a steady, regular patronage, commercial operations would not survive lean times and might barely make a profit during good times. Continued, repeat patronage or the kind that can reflect a steady profit margin is not an automatic right of anyone entering the foodservice business. It is earned. Further, good customers are the best form of advertising. They tell other customers.

MARKETING SANITATION

Sanitation should be marketed on both an internal and an external level. Marketing sanitation requires that management address both employees and the public, making sure that each is aware of the high priority that the operation assigns to sanitation.

Internal Marketing

To effectively market sanitation to employees, management must help them realize its importance. Appropriate emphasis must be placed on the need to practice sanitary work habits.

Some large chains provide excellent examples, by ensuring that sanitation is made a priority from the top of the company down. In other words, the emphasis on sanitation begins with the president of a company. At the unit level, the operator or manager of an operation should display his or her commitment to sanitation.

Several large chains have special training programs for managers. Those chains share the corporate philosophy that managers cannot be held accountable for sanitation unless they are properly trained. Trained managers in turn conduct sanitation training for their employees.

There are several ways to keep the momentum going once sanitation has been assigned a high priority. Some operations use award systems to encourage their employees. Identifying and recognizing the completion of training proves effective; certificates might be awarded in such cases.

The importance of education and training cannot be stressed enough. Management must make sure that employees are continually trained and retrained when necessary. Management should conduct follow-up evaluations, guaranteeing that training is adequate and that proper procedures are being carried out.

Good supervision is required to make internal marketing successful. It is critical that employees receive explicit instructions. Management should observe employees as they perform their jobs and offer feedback—both positive and negative. Remember, it is extremely important to compliment a job well done; merely criticizing lapses is considerably less successful.

External Marketing

Your commitment to sanitation needs to be communicated to your customers. They should be made aware of all the steps that you follow to ensure that food is served safely. Communication of your commitment can result in instilling customer confidence in your operation. Some relatively simple practices can inform your customers of your operation's emphasis on sanitation.

You might use employee buttons or display signs. One major foodservice chain uses place mats to describe its commitment to sanitation. Posting or distributing information about food safety can be another effective method. Carry-out establishments can offer information about keeping food safe after the customer removes it from the operation. External marketing can be accomplished through many ways, so devise your own creative ideas.

Use training to ensure that your employees are well-informed and can answer any questions that customers might have about how food is handled or prepared in your operation.

All the efforts that you make to market sanitation can help build your reputation and patronage, because customers will recognize that you care about them and their safety.

SUMMARY

People are eating out more frequently than ever before. When people eat out they have a number of expectations: good, safe food; clean surroundings; and pleasant service. Foodservice managers need to recognize the customer's desire for safe food and implement procedures to ensure food safety.

Sanitation is the creation and maintenance of healthful conditions to prevent food contamination and meet customers' expectations. The sanitation program is directed at the elimination of foodborne illness through reducing opportunities for food contamination, and counteracting contamination that does occur. This task is complicated by the fact that food contaminants—particularly disease-causing microorganisms—are everywhere. But difficult or not, the foodservice manager has a legal and a moral responsibility to see that this job is done. In addition, sanitary practices help maintain the bottom line in profits and a balanced budget.

A *foodborne illness* is a disease or injury caused by the ingestion of contaminated food. Since any type of foodservice can be responsible for an outbreak of foodborne illness, sanitary procedures are necessary in every type of facility.

We will analyze the solution to the problem of safe food in three parts. First is the food: how to obtain safe food and how to keep it safe with proper time and temperature controls and other measures. Second is the people: how to train employees to learn and apply hygienic foodhandling practices. Third is the facility: how to build in sanitation, how to manage the operation so the foodservice environment does not hinder the sanitation effort, and how to choose equipment with sanitation features. At the end of each chapter, there will be a case study to draw on your knowledge of the material covered to date.

Most of the opportunities for contamination are controllable by an alert management with well-trained employees, using a planned and thorough sanitation program. One such management system is the S.A.F.E. or HACCP-based system.

It is the foodservice manager's attitude that will cause the success or failure of an operation's sanitation program. When the foodservice manager is convinced of the importance of sanitation, well-informed on the subject of food protection, and actively interested in preventing foodborne illness, this effort will succeed.

A CASE IN POINT

Nearly 200 passengers who had been on an excursion train that traveled through the South experienced foodborne illness on June 14 and 15. At least 55 required hospitalization. The train had stopped to pick up box lunches prepared by a restaurant in Tennessee. An analysis of the outbreak implicated ham served in the lunches as the most likely vehicle for transmitting the illness.

Investigation revealed that on June 11, three days before the lunches were served, 40 hams were delivered to the restaurant and stored in an improperly operating walk-in refrigerator. The next day, June 12, the hams were deboned. They were then refrigerated in stainless steel pans until June 13, when they were sliced. On the morning of June 14, the ham portions were boxed with other food items for the lunch. The boxes were closed and delivered to the railroad station. The box lunches had been unrefrigerated for three hours before distribution to passengers.

All of the ill passengers had eaten lunches containing ham, baked beans, potato salad, rolls, and coffee or tea. A sample from the ham eaten by the passengers was tested and bacteria in sufficient numbers to cause the illness were identified. A fingernail culture of a foodhandler yielded bacteria of a type identical to that found in the implicated ham.

What do you think happened here?

STUDY QUESTIONS

1. What is a foodborne illness?
2. What is an outbreak?
3. What is the difference in meaning between the terms *clean* and *sanitary?*
4. Describe how food, people, and facilities can contribute to the problem of foodborne illness.
5. What are the legal reasons foodservice managers should provide safe food?
6. How can a foodservice manager market sanitation?

ANSWER TO A CASE IN POINT

People contributed to this outbreak, and people could have prevented it. In most operations, the person who is the most responsible for food safety is the foodservice manager.

A contributing factor was a foodhandler who didn't wash his or her hands before slicing the cooked ham. Another factor was inadequate cooling of the hams. Following preparation, the ham should have been refrigerated because it is a potentially hazardous food, but since it wasn't, the bacteria continued to grow.

Although you might not have been able to guess everything that went wrong in this case study, in the following chapters, you will be given the information to identify individual problems, and prevent them from happening.

MORE ON THE SUBJECT

THE EDUCATIONAL FOUNDATION OF THE NATIONAL RESTAURANT ASSOCIATION. Chicago: The Educational Foundation, 1990. The Foundation offers the following employee sanitation training aids:

- *Serving Safe Food: A Guide for Foodservice Employees.* Written for foodservice employees, this easy-to-read guide reinforces the key principles of foodservice sanitation, such as personal hygiene, receiving and storing food, safe foodhandling, and cleaning and sanitizing. Includes learning exercises and a certificate of completion.
- *Serving Safe Food Video Series.* This four-part employee training program includes "Introduction to Food Safety: Employee Health and Hygiene," "Safe Foodhandling:

Receiving and Storage," "Safe Foodhandling: Preparation and Service," and "Cleaning and Sanitizing." Each comes with a leader's guide containing learning objectives and questions for discussion.

• *SERVSAFE Manager Kit.* This kit provides a comprehensive training package for foodservice managers and includes a step-by-step leader's guide, five *Serving Safe Food* guides, five certificates of completion, an employee wall chart for tracking progress through the training program, and nine posters for strategic placement in a foodservice operation.

GUTHRIE, RUFUS K. *Food Sanitation.* Westport, CN: AVI, 1980. 326 pages. This reference text presents an overview of food sanitation, covering common diseases and their modes of transmission. Representative sanitation laws and ordinances and recommended health codes are included. The purpose of the book is to give the food industry worker an understanding of the biological principles involved in air, land, and water pollution.

LONGRÉE, KARLA, and GERTRUDE G. BAKER. *Sanitary Techniques in Food Service.* 2nd ed. New York: Macmillan, 1982. 271 pages. This text provides practical guidance in culinary sanitation for the foodhandler. Designed as a teaching aid for vocational training, the book contains many charts and tables depicting safe operating procedures. It includes new material on foods prepared away from the serving premises, new equipment, changes in the food supply, and changes in microbiological problems.

U.S. PUBLIC HEALTH SERVICE. *Food Service Sanitation Manual.* Washington, DC: U.S. Government Printing Office, 1978. 96 pages. These official federal guidelines contain the 1976 Public Health Service recommendations for "A Model Food Service Sanitation Ordinance and Code" presented for adoption by state and local jurisdictions or for incorporation into their laws and regulations. The manual is annotated with descriptive material concerning reasons for the code's provisions and compliance criteria.

2

The Microworld

Of the three hazards to safe food—biological, chemical, and physical—the biological hazard causes the greatest percentage of foodborne-illness outbreaks. The biological hazard is a challenge to control, because it mainly consists of tiny living creatures called *micro-organisms,* from the Greek words meaning *small* and *living beings.* They are so small that they can only be seen with the aid of a microscope.

The existence of this invisible world was first discovered in 1693 by the Dutch lens maker, Anton van Leeuwenhoek. However, it was another 200 years before the French scientist Louis Pasteur, and a small number of his associates, became convinced that certain types of micro-organisms cause disease. Since that time, the study of micro-organisms has led to the control of the spread of many infectious diseases. Eventually this research revealed the connection between disease and the contamination of food and water.

One of the primary responsibilities of a foodservice manager is to protect the consumer by serving safe and wholesome food. One way to do this is to follow the adage, "Know thine enemy"; in this case, *bacteria, viruses, parasites,* and *fungi—yeasts,* and *molds.* We will examine these principal micro-organisms that threaten food safety in storage, preparation, holding, and serving. This chapter will:

- Review the ways in which bacteria multiply.
- Examine what kind of environment bacteria thrive in and how nourish-

ment, moisture, temperatures, and time affect their survival and growth.

- Consider how certain parasites may enter foods and cause foodborne illness.
- Discuss yeasts and molds and the conditions that influence their growth in food.
- Show that micro-organisms can be both harmful and helpful to humans.

BACTERIA

Of all micro-organisms, bacteria are of the greatest concern to the foodservice manager. Bacteria are more commonly involved in foodborne illness than are any other biological forms, including viruses, yeasts, and molds. Knowing what bacteria are, and particularly knowing the environment they flourish in, is a first step in controlling them.

A *bacterium* is a living organism made up of a single cell. Like all living things, bacteria need nutrients to maintain their functions. They take in these nutrients through their cell walls. Bacteria in food causes illness in two ways. Some bacteria are infectious disease-causing agents called *pathogens,* which feed on nutrients in potentially hazardous foods and multiply very rapidly at favorable temperatures. These micro-organisms are using the food as a medium for growth and also as transportation to the human body. Other bacteria are not infectious in themselves, but as they multiply in food, they discharge toxins that poison humans when the food is ingested. (See Chapter 3.)

Spore-Forming and Vegetative Stages

A *vegetative organism* is capable of growth and reproduction. Vegetative bacteria may be killed by high temperatures, but are more resistant to low temperatures and may survive freezing. A bacterium in a developing, nonspore stage is in a *vegetative stage.*

Certain bacteria have the ability to produce a special structure called a spore that serves as a means of protection against an unfavorable environmental condition. A *spore* is a thick-walled formation within the bacterial cell that is capable of becoming a vegetative organism when conditions again become

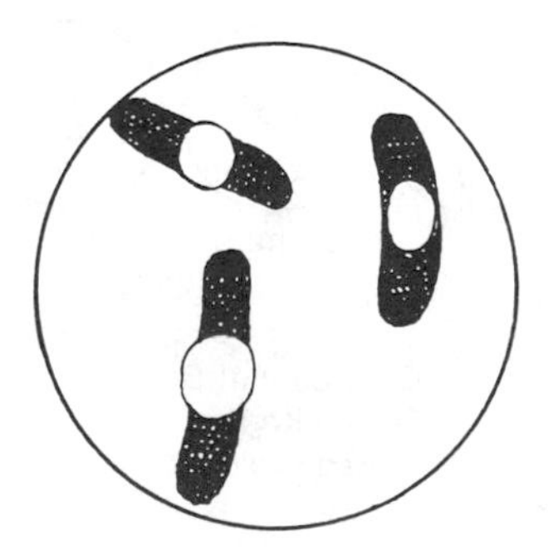

Exhibit 2.1 Bacteria: *left,* vegetative; *right,* with spores

favorable. The spore itself does not reproduce. The thick wall of the spore makes it more resistant to heat, cold, and chemicals than the vegetative bacterial cell (see Exhibit 2.1). What this means to the foodservice manager is that a spore may survive some cooking temperatures and can cause foodborne illness.

For example, a spore may survive boiling water for an hour or more. It also holds up well under freezing and may resist some sanitizing solutions. The fact that spores are so difficult to destroy is one of the reasons for careful procedures to keep harmful numbers of bacteria out of the food in the first place.

Reproduction of Bacteria

Given the right conditions, bacteria reproduce in a very simple manner. The vegetative cell enlarges and then divides in two. Each of these two bacteria may divide into two more cells, and so on. The result of this kind of growth is a *tremendous increase in the numbers of bacteria over a relatively short period of time.* The offspring of a single bacterium will double with each division—2, 4, 8, 16, 32, and so on (see Exhibit 2.2). This rapid reproduction increases the risk of bacteria causing a foodborne illness. The conditions important to the reproductive process are food, acidity (pH), time-temperature, oxygen, and moisture (FAT-TOM).

Growth Pattern of Bacteria

Under ideal conditions, the growth rate for bacteria follows a distinct pattern (see Exhibit 2.3). If you were to touch a slice of ham with your thumb, you could plant several thousand bacteria on its surface. Some of these organisms will not survive the change in environment, so for a time there will be fewer organisms. This period of adaptation to a new environment is the *lag phase* of bacterial growth.

Exhibit 2.2 Bacteria reproduce by dividing. Under ideal conditions, bacteria multiply at an explosive rate, a single cell becoming billions in 10 to 12 hours.

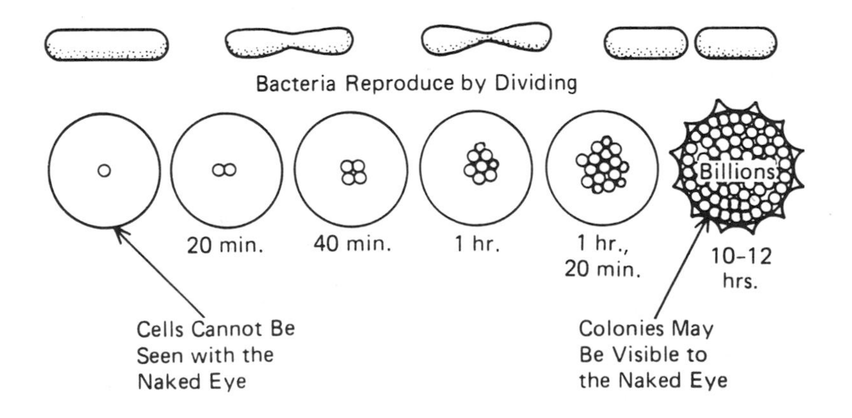

Exhibit 2.3 Bacterial growth curve

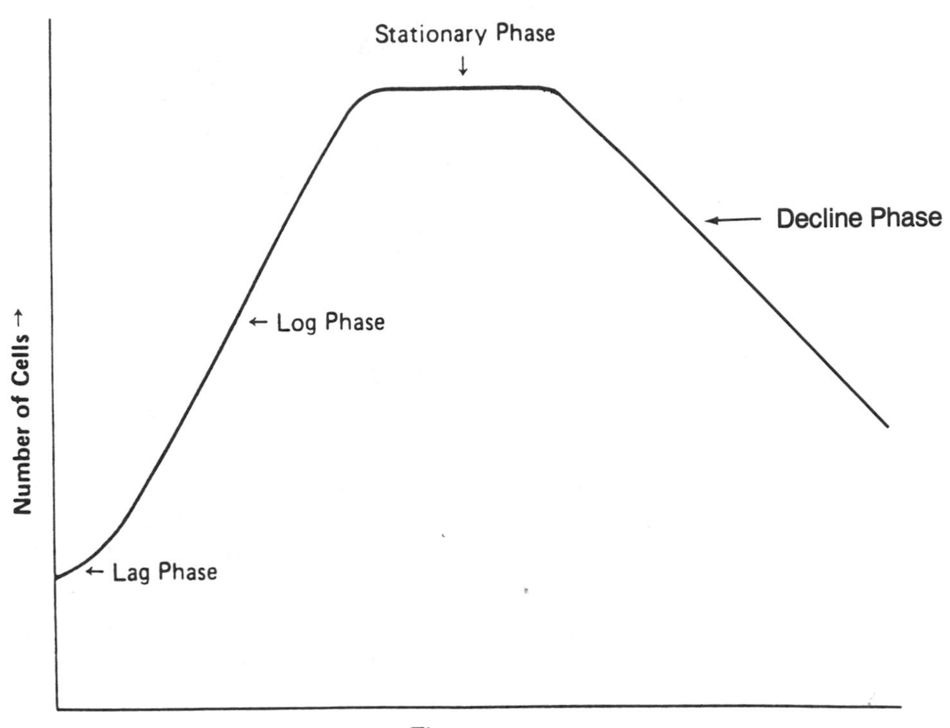

After a longer period of time (from an hour to several days, depending on conditions), the bacteria would start to multiply very rapidly. This accelerated growth phase is the *log phase.* When the bacteria have increased to such large numbers that they compete for space and nourishment, they no longer multiply so rapidly and some may start dying. The period of competition is the *stationary phase.* The last phase, or *decline phase,* is when bacterial cells begin to die more quickly due to lack of nutrients or because of their own waste products.

The development and decline of bacterial cells depends on temperature, food supply, the species of the bacteria, the amount of initial contamination, and the age of the organisms (see Exhibit 2.4).

Bacteria should not be allowed to grow past the lag phase. During the lag phase, the foodservice manager has *the most control.* Allowing the product to remain in the *temperature danger zone,* 45° to 140°F (7.2° to 60°C) for four hours or more provides conditions favorable for growth where bacteria can expand in sufficient number to cause illness. Some

Exhibit 2.4 Effect of temperature in *Salmonella* growth

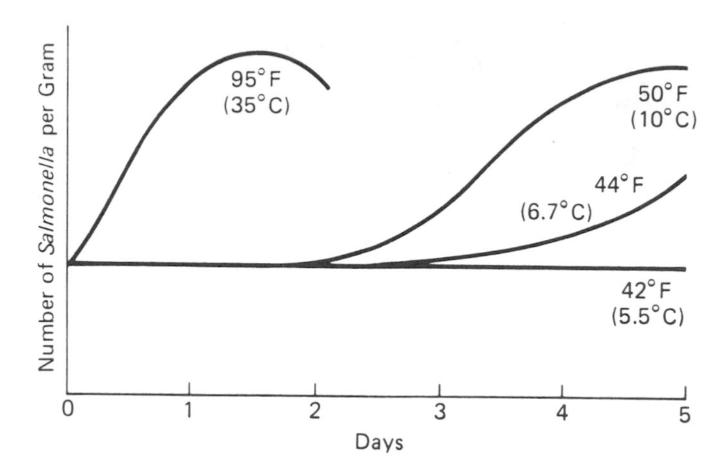

health codes specify 40° to 145°F (4.4° to 62.8°C) as the temperature danger zone for potentially hazardous foods. *Temperature control and time* are the key factors in preventing bacterial growth.

Toxins

Chemical changes occur in bacteria, just as they do in living human cells. Like humans, bacteria discharge wastes and decompose when they die. The wastes and decomposed materials are *toxins,* which are poisonous to humans. The toxins can themselves cause foodborne illness. So even though the disease-causing bacteria may be dead, the toxins they leave behind make people ill.

Mobility

Bacteria are moved from place to place by water, wind, food, insects, rodents, or other animals (including human beings). Bacteria are notorious hitchhikers on human hosts. They occur in the hair, on the skin, and in the clothing. They are found in the mucous membranes in the mouth, nose, and throat; in the intestinal tract; and on scabs or scars from skin wounds or lesions. They frequently end up on the hands, and potentially in food.

Environmental Needs of Bacteria

Bacteria can live anywhere a human being can live. In fact, they survive hotter and colder temperatures and a wider range of atmospheric conditions than humans. Generally speaking, bacteria thrive in a warm, moist, protein-rich environment (or medium) that is neutral or low acid. Some species, however, tolerate—and even prefer—extremes of heat or cold; and a few can even survive in a dry medium or one of high acid or salt content.

Time and Temperature Requirements

Most bacteria can grow well in the temperature range between 60° and 110°F

(15.6° and 43.3°C). These temperature limits coincide closely with the range between normal room temperature and normal body temperature for humans. Therefore, many types of the bacteria that invade human beings find an ideal temperature for growth. Other types of bacteria live and multiply better between 110° and 130°F (43.3° and 54.4°C).

Both of these types of bacteria thrive within the temperature danger zone for foods, between 45° and 140°F (7.2° and 60°C) (see Exhibit 2.5). Given the right medium (moist environment and enough time), they will multiply in this temperature range.

Other bacteria seem to prosper between 32° and 45°F (0° and 7.2°C), and some can grow at temperatures as low as 19°F (−7.2°C). Such organisms can continue to multiply at refrigerator temperatures. Placing food in a refrigerator will not provide absolute protection against microbial growth. For this reason, the foodservice manager must monitor the length of time a product is kept in refrigerated storage. (Proper refrigeration of foods will be discussed in Chapter 7.)

Exhibit 2.5 Temperature and bacterial growth

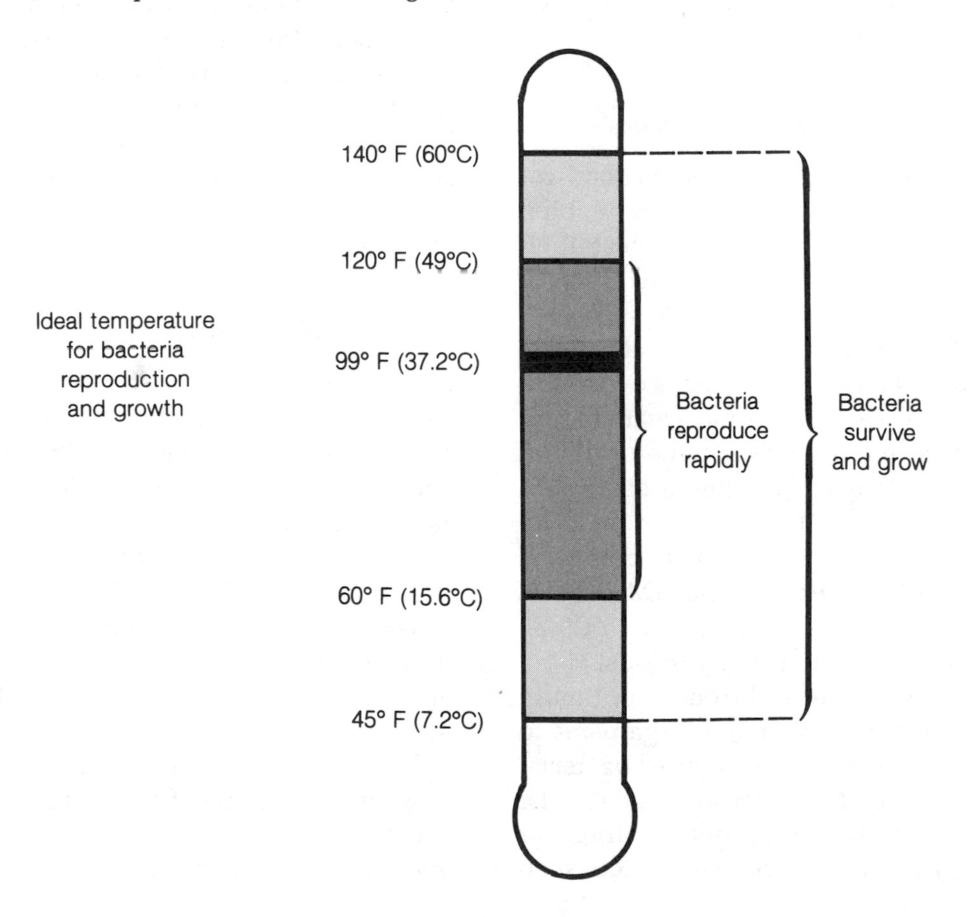

Knowledge of the effects of temperature on bacteria can be a valuable tool in keeping food safe. Most bacteria grow very slowly at lower temperatures. They are not killed by freezing, but their growth slows substantially. Potentially hazardous foods, those items most often implicated in foodborne illness, that spend a total of over four hours in the temperature danger zone provide ample opportunity for bacterial contamination to increase. In order to multiply rapidly, bacteria need *both* time and temperature.

At about 110°F (43.3°C), many vegetative bacteria will start dying, and by the time the temperature reaches 140°F (60°C), most will be killed, although spores will survive.

Food and Moisture Requirements

Bacteria need water as well as food for growth and development (see Exhibit 2.6). Since bacteria cannot ingest their food in solid form, they must receive their nutrients in some kind of water solution. The availability of the water is described as *water activity,* and is expressed as A_w. Most fresh foods have water activity values that are close to the optimum level for growth of micro-organisms (0.97–0.99). The lowest water activity value at which bacterial pathogens will occur is 0.85. Therefore, the FDA defines those foods with a water activity of 0.85 or higher as potentially hazardous.

As the A_w is lowered from an optimum value, the ability for micro-organisms to grow is reduced. To reduce bacterial growth in foods, water activity can be lowered by freezing, dehydrating, or mixing with a dissolved substance, such as sugar or salt. For example, when bacon, which is a potentially hazardous food, is cooked crisp, it has a water activity level that is less than 0.85. Cooking removes much of the available water, and the sugar and salt included in the chemical composition of the bacon tie up (or bond) the available moisture, thereby limiting bacterial growth. For the foodservice manager, this means that crisp, cooked bacon need not be subjected to strict time-temperature controls.

Although bacterial growth is halted in foods that are very low in available moisture, the bacteria remain alive. This fact is a key to food preservation. Many food products are handled dry, such as sugar, flour, some meats, and some fruits. So long as they are kept dry, these supplies will be safe in storage, even though some bacteria are present. Some dry foods, such as rice or beans, become potentially hazardous in the cooking process once moisture is added.

Acidity (pH) Requirements

Bacteria grow best in a medium that is neutral or slightly acidic. The growth of most bacteria is greatly inhibited in a very acidic medium. Acidic foods, such as vinegar and most fresh fruits, especially citrus fruits, are very seldom vehicles for pathogenic bacteria.

The *pH* of a food or other medium is a measure of its acidity or alkalinity. pH is measured on a scale from 0 to 14.0. A solution with a pH of 7.0 is exactly neutral, neither acidic nor alkaline. Distilled water is an example of a neutral solution. A food with a pH *below* 7.0 is *acidic,* and one with a pH *above* 7.0 is *alkaline.* The

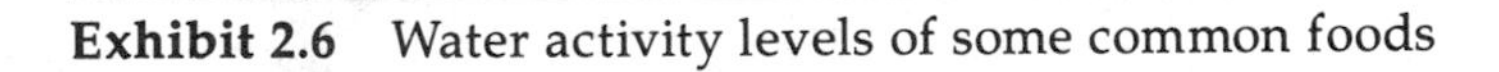
Exhibit 2.6 Water activity levels of some common foods

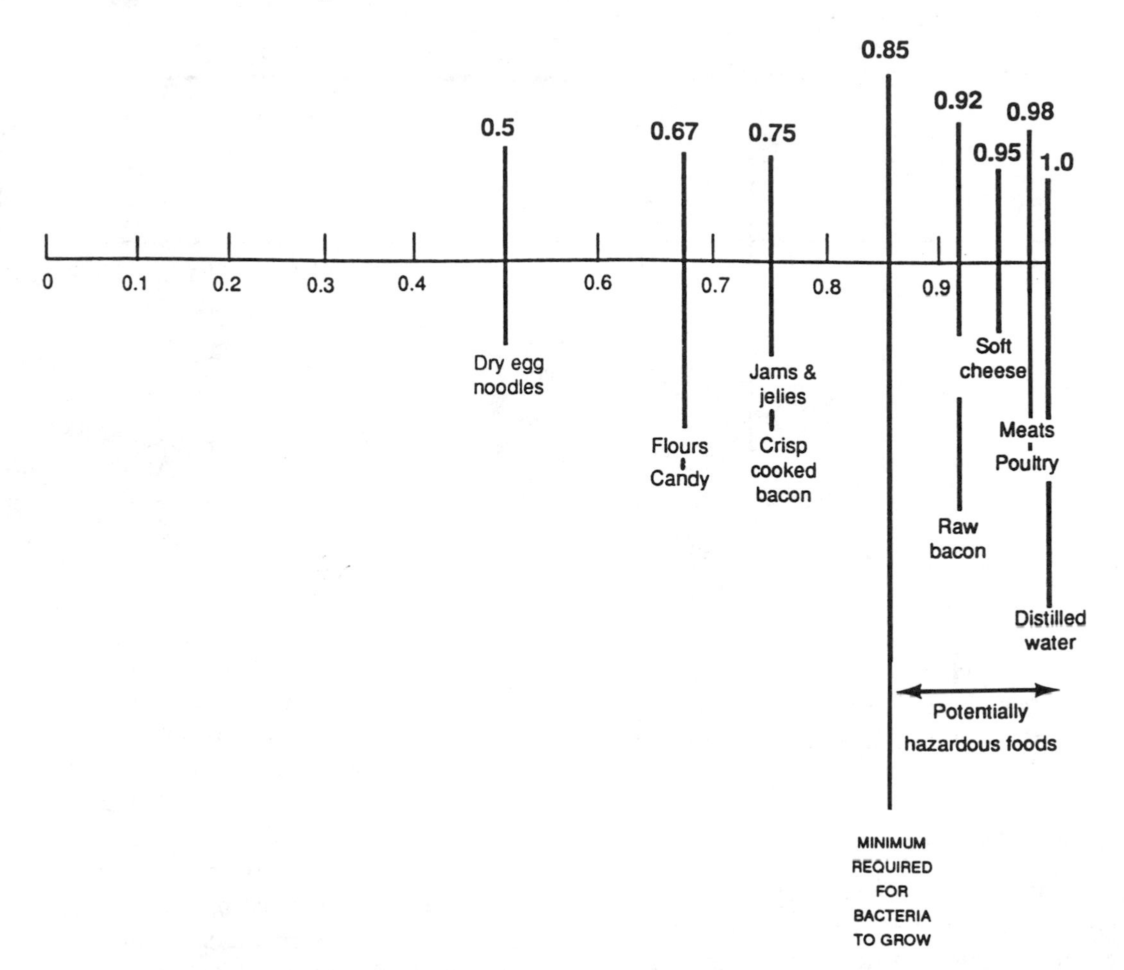

lower the pH value, the higher the acid content of the food; the higher the pH value, the lower the acidity. For example, limes have a pH of 2.0, which is highly acidic, while fresh meat has a pH of 6.4, which is almost neutral. Most bacteria will not grow well at pH levels *below* 4.6. Although many micro-organisms can survive in the pH range between 4.6 and 9.0, most grow best at a pH between 6.6 and 7.5. Many foods, especially meats, have a pH that is very favorable for the growth of bacteria (see Exhibit 2.7).

While adding an acidic substance to a food item can be a protection against bacterial growth, the foodservice manager should not rely only on pH to stop bacterial growth. For instance, commercially prepared mayonnaise has a pH below 4.6; however, using it to prepare ham salad may not stop the growth of bacteria already in the ham. Instead, the

Exhibit 2.7 pH of some common foods

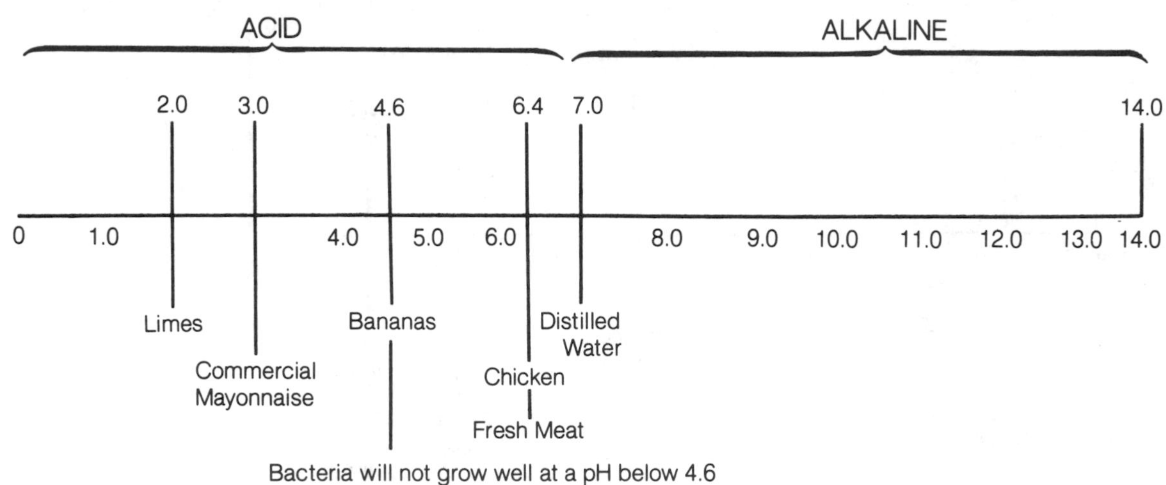

favorable pH and moisture of the ham will most likely raise the pH of the ham salad to a level favorable to bacterial growth.

Oxygen Requirements

Bacteria vary in their requirements for oxygen. Some bacteria will grow only when supplied with free oxygen; these are called *aerobes.* Other bacteria will grow only when free oxygen is excluded or absent; these are called *anaerobes.* These bacteria grow in a vacuum-sealed jar or pouch, in a can, or in a large pot of food. Most of the bacteria that cause foodborne illness are *facultative,* meaning they can grow either with or without the presence of free oxygen.

VIRUSES

Another type of micro-organism of concern to the foodservice manager is the virus (from the Latin word for *poison*). *Viruses* are the smallest and perhaps the simplest form of life known; there is even some doubt as to whether they are alive at all. Unlike bacteria, viruses are not complete cells. They are genetic material in a protein wrapper or overcoat.

Because of their primitive structure, viruses cannot reproduce outside of a living cell. Once viruses gain entrance to a cell they stop its life processes and force the cell to assist in producing more viruses. Different kinds of viruses require different hosts, ranging from bacteria and larger plants to insects, higher animals, and humans.

Viruses vary in their reaction to heat and cold. One type of virus has been shown to survive in food even at temperatures as high as 176°F (80°C). This same virus, which is so resistant to heat, can also remain active after storage in a refrigerator for over a year.

Viruses do not require potentially hazardous foods to survive. Also, viruses do not *increase* in number while they are in food. The food and food-contact surfaces merely serve to transport the viruses, which then may lodge themselves in the human host and reproduce abundantly. Viruses have a built-in capacity to cause disease in a susceptible person, so *they must not be allowed to contaminate food in the first place.*

Outbreaks of foodborne or waterborne viral diseases are most often attributed to poor personal hygiene or a contaminated water supply. (For a detailed discussion of personal hygiene refer to Chapter 4.) Viruses are often found in drinking water that has not been filtered and disinfected. (Methods of preventing contaminated water from coming into contact with usable water in the foodservice facility are discussed in Chapter 9.) Molluscan shellfish (oysters, mussels, and clams) illegally harvested from polluted water, which are eaten raw or slightly cooked, are often found to be a source in foodborne viral illnesses.

PARASITES

Parasites are small or microscopic creatures that need to live on or inside a host to survive. *Trichinella spiralis* is the best known of the parasites that contaminate food. This parasite is a roundworm (see Exhibit 2.8) that prefers mammals, including humans, pigs, rats, and wild game (e.g., bears and walrus). The larvae from this worm cause trichinellosis, commonly called *trichinosis.* Other foods, such as ground beef, have been linked to trichinosis outbreaks because of the addition of contaminated pork products to them.

Exhibit 2.8 *Trichinella spiralis*

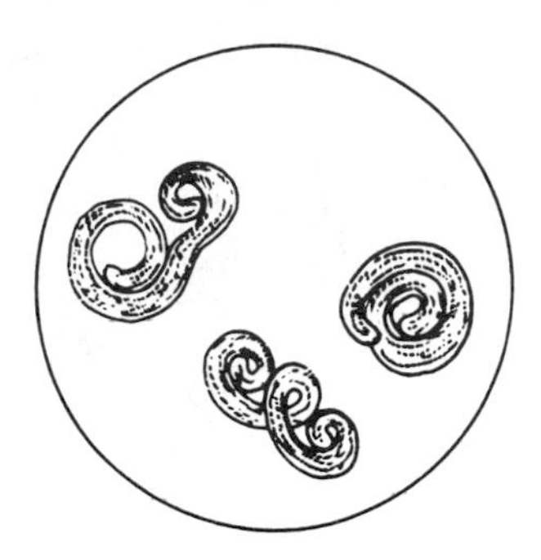

With a growing interest in exotic foods, more people in the United States are eating raw, marinated, or partially cooked fish. Fish-related parasite infections are an increased risk for these consumers. One parasite that affects fish is the *Anisakis* roundworm.

FUNGI

Fungi range in size from microscopic, single-celled plants to giant mushrooms. They are found naturally in air, soil, plants, animals, water, and some food. The members of this group of organisms that are of most concern to the foodservice manager are molds and yeasts (see Exhibit 2.9).

Molds

Molds grow quickly, so the tangled mass of mold plants can easily be seen as a fuzzy growth. Individual mold cells can usually be seen only with a microscope; mold colonies that are visible to the naked eye consist of large numbers of

Exhibit 2.9 Fungi: *left,* yeast; *right,* mold

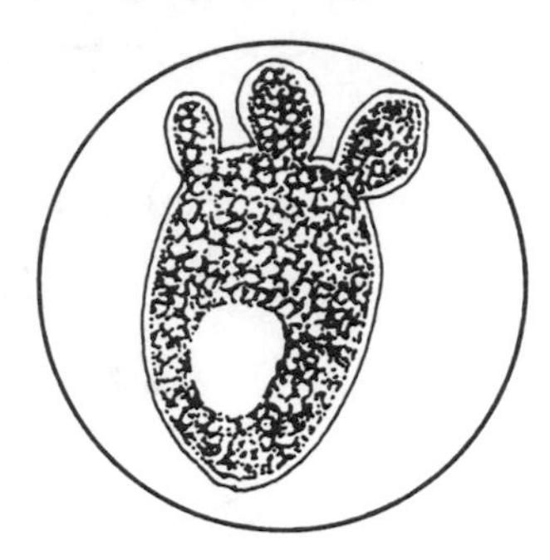

cells. *Spores* formed by molds are not survival devices, as with bacteria, but are for the purpose of reproduction.

It was once thought that food molds were not dangerous to humans, but we now know that certain molds produce toxins, such as aflatoxin, that have been shown to cause cancer in animals. These same toxins have been linked to rare, isolated incidents of foodborne illness. Also, certain other types of molds can cause serious infections and allergies (often contracted by sniffing mold in spices and other foods).

Mostly, molds affect the quality of foods. They cause spoilage with discoloration, formations of musty odors, and off-flavors in foods. They can grow on almost any food, at almost any storage temperature, and under almost any conditions: moist or dry, high or low pH, salty or sweet.

Molds commonly grow on fruits and breads. They also grow on vegetables, refrigerated meats, and cheese that has been exposed to the air. They are the primary spoilage organism in dried foods, such as spices, nuts, and popcorn, because they can survive and grow with a low A_w.

Although the cells and spores of most molds can be killed by heating to 140°F (60°C) for ten minutes, the toxins are heat-stable and are not destroyed by normal cooking methods. Freezing prevents the growth of molds, but will not kill those already present in the food. Therefore, foods with molds that are not a natural part of the items should be discarded.

Mold is a natural part of certain cheeses (gorgonzola, bleu cheese, brie, and camembert). For other cheeses, it is unacceptable merely to cut away the portion of a cheese that has become moldy. According to the FDA, mold on certain types of cheeses, for example, hard cheeses (cheddar, colby, romano, Monterey Jack, and parmesan) can be cut away if:

1. The cheese does not have holes.
2. Unrefrigerated cheese has one inch beyond the molded cheese removed.
3. Refrigerated cheese has at least one-half inch beyond the molded cheese removed.

These measures will help ensure quality control as well as food safety.

Yeasts

Yeasts are fungi. Since they require sugar and moisture for survival, they often consume these ingredients from food products, such as jellies and honey, while spoiling the food in the process. Yeast contamination sometimes creates a visible slime on pickle brine or fruit juices. In addition, the pink discoloration in cottage cheese may be due to the presence of yeasts. In general, yeast spoilage is most readily recognizable by the presence of bubbles and an alcoholic smell or taste.

Yeast organisms can be killed by heating food products to 136°F (57.8°C) for 15 minutes. While yeasts are responsible for a few diseases in human beings, no evidence suggests that these diseases are transmitted by food or that the yeasts occurring naturally in foods are harmful to human beings. However, since yeasts spoil food, they still need to be controlled.

BENEFITS FROM MICRO-ORGANISMS

Certain micro-organisms are native to our skin, our mouths, and our digestive tracts. They are present in our food and water. They are, quite literally, found everywhere life can exist.

Bacteria aid in the digestion of food and help to break down wastes in our bodies. Some molds are useful (as sources of antibiotics, such as penicillin, and in the ripening of some cheeses). The beneficial uses of yeasts are well known; they aid in leavening bread and in fermenting beer and other products. Our concern in this book, however, is with the micro-organisms that cause illness.

SUMMARY

There are five main classes of micro-organisms with which the foodservice manager is concerned. The greatest threat to food safety comes from *bacteria.* While bacteria can survive a wider range of environmental conditions than human beings can, they flourish at temperatures that very nearly coincide with those most favorable to human life. High temperatures, for the most part, kill bacteria; low temperatures and stable, dry conditions slow their growth. Time and temperature are critical factors in controlling the growth of bacteria.

The smallest of the micro-organisms, *viruses,* can multiply only in living cells; food items serve largely as a means of transportation to the potential disease victim.

Parasites are also dangerous to humans, particularly the larvae of *Trichinella spiralis,* which causes trichinosis. These larvae usually are transported to humans via undercooked meat, mainly pork. *Anisakis,* another parasite, is harbored in raw fish.

Mold and *yeast* growth occurs commonly on certain foods. Molds are highly adaptable organisms that are of value in medicine and cheese-making, but some varieties contribute to disease. Although yeasts are helpful as agents in fermentation and leavening, they detract from the flavor of some foods.

Micro-organisms are found everywhere. They are found in our bodies, in the soil, in water, and in the air. They perform useful functions, such as decomposing debris and producing certain foods; but their effects are obviously harmful when they serve as agents of foodborne disease and food spoilage.

A CASE IN POINT

The Department of Health in a midwestern state began an investigation when it received reports that three unrelated patients hospitalized with abdominal cramps and diarrhea had stool cultures positive for *Salmonella* bacteria. Investigation revealed that one of these patients worked as a cook at a suburban restaurant and that the other two had eaten at the same restaurant four days apart. After three days, seven more cases of illness were reported. The management voluntarily closed the restaurant.

Forty-seven restaurant patrons were interviewed. Twenty-five of them were found to be ill. When questioned, the employees identified incomplete cooking and poor cooling practices. *Salmonella* was isolated from the cooked prime rib, the cooked roast beef, the cooked ham, the lettuce, and the coleslaw, as well as from the surface of a wooden cutting board.

Why do you think the meat was implicated as the cause of this outbreak?

STUDY QUESTIONS

1. Why are micro-organisms a concern of foodservice managers?
2. Which type of micro-organism is most associated with foodborne illness?
3. What is the temperature danger zone for bacterial growth?
4. What is the function of a bacterial spore?
5. Name four environmental needs for the growth of bacteria.
6. Why is pH an important factor in the control of bacteria?
7. How are viruses different from bacteria?
8. Define a *parasite.*
9. Can moldy cheese still be used? Explain.
10. How can yeasts be controlled?

ANSWER TO CASE IN POINT

The meat was implicated as the source of the bacteria that caused the outbreak because meats are potentially hazardous foods, which require very careful time and temperature controls. Meats are considered potentially hazardous because they are a favorable medium to support the growth of bacteria.

Meats have a pH of about 6.4, which is near the neutral level that bacteria prefer. Meats also have a high water activity level since they are moist. The other requirements for bacteria survival and reproduction are oxygen, time, and temperature. The bacteria also had a favorable temperature over time to grow as the foodhandlers interviewed said the meats were not adequately cooked or cooled. Cross-contamination of the other foods (the lettuce and cole slaw) occurred because the cutting board had not been cleaned and sanitized after the meat had been sliced on it.

In order to control bacteria in potentially hazardous foods, you need to practice careful time and temperature control. Keep foods out of the temperature danger zone during all steps of preparation, storage, and serving. Also be sure not to contaminate ready-to-eat foods with knives or cutting boards that have come into contact with potentially hazardous foods. To learn more control measures—keep reading!

MORE ON THE SUBJECT

BANWART, GEORGE J. *Basic Food Microbiology.* Westport, CN: AVI, 1981. 519 pages. Chapters 1–4 on micro-organisms associated with food provide a very useful source on factors influencing microbial growth. The book presumes some background in biology and chemistry.

LONGRÉE, KARLA and GERTRUDE ARMBRUSTER. *Quantity Food Sanitation.* 4th ed. New York: Wiley, 1987. 452 pages. The introduction to bacteria, yeasts, molds, and viruses in Chapter 2, "Some Basic Facts on Micro-organisms Important in Food Sanitation," is complete and especially useful for the drawings it presents. This book is written for students with some knowledge of elementary biology and chemistry.

3

Contamination and Foodborne Illness

With the scientific advances of recent decades, one might think that the biological, chemical, and physical hazards that contaminate our food and compromise our health would long since have been conquered. *Contamination* is the unintended presence of harmful substances or organisms in food. While it is true that advances have resulted in safer foods, better methods of preservation, and improved storage practices, it is still necessary to guard against the practices that can increase the likelihood of contamination, or add to the growth of micro-organisms that are already present in the product. Preventing contamination of safe food needs to be a prime objective of every foodservice manager.

This chapter will cover the following three categories of hazards responsible for outbreaks of foodborne illness:

- *Biological hazards:* Harmful bacteria, viruses, or parasites

- *Chemical hazards:* Harmful substances, such as cleaning solutions, sanitizers, or metals
- *Physical hazards:* Foreign particles, such as glass or metal particles.

Although the biological hazard represents the most widespread problem, contamination of food by chemical and physical hazards should be considered dangerous as well.

ESTABLISHING MULTIPLE BARRIERS TO BACTERIAL GROWTH

In Chapter 2, we examined the five conditions bacteria need in order to grow: **f**ood, **a**cidity, **t**ime-**t**emperature, **o**xygen, and **m**oisture (FAT-TOM). You can develop preparation and handling procedures that establish barriers to these growth conditions when you determine your menu items. These procedures will help prevent foodborne illness.

Let's look at the preparation of tuna salad from a recipe that includes tuna, mayonnaise, celery, onion, and salt. Canned tuna and commercial mayonnaise do not normally require refrigeration until opened, but by moving the cans of tuna and mayonnaise from dry storage to the refrigerator one day in advance, your main ingredients are prechilled before preparation. At the time of preparation, the temperature of these items should be at or below 45°F (7.2°C). A temperature level that slows the growth of bacteria is the first barrier in preparation.

The next barrier directly involves the cook. Properly washing his or her hands and using cleaned and sanitized utensils greatly reduces the risk of contaminating or cross-contaminating the food itself during preparation. In addition, the cook must thoroughly wash the celery to keep the dirt found on the celery from contaminating the tuna salad with bacteria. The cook might even blanch the celery in boiling water and refrigerate it to further reduce any contamination that may be added to the tuna salad.

The third barrier is the adjustment of the pH, or acidity, of the product. By adding chilled *commercial* mayonnaise, rather than homemade, the acidity of the tuna salad has been made less favorable for bacterial growth. This barrier can be further strengthened by adding lemon juice to the recipe, which will further increase the acidity out of the ideal range for growth. It is important to realize that the adjustment of the pH of a product does not mean that the product can be time or temperature abused. *Adjusting the pH* means that another barrier has been established to slow bacterial growth.

Serving involves a fourth barrier, both temperature and time. Keeping the tuna salad cold at 45°F (7.2°C) or lower while serving it, by limiting the serving time to four hours or less, and using strict procedures for handling leftovers, greatly reduces the opportunity for bacterial growth.

Taken in combination, these properly designed and well-executed preparation procedures operate as multiple barriers to control the conditions that could result in contamination and bacterial growth. The more points in the preparation process where there are established barriers to bacterial growth, the lower the risk that a single foodhandling error will cause a foodborne illness. Understanding foodhandling errors, the hazards they involve, and the role of multiple barriers are keys in establishing an effective food safety system.

THE BIOLOGICAL HAZARD

Biological hazards are dangers to food from pathogenic (disease-causing) micro-organisms (bacteria, viruses, parasites, and fungi) and from toxins that occur in certain plants and fish. When biological hazards result in foodborne illnesses, these illnesses are generally classified as either infections or intoxications.

Infection, Intoxication, and Toxin-Mediated Infection

A *foodborne infection* is a disease that results from eating food containing living harmful micro-organisms.

A *foodborne intoxication* results when toxins, or poisons, from bacterial or mold growth are present in ingested food and cause illness in the host (the human body). These toxins are generally odorless and tasteless, and are capable of causing disease even after the micro-organisms have been killed. Toxins may be present in foods naturally, as in the case of certain plants, such as mushrooms, and certain animals, such as puffer fish. When a human being consumes these plants and animals, illness may result.

A *foodborne toxin-mediated infection* is a disease that results from eating a food containing a large amount of disease-causing micro-organisms. Once these micro-organisms are ingested, the human intestine provides the perfect conditions for the micro-organisms to produce toxins.

The term *food poisoning* is commonly misused to refer to various food-related illnesses caused by micro-organisms. The best term to use is foodborne illness.

Foodborne Infections of Bacterial Origin

We will discuss in some detail the principal bacterial micro-organisms, their reservoirs, and the diseases they cause. A *reservoir* is a host or carrier of disease-causing organisms. A *host* is a person, animal, or plant on which another organism lives and takes nourishment. A *carrier* is a person or animal who harbors disease-causing micro-organisms in their body without being noticeably affected. Humans and animals are a common reservoir for many pathogens (see Exhibit 3.1).

Salmonellosis

Salmonellosis results from the consumption of food contaminated with live pathogenic *Salmonella.* It is one of the most frequently reported foodborne infections, and its occurrence is on the rise.

Symptoms. The symptoms of salmonellosis are diarrhea, fever, chills, abdominal pain, and possibly a headache or vomiting. The incubation period is between 6 and 48 hours after ingestion of contaminated food. *Incubation period,* or *onset,* means the length of time it takes

Exhibit 3.1 Major foodborne diseases of bacterial origin

	Salmonellosis Infection	Shigellosis Infection	Listeriosis Infection	Staphyloccal Intoxication	Clostridium Perfringens Toxin Mediated Infection	Bacillus Cereus Intoxication	Botulism Intoxication
Bacteria	*Salmonella* (facultative)	*Shigella* (facultative)	*Listeria monocytogenes* (reduced oxygen)	*Staphylococcus aureus* (facultative)	*Clostridium perfringens* (anaerobic)	*Bacillus cereus* (facultative)	*Clostiridium botulinum* (anaerobic)
Incubation Period	6–72 hours	1–7 days	1 day to 3 weeks	1–6 hours	8–22 hours	$1/2$–5 hours; 8–16 hours	12–36 hours + 72
Duration of Illness	2–3 days	Indefinite, depends on treatment	Indefinite, depends on treatment, but has high fatality in the immunocompromised	24–48 hours	24 hours	6–24 hours; 12 hours	Several days to a year
Symptoms	Abdominal pain, headache, nausea, vomiting, fever, diarrhea	Diarrhea, fever, chills, lassitude, dehydration	Nausea, vomiting, headache, fever, chills, backache, meningitis	Nausea, vomiting, diarrhea, dehydration	Abdominal pain, diarrhea	Nausea and vomiting; diarrhea, abdominal cramps	Vertigo, visual disturbances, inability to swallow, respiratory paralysis
Reservoir	Domestic and wild animals; also humans, especially as carriers	Human feces, flies	Humans, domestic and wild animals, fowl, soil, water, mud	Humans (skin, nose, throat, infected sores); also, animals	Humans (intestinal tract), animals, and soil	Soil and dust	Soil, water
Foods Implicated	Poultry and poultry salads, meat and meat products, milk, shell eggs, egg custards and sauces, and other protein foods	Potato, tuna, shrimp, turkey and macaroni salads, lettuce, moist and mixed foods	Unpasteurized milk and cheese, vegetables, poultry and meats, seafood, and prepared, chilled, ready-to-eat foods	Warmed-over foods, ham and other meats, dairy products, custards, potato salad, cream-filled pastries, and other protein foods	Meat that has been boiled, steamed, braised, stewed or roasted at low temperature for a long period of time, or cooled slowly before serving	Rice and rice dishes, custards, seasonings, dry food mixes, spices, puddings, cereal products, sauces, vegetable dishes, meat loaf	Improperly processed canned goods of low-acid foods, garlic-in-oil products, grilled onions, stews, meat/poultry loaves
Spore Former	No	No	No	No	Yes	Yes	Yes
Prevention	Avoid cross-contamination, refrigerate food, cool cooked meats and meat products properly, avoid fecal contamination from foodhandlers by practicing good personal hygiene	Avoid cross-contamination, avoid fecal contamination from foodhandlers by practicing good personal hygiene, use sanitary food and water sources, control flies	Use only pasteurized milk and dairy products, cook foods to proper temperatures, avoid cross-contamination	Avoid contamination from bare hands, exclude sick foodhandlers from food preparation and serving, practice good personal hygiene, practice sanitary habits, proper heating and refrigeration of food	Use careful time and temperature control in cooling and reheating cooked meat dishes and products	Use careful time and temperature control and quick chilling methods to cool foods, hold hot foods above 140°F (60°C), reheat leftovers to 165°F (74°C)	Do not use home-canned products, use careful time and temperature control for sous-vide items and all large, bulky foods, keep sous-vide packages refrigerated, purchase garlic-in-oil in small quantities for immediate use, cook onions only on request

before the symptoms of the illness appear. Milder cases of salmonellosis usually last two to three days, while severe infections may last longer and can, in rare cases, be fatal.

Causative Agent. The infection Salmonellosis is caused by the *Salmonella* genus of bacteria. There are more than 2,000 types of this bacteria that cause illness in human beings. One type of *Salmonella*, the *Salmonella typhi*, causes typhoid fever (see Exhibit 3.2).

Salmonella are facultative bacteria, meaning they grow with or without the presence of oxygen. They do not form spores. They grow best at human body temperature 98.6°F (37°C), but will survive at somewhat higher and lower temperatures. They will be killed by temperatures of 130°F (54.4°C) and higher after two hours, or 165°F (73.9°C) in a few seconds. They can also be killed with an extended time exposure at lower temperatures.

Exhibit 3.2 *Salmonella*

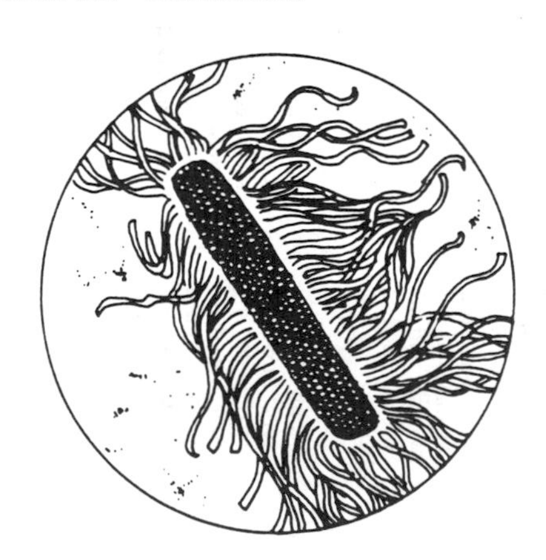

Source. *Salmonella* are found in domestic and wild animals, especially poultry; in eggs and on their shells; in pets, such as puppies, kittens, turtles, and ducklings; and in human beings. However, *Salmonella enteritidis* has also been found in uncracked shell eggs from chickens that carry the bacteria in their ovaries. Persons who eat food contaminated by *Salmonella* may not necessarily become ill, but they may become carriers and transmit the bacteria to food by not washing their hands and then handling the food.

Foods Involved. A multitude of foods are implicated in outbreaks of salmonellosis. The main vehicles for transmission are meat and poultry. Others include foods containing raw or not thoroughly cooked eggs or egg products, and unpasteurized milk or dairy products.

A Typical Case. A cook takes raw chicken out of the freezer and raw beef out of the refrigerator. The cook then sets the chicken out to thaw at a room temperature of 70°F (21.1°C) and begins to prepare and cook the beef.

A few hours later, when the beef has cooked to an internal temperature of 150°F (65.6°C), it is set aside on a preparation table to cool at room temperature.

The cook now begins to prepare the chicken. It is cut up for cooking and put in the oven. Using the same knife and cutting board, the cook cuts the beef, putting *Salmonella* from the raw chicken onto the cooked roast beef. The roast beef is cut up to await orders for sandwiches. When the chicken is taken out of the oven, it is cut up with the same unwashed knife and cutting board used on the roast beef and is recontaminated.

Cross-contamination or recontamination of cooked foods from raw foods is a frequent factor in salmonellosis outbreaks (see Exhibit 3.3). *Cross-contamination* is the transfer of harmful micro-organisms from one item of food to another by means of a nonfood surface (equipment, utensils, human hands), or from storing and thawing raw meat and poultry (which can drip contaminated fluids) above other foods that will receive no further cooking.

Control Measures. Since food producers have found it difficult to eliminate *Salmonella* contamination from raw foods of animal origin, for example meat and poultry, the burden of protecting the customer is on the shoulders of the foodservice manager. The following steps are control measures.

- Be sure to cook foods adequately to the minimum internal temperature appropriate for each item, to chill foods within four hours and to hold foods at proper temperatures during display. (Cooking, chilling, and holding

Exhibit 3.3 Typical events that lead to outbreaks of salmonellosis

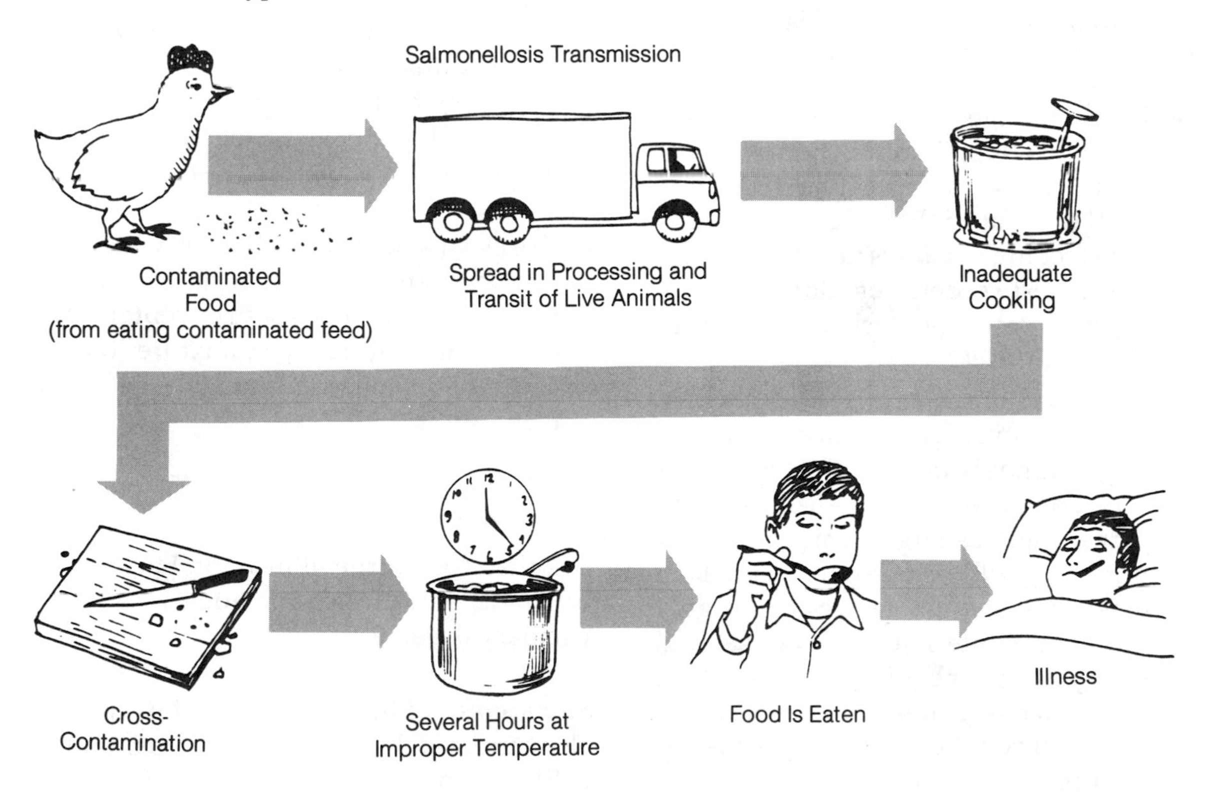

procedures will be discussed in detail in Chapter 8.)

- Guard against cross-contamination. Do not use the same utensils for both raw and cooked product without cleaning and sanitizing the utensils between uses. Separate cutting boards should be maintained for raw and cooked products as well.
- Make sure that all foodhandlers follow the rules of personal hygiene, particularly hand washing after using the toilet. Persons contaminated with *Salmonella* may carry the bacteria in their fecal matter long after they are first contaminated and seemingly recovered. (A complete discussion of personal hygiene is found in Chapter 4.)

The following steps are added precautions for handling shell eggs:

- Purchase eggs only from approved, reputable suppliers. Request refrigerated delivery. Keep eggs refrigerated at all times at 45°F (7.2°C). This means not stacking egg trays near the grill for on-request orders or during mealtimes. Rotate egg stock using First In, First Out (FIFO).
- Cook and break eggs to order whenever possible. Do not pool and hold large numbers of eggs. *Pooling* means breaking quantities of shell eggs together into one large storage container until they will be used, for example in omelets. Cook whole eggs thoroughly until the white is completely set and the yolk is beginning to thicken. Scrambled eggs need to be cooked until firm throughout with no visible liquid product remaining.
- Do not combine freshly prepared eggs with a batch of eggs previously held on the steam table.
- When preparing products containing eggs that will not be thoroughly cooked, such as hollandaise sauces, Caesar salad dressings, meringues, and mousses, use frozen or liquid pasteurized eggs, or commercially prepared mixes.
- Use frozen or liquid pasteurized eggs when serving high-risk populations that are especially susceptible to foodborne illness, such as the aged, the ill, pregnant women, infants, or other individuals with suppressed immune systems.
- When breaking eggs, do not allow shells to come into contact with the internal contents of the egg. Any procedures that grind or break whole shell eggs and then separate the shell from the liquid part of the egg by using a centrifuge or china cap are prohibited.
- Use clean and sanitized bowls, whisks, blenders, and other utensils for each new order. Do not reuse a container after it has had raw egg mixture in it. Clean and sanitize thoroughly before using again.

Shigellosis

Another foodborne illness of bacterial origin is shigellosis, sometimes called bacillary dysentery.

Symptoms. Shigellosis is an infection characterized by diarrhea, cramps, and chills, often accompanied by fever. The onset of the symptoms is between 1 to 7 days after ingestion of the bacteria.

Causative Agent. The bacteria responsible for shigellosis belong to the *Shigella* genus. The bacterium is facultative.

Source. Humans themselves are the prime reservoir for *Shigella.* People may carry this pathogen for periods of several weeks. Carriers excrete *Shigella* in their feces. When they subsequently fail to wash their hands properly, they transmit the bacteria to food. Flies are also thought to be responsible for the transmission of this bacteria.

Foods Involved. Implicated foods include raw produce and moist-prepared foods, such as potato, tuna, turkey, and macaroni salads that have been handled with bare hands during preparation. In addition, water has also been implicated.

Control Measures. Having a potable water supply, paying attention to washing hands after using the toilet, rapid cooling, and fly control are important weapons against shigellosis.

Listeriosis

Another foodborne infection is listeriosis.

Symptoms. Listeriosis is characterized by meningitis in immuno-compromised individuals. *Meningitis* is inflammation of the brain and spinal cord. Symptoms in otherwise healthy adults include nausea, vomiting, and headache. If a woman is pregnant, the illness can cause a miscarriage or result in the birth of a stillborn child. Listeriosis can cause severe retardation, meningitis, and death in newborn infants. The incubation period ranges from one day to three weeks.

Causative Agent. The bacterium responsible for listeriosis is *Listeria monocytogenes.* It is facultative. It also grows well in a damp environment.

Source. Infected wild and domestic mammals and fowl are carriers for these bacteria. The organism is frequently found free-living in soil, water, and plant matter, which animals might ingest. Carriers excrete the bacteria in their feces.

Foods Involved. Raw, soil-grown vegetables can be contaminated. Dairy products, especially unpasteurized milk and certain soft cheeses, raw meats, and poultry have been implicated in outbreaks of listeriosis.

Control Measures. *Listeria* can grow slowly at refrigeration temperatures between 32° and 34°F (0° and 1.1°C), so refrigerating implicated foods is not a complete protection against its growth. However, thorough cooking of food will kill the bacteria. Procedures to avoid cross-contamination should also be followed. Although research is still being conducted, the bacteria do not appear to survive the pasteurization process.

The foodservice manager should also keep facilities clean and dry, because *Listeria* can grow on wet floors, in drains, in ceiling condensates (liquid that forms and drips off of surfaces), and on sponges. Keeping sponges and cloths in sanitizing solutions and sanitizing floor drains will prevent its growth in these areas.

Foodborne Intoxications of Bacterial Origin

Staphylococcal Food Intoxication

Staphylococcal food intoxication is one of the most common types of foodborne illness reported in the United States. The foodservice operator who understands the characteristics of this disease and its causative agent can readily understand the control measures needed for its prevention.

Symptoms. The symptoms of staphylococcal food intoxication are nausea, vomiting, diarrhea, dehydration, prostration, and cramps. The signs appear suddenly, usually within 1 to 6 hours after the contaminated food is eaten, and some symptoms last for 24 to 48 hours. Most affected individuals recover from this foodborne illness without any complications. However, it can cause dehydration or other serious complications.

Causative Agent. The cause of staphylococcal food intoxication is a bacterium known as *Staphylococcus aureus* or staph, for short. (Exhibit 3.4). As they grow, these bacteria secrete their toxins into food, contaminating it. Even after the bacteria are killed or die, the toxins remain to cause illness. The toxin is not destroyed or inactivated by cooking. Staph are facultative bacteria. Staphylococci do not form spores.

Exhibit 3.4 *Staphylococcus aureus*

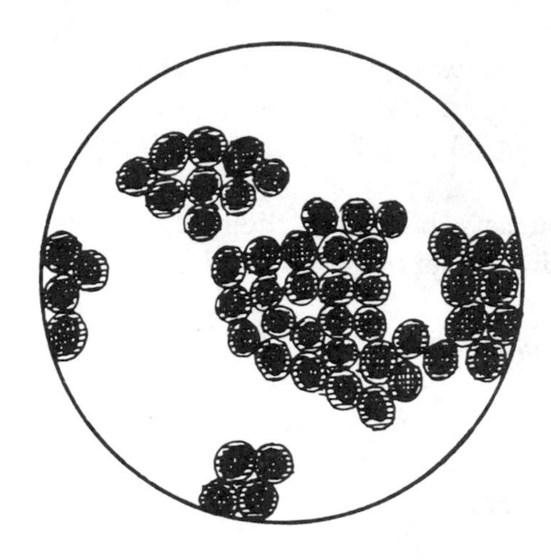

Source. Though this pathogen is sometimes found in meat and poultry, human beings are considered to be the most important reservoir of *Staphylococcus aureus* that reach foods. It is estimated that at any given time, 40 to 50 percent of all healthy human adults carry this bacteria. *Staphylococcus* is most commonly found in the nasal passages, in the throat, on the hands and skin, and in infected cuts, abrasions, burns, boils, and pimples.

Foods Involved. Staph bacteria do not compete well with other microorganisms, so illness is most commonly associated with cooked foods that are contaminated after cooking kills other micro-organisms. Food products frequently involved in staph intoxications include cooked meat products (especially ham and sliced meats), potato salad, custards, pastry fillings, cheeses, and other moist, high-protein and salty foods. Foods contaminated with staph bacteria or toxins have no particular odor or taste that might make the contamination obvious.

A Typical Case. Imagine that a chef has a skin infection on his neck. When it hurts or itches, he picks or scratches it.

One of his assignments is to prepare leftover ham for use in ham salad, an item usually held for some time before being served. The cooked ham is sliced by the chef, who plants staph bacteria from his neck everywhere he touches the ham with his hands. The sliced ham should be quickly chilled, which would stop or slow bacterial growth. Instead, it is put aside on a work table at room temperature.

Next, the ham is chopped, diced, and mixed into the salad base, which was prepared separately and also kept unrefrigerated. All the while—from slicing to mixing—the population of staph bacteria is producing harmful toxin. The completed ham salad mixture is once more set aside, unrefrigerated, to await customers' calls for ham salad sandwiches. (If you study Exhibit 3.5, you can see the effect of temperature on the growth patterns of most bacteria.)

This description is typical of cases actually reported to the U.S. Centers for Disease Control (see Exhibit 3.6).

Control Measures. Preventive measures begin with the initial storage and handling of the product. Staph bacteria will be killed under proper cooking conditions, but the toxins produced by the bacteria are heat resistant. For this reason, the foodservice manager cannot rely on proper cooking and reheating to prevent staph contamination. The cooling of the product should have specific procedures. (Proper preparation and handling procedures will be covered in Chapter 8.) The following control measures in the prevention of staphylococcal intoxication should be used:

- Cool and refrigerate cooked food rapidly to 45°F (7.2°C) or below and maintain the temperature.
- Throw out leftover custards, gravies, sauces, and other potentially hazardous foods. Other leftover foods should only be reheated once, and then thrown out.
- Protect cooked products from cross- or re-contamination.
- Exclude employees with respiratory infections, pimples, boils, and infected cuts and burns from foodhandling duties.
- Avoid hand contact with hair, face, nose, mouth, and other parts of the body.
- Use cleaned and sanitized utensils and disposable gloves to minimize hand contact with foods. (This is in addition to the guidelines for proper hand washing that will be discussed in Chapter 4.)
- Clean and sanitize equipment, utensils, and food-contact surfaces.
- Cover the product.

Bacillus cereus

Bacillus cereus foodborne illness is of growing concern to foodservice managers. The disease is generally classified

Exhibit 3.5 Temperature of food for the control of bacteria

°F		°C
250		121.1
	Canning temperatures for low-acid vegetables, meat, and poultry in pressure canner.	
240		115.5
	Canning temperatures for fruits, tomatoes, and pickles in water-bath canner.	
212		100.0
	Cooking temperatures destroy vegetative forms of most bacteria. Time required to kill bacteria decreases as temperature is increased.	
165		73.8
	Hot-holding temperatures prevent growth of bacteria.	
140		60
130	Some bacterial growth may occur. Many bacteria survive.	55
120		48.8
	Temperatures in this zone allow rapid growth of bacteria and production of toxins by some bacteria.	
60		15.5
	Some growth of food-poisoning bacteria may occur.	
45		7.2
40		4.4
32	Cold temperatures permit slow growth of some bacteria that cause spoilage and illness.	0
14		−10
	Freezing temperatures stop growth of bacteria, but allow many bacteria to survive.	
0		−17.7

Exhibit 3.6 Typical events that lead to outbreaks of staphylococcal intoxication

Staphylococcus Transmission

Foods Containing Protein

Cooked (Bacteria Usually Killed)

Contaminated by Foodhandler

Room Temperature Cooling 80°F (27°C)

Cooled in Large Masses

Time

Staphylococcal Intoxication

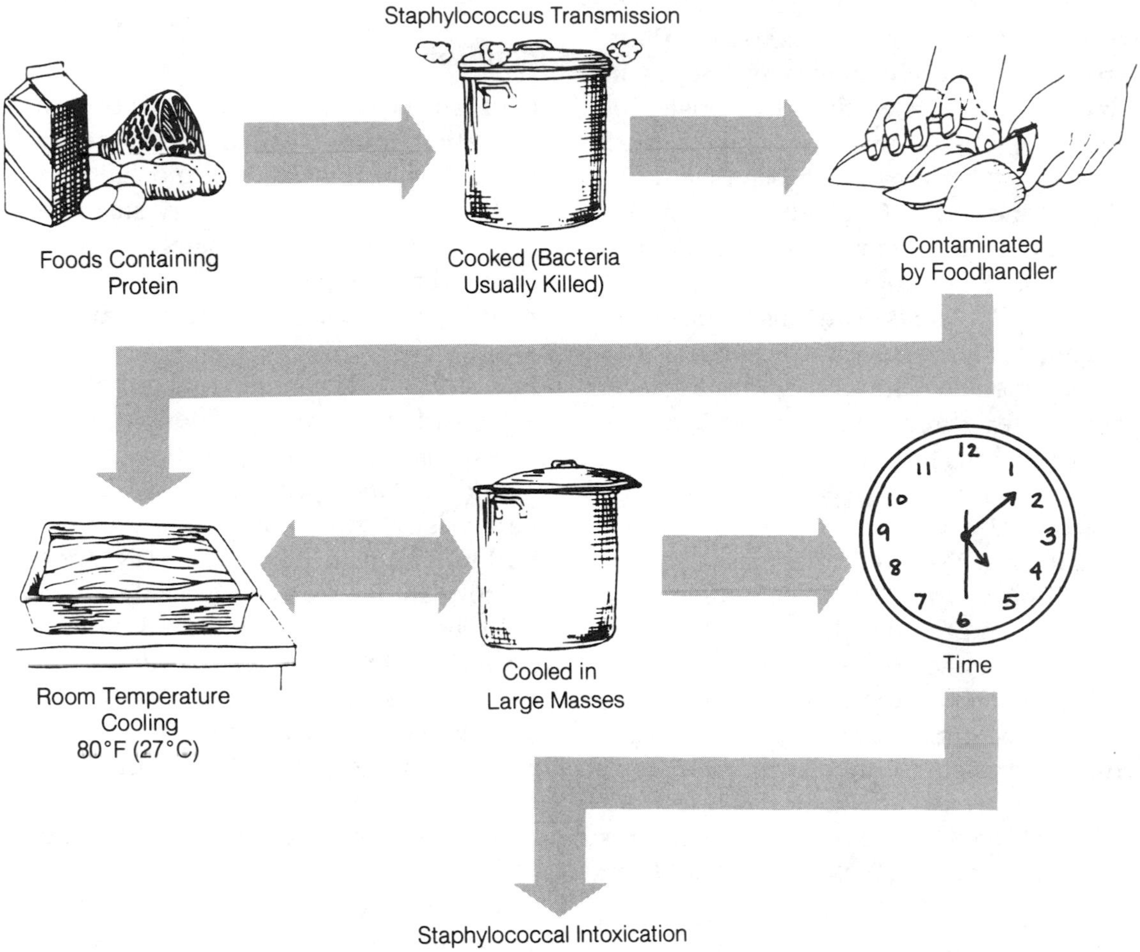

Source: Centers for Disease Control, Atlanta, Georgia

as an intoxication. One of the problems of diagnosis of the disease is there are actually two different toxins formed by *Bacillus cereus* with differing times of onset and symptoms. The bacteria can be found in a wide variety of foods.

Symptoms. The symptoms for one form of *B. cereus* toxin are diarrhea and abdominal pain with an onset time of 8 to 16 hours. The primary symptom for the other form of *B. cereus* toxin is vomiting. The onset time for this form is usually 30 minutes to 5 hours after ingestion of the contaminated food.

Causative Agent. *B. cereus* is facultative and form spores.

Source. *B. cereus* is found in soil.

Foods Involved. The bacteria are frequently found in grains, rice, flour, spices, and starch; they may also be in dry-mix products, such as those used for soups, gravies, puddings, and dried potatoes. They have also been found in meats and milk. Cooked rice that has been allowed to remain at room temperature has been implicated in many outbreaks. *B. cereus* can also grow on vegetable sprouts. Alfalfa sprouts and wheat sprouts can harbor the bacteria even after three washings and subsequent cooking.

Control Measures. As with other types of bacteria, time and temperature abuse cause rapid increase in the vegetative bacteria and the development of spores. Since these bacteria come into the establishment on the food, there is little the foodservice manager can do to keep them completely out of food, but they must be prevented from growing once they get in food. The following measures should prevent *Bacillus cereus* from growing in food:

- Do not hold foods at room temperature for any period of time.
- Chill foods to at least 45°F (7.2°C) within four hours after preparation.
- Store food in the refrigerator in shallow pans that are less than four inches deep rather than in large, deep pots, so the food will cool more rapidly.
- If the food you are storing is a thick substance, such as refried beans, you should store it in shallow pans in amounts less than three inches.
- Use food items as quickly as possible after preparation.

Botulism

Botulism is probably the foodborne intoxication most familiar to the general public because of its high mortality rate and the sensational publicity surrounding any outbreak. Fortunately, botulism outbreaks originating in commercial foodservice establishments are rare.

Symptoms. The symptoms of botulism are headache, vertigo, double vision, weakness, difficulty swallowing and speaking, and progressive respiratory paralysis. Symptoms usually appear within 12 to 36 hours but may appear up to a week after eating contaminated food. Unlike most foodborne illnesses of bacterial origin, botulism attacks the nervous system. Depending on the age and the condition of the individual and the promptness of treatment, the victim of botulism may face a long convalescence with respirator assistance or death.

Causative Agent. The bacterium that causes botulism is *Clostridium botulinum,* which is spore-forming. Because it is an anaerobic organism, it prefers environments like those found in a sealed container, like a can or pouch, or in a large container of thick food in which the freely available oxygen has been driven off through heating. Only small amounts of the toxin are needed *to cause illness and death.*

Source. *Clostridium botulinum* vegetative bacteria and spores are found in soil, water, and in the intestinal tracts of animals, including fish.

Foods Involved. Foods implicated in botulism outbreaks are improperly processed, usually home-canned, low-acid foods (for example, green beans, asparagus, peppers, corn, beets, spinach, and mushrooms); smoked, vacuum-packed fish; garlic products packed in oil; grilled onion; baked potatoes; turkey loaf; and stew. Modified atmosphere packaging and sous vide products offer a potential risk because they are vacuum packaged.

Control Measures. Preventive measures include adherence to the following five rules:

1. Never use home-canned foods at any time in a commercial foodservice establishment.
2. Never accept commercial canned goods if the cans are swollen or show signs of internal pressure or if the contents are foamy, foul-smelling, or give some other indication of being spoiled. Do not even *taste* suspect goods. Death can result from a single taste of food contaminated with *C. botulinum.*
3. Prepare garlic-in-oil and similar products fresh and in small batches that can be used quickly, or use commercially prepared garlic-in-oil products and keep the product refrigerated.
4. Prevent foods that have been heated from remaining in the temperature danger zone, 45° to 140°F (7.2° to 60°C).
5. Store vacuum-packaged/sous vide food below 45°F (7.2°C).

Clostridium Perfringens: A Foodborne Toxin-Mediated Infection

Clostridium perfringens illness is not easily classified because it has characteristics of both an intoxication and an infection. It is a toxin-mediated infection. Once the micro-organisms are ingested, the human intestine provides the necessary conditions for the micro-organisms to produce their toxins. *C. perfringens* is suspected of causing many of the mild (and unreported) cases of foodborne illness.

Symptoms. The symptoms of clostridium perfringens illness are usually milder than those of staphylococcal intoxication and usually clear up within 24 hours of their first appearance. This illness is marked by abdominal pain and diarrhea. These signs show up from 8 to 22 hours after the contaminated food has been eaten.

Causative Agent. The bacterium *Clostridium perfringens* (Exhibit 3.7) is the cause of the foodborne illness. These bacteria are spore-forming organisms that are anaerobic. The vegetative cells of *C. perfringens* on surfaces of food are normally killed by cooking, but heat-stable spores survive, and some of these spores are able to withstand boiling temperatures and can even resist freezing.

Source. *C. perfringens* are found in soil, dust, and human and animal feces. These pathogens are therefore likely to accom-

Exhibit 3.7 *Clostridium perfringens*

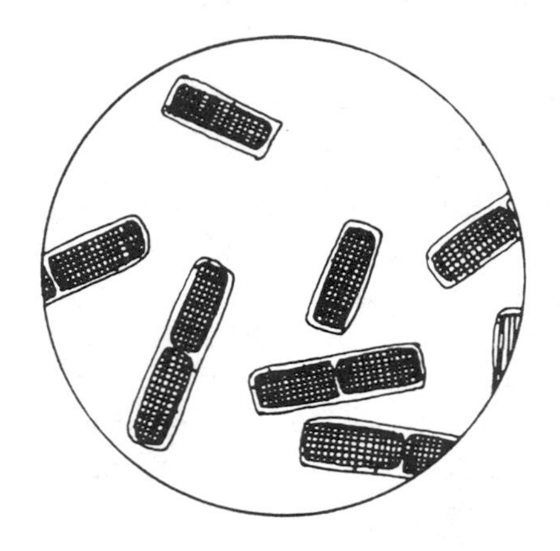

pany any raw food products brought into a foodservice establishment.

Foods Involved. Occurrences of clostridium perfringens illness are associated with cooked meats, poultry, gravy, and beans. Inadequate heating or cooling is a particular problem in the interiors of large food masses because of anaerobic conditions in deep pots.

A Typical Case. A cook is preparing a large pot of beef stew that will contain meat, onions, potatoes, carrots, herbs, and spices. The stew is cooked in a 20-gallon stockpot on Wednesday afternoon for serving at lunch on Thursday. After cooking, without any pre-cooling, the stockpot is placed directly in the refrigerator. The soup is removed from the refrigerator at 10:30 A.M. on Thursday and brought to the steam table on the serving line, where it is slowly heated to a palatable temperature. The carrots and other vegetables are grown in soil. *C. perfringens* vegetative bacteria and spores, which live in the soil, may collect on the vegetables. In addition, *C. perfringens* are also commonly found in the intestinal tracts of cattle and on beef carcasses. Normal cleaning methods do not remove all the bacteria and spores. The process of heating the stew may kill the vegetative cells of the *C. perfringens,* but it would not kill the spores.

Because it was refrigerated in a large container, the stew was allowed to remain in the temperature danger zone for several hours during cooling. Large volumes of stew will not cool quickly without using quick-chill methods; therefore, the spores of the bacteria had favorable conditions to become vegetative cells and had plenty of time to multiply. Later, the stew was brought directly from the refrigerator to the steam table. A steam table is designed to hold food that is already hot, so it cannot heat large amounts of food to temperatures high enough to kill the bacteria. Therefore, the *C. perfringens* was able to continue to multiply during the entire time that the stew was in the steam table.

Control Measures. Since *C. perfringens* are present naturally in so many of the raw foods available to the foodservice operator, the total prevention of their entry into the establishment cannot be expected. And since the spores of these bacteria are so heat-stable, the operator cannot expect to kill all of the spores.

Effective control of *C. perfringens* can be achieved through the following measures:

- Rapidly cool meat dishes that will be eaten later. (Cooling methods will be discussed in Chapter 8.)

- Serve meat and poultry dishes hot or as soon as they are cooked; avoid preparing food a day in advance.
- Reheat food to 165°F (73.9°C) or higher within two hours. Do not reheat foods on a steam table or on other hot-holding equipment.
- Hold cooked foods during display at 140°F (60°C) or higher.

Hepatitis A: A Viral Disease

Hepatitis is a contagious viral disease which causes inflammation of the liver.

Symptoms. The symptoms of hepatitis A are fever, nausea, abdominal pain, fatigue, and possibly jaundice (yellowing of the skin). The symptoms appear 15 to 50 days after infection. A mild case may last several weeks, whereas a severe infection may linger for several months.

Causative Agent. Hepatitis A is caused by a viral agent. Although there are several different types, hepatitis A is the form that most often affects the foodservice industry through poor personal hygiene practices.

Source of Virus. Hepatitis A is caused by a virus found in the feces and urine of infected persons. This disease is most often transmitted by individuals who have not washed their hands after using the restroom and then contaminate food or water served to others. Hepatitis A can also be transmitted through contaminated water.

Foods Involved. The foods most often implicated in outbreaks are prepared salads, items from salad bars, and raw or lightly cooked oysters and clams (from sewage-contaminated waters). Other foods implicated are those foods that will not receive cooking or reheating before service but are subject to handling by an infected individual who practices poor hygiene.

Control Measures. The most important factor in the control of outbreaks is making sure that foodhandlers practice good personal hygiene by washing their hands thoroughly and often. Poor personal hygiene is the leading cause for hepatitis A. Other controls include obtaining shellfish from safe, certified sources, and using safe water supplies.

Emerging Pathogens

An *emerging pathogen* is a disease-causing agent that has been increasingly identified as causing foodborne disease. Three emerging diseases of concern to the foodservice manager are campylobacteriosis, gastroenteritis caused by *Escherichia coli,* and Norwalk virus (see Exhibit 3.8).

Campylobacter Jejuni

Campylobacteriosis is a foodborne infection of bacterial origin.

Symptoms. Symptoms of this infection include fever, headache, and fatigue, which appear from 12 to 48 hours after ingestion. Abdominal pain, diarrhea, and sometimes vomiting follow. The disease ordinarily lasts from one to four days; in approximately 20 percent of

Exhibit 3.8 Emerging pathogens that cause foodborne illness

	Campylobacteriosis Infection	**E. Coli 0157: H7 Infection/Intoxication**	**Norwalk Virus Illness**
Pathogen	*Campylobacter jejuni*	*Escherichia coli*	Norwalk and Norwalk-like viral agent
Incubation period	3–5 days	12–72 hours	24–48 hours
Duration of illness	1–4 days	1–3 days	24–48 hours
Symptoms	Diarrhea, fever, nausea, abdominal pain, headache	Bloody diarrhea; severe abdominal pain, nausea, vomiting, diarrhea, and occasionally fever	Nausea, vomiting, diarrhea, abdominal pain, headache, and low-grade fever
Reservoir	Domestic and wild animals	Humans (intestinal tract); animals, particularly cattle	Humans (intestinal tract)
Foods implicated	Raw vegetables, unpasteurized milk and dairy products, poultry, pork, beef, and lamb	Raw and undercooked beef and other red meats, imported cheeses, unpasteurized milk, raw finfish, cream pies, mashed potatoes, and other prepared foods	Raw vegetables, prepared salads, raw shellfish, and water contaminated from human feces
Spore former	No	No	No
Prevention	Avoid cross-contamination, cook foods thoroughly	Cook beef and red meats thoroughly, avoid cross-contamination, use safe food and water supplies, avoid fecal contamination from foodhandlers by practicing good personal hygiene	Use safe food and water supplies, avoid fecal contamination from foodhandlers by practicing good personal hygiene, thoroughly cook foods

reported cases, symptoms recur for up to several weeks.

Causative Agent. Campylobacteriosis is caused by *Campylobacter jejuni,* a nonspore-forming, anaerobic pathogen.

Source. These bacteria are widely distributed in nature. Found in the intestines of sheep, pigs, cattle, and poultry, the bacteria are spread in the slaughtering and processing operations.

Foods Involved. The most common foods implicated are uncooked or insufficiently cooked meat and poultry products, unpasteurized dairy products, and foods that have been cross-contaminated.

Control Measures. Since *C. jejuni* are not hardy survivors out of the environment of their carrier, the bacteria are sensitive to freezing, drying, and acidic conditions, which can be used as barriers. Proper sanitation practices, including good personal hygiene, conscientious temperature control, and the avoidance of cross-contamination are also effective control measures.

Escherichia Coli

Escherichia coli 0157:H7 was first recognized as a foodborne pathogen in 1982. It is of great concern because of the severity of the illness it causes.

Symptoms. *E. coli* 0157:H7 causes symptoms similar to those of shigellosis, but it can cause either an infection or an intoxication. The symptoms for the intoxication are severe abdominal pain, cramps, nausea, vomiting, diarrhea, and occasionally fever. The symptoms for the infection are bloody diarrhea and colitis. The onset of the illness is between 12 and 72 hours, and it usually lasts from one to three days. In children, the illness can be more severe, possibly leading to kidney failure and blood poisoning.

Causative Agent. *E. coli,* a nonspore-forming, facultative bacterium, is responsible for this variety of gastroenteritis, which is an inflammation of the linings of the stomach and intestine.

Source. Human intestinal tracts are the primary source of this bacteria. The reservoir for *E. coli* 0157:H7 is the intestinal tract of cattle. *E. coli* can also be present in water contaminated from sewage.

Foods Involved. Raw or undercooked ground beef and red meats (lamb and pork) are the main vehicles of transmission. However, it has also been found in prepared foods like cream pies and mashed potatoes, untreated water, unpasteurized milk, imported cheeses, and finfish.

Control Measures. Cooking certain cuts of beef to at least 130°F (54.4°C) and holding it there for two hours or cooking throughly is the best control measure to use, along with an avoidance of cross-contamination. Reheat leftover red meat to 165°F (73.9°C) to kill the bacteria. Proper hand washing and personal hygiene are also good precautions against this illness.

Norwalk Virus

Norwalk is a viral illness caused by poor personal hygiene among infected food-handlers.

Symptoms. Norwalk and Norwalk-like viruses are characterized by nausea, vomiting, diarrhea, abdominal pain, headache, and low-grade fever. The incubation period ranges from 24 to 48 hours, with the symptoms then persisting for an additional 24 to 48 hours.

Causative Agent. This illness is caused by a viral agent named for the city in Ohio where it was first implicated. Because Norwalk is a virus, it does not reproduce in food and it remains viable for an extended amount of time until the food is eaten.

Source. The Norwalk virus is found in the feces of infected persons. Humans are the only known reservoir of the virus.

Foods Involved. Although Norwalk virus is mainly transmitted through contaminated water supplies and human contact, raw vegetables fertilized by manure, cole slaw and other salads, raw shellfish, eggs, and even icing on baked goods have been implicated. Manufactured ice cubes and frozen foods have been vehicles for transmission as well.

Control Measures. Since Norwalk virus can survive freezing temperatures and chlorine sanitizing solutions, it makes control difficult. Emphasis should be placed on good personal hygiene; a potable water supply for drinking and ice-making; reliable, certified seafood suppliers; avoiding cross-contamination; and thoroughly cooking foods. Norwalk virus is susceptible to high heat and cooking temperatures.

Parasitic Diseases

Trichinosis

Trichinosis is a foodborne illness caused by parasites. A *parasite* is a tiny organism that depends on a living host, in this case an animal, to provide certain requirements for growth and nourishment.

Symptoms. The initial symptoms of trichinosis, which appear between 2 to 28 days after eating the contaminated food, are vomiting, nausea, abdominal pain, and swelling of the tissues surrounding the eyes. Later, fever and muscular stiffness develop.

Causative Agent. Trichinosis involves a coiled roundworm called *Trichinella spiralis.* A person who eats undercooked pork or wild game meat that is infested with the larvae of this parasitic worm will become a host, then become ill when the larvae develops.

Source. The most common reservoir of this parasite in the United States is undercooked pork. Despite improvements in methods of raising pork for food in the past 20 years, the foodservice manager can still expect that some raw pork, including government-inspected pork, is infected. Trichinae also infest certain game animals, including bear and walrus.

Foods Involved. Undercooked or raw pork and game animals are most commonly implicated. Another concern is cross-contamination resulting from equipment like sausage grinders, which may mix raw pork into nonpork foods.

Control Measures. The only sure safeguard against trichinosis is the cooking of pork until it reaches an internal temperature of 150°F (65.6°C) or 170°F (76.7°C) if cooked in a microwave oven. Freezing will also kill the larvae found in pork products that are less than six inches thick if it is stored at 5°F (−15°C) for 30 days, −10°F (−23.3°C) for 20 days, or −30°F (−34.4°C) for 12 days. Since freezing does not kill other bacteria, pork should never be eaten raw or rare. Equipment, such as sausage grinders, and utensils should be washed, rinsed, and sanitized after working with raw pork.

Anisakiasis

Anisakiasis is another illness caused by parasitic roundworms.

Symptoms. The symptoms are abdominal pain and vomiting if the worms attach themselves in the stomach, and coughing if they attach in the throat. The symptoms caused when they attach in the intestines are identical to those produced by appendicitis—sharp abdominal pain and fever. In rare instances, they can require surgical removal.

Causative Agent. The *Anisakis* roundworm, which lives in the vital organs of some finfish, is the parasite that causes this illness.

Source. Bottom-feeding fish taken from contaminated waters inhabited by larger marine mammal populations are often the source since marine mammals are carriers of the adult worms. In addition, the failure to clean fish soon after catching them allows the worms to move from internal organs to muscle tissues.

Foods Involved. Raw fish items, including herring, sushi and sashimi, ceviche, salmon, and *lightly*-cooked filets, are foods most likely to be implicated. It is important to remember parasites found in raw fish can survive marinades that ordinarily would lower the pH or acidity of a food item to a level where bacteria cannot grow.

Control Measures. One method for making sure that fish products are free of parasites is cooking fish to a minimum of 140°F (60°C). Freezing at −31°F (−25°C), or lower, for 15 hours will also kill these parasites. Freezing at −10°F (−23°C) can be effective, but requires seven days in order to kill the parasites. The most reliable method for purchasing safe fish products intended for raw consumption is using a reliable supplier and obtaining a guarantee that the product has been properly frozen in the manner described above.

Fish Toxins

Fish and shellfish can be designated as poisonous for one of two reasons: either because toxins are produced in the fish or because certain circumstances encourage the fish to collect toxins from their food supply or marine environment. Some fish that contain toxins include the puffer, blowfish, moray eel, and freshwater minnow. Other fish can collect toxins from their diet during certain seasons or in certain waters. For example, the barracuda on one side of an island can be toxin-free, while those on the other can be carrying toxins, depending on their food source.

Two illnesses that are linked to seafood are *ciguatera* (from ciguatoxin), and *scombroid* intoxication (from histamine poisoning). Ciguatoxin collects in predatory, marine reef fish that have eaten smaller reef fish that have eaten toxic algae. Ciguatera causes vomiting, severe itching, nausea, dizziness, hot and cold flashes, temporary blindness, and sometimes hallucinations. Recovery often takes weeks or months. There is not yet a commercially feasible method to determine if ciguatoxin is present in a particular fish. The toxin is not destroyed by cooking.

Scombroid poisoning is most often found in tuna, bluefish, mackerel, bonito, and mahi-mahi (dolphin fish), which have been allowed to decompose through time-temperature abuse. While histamine is produced during this decomposition process, decomposition might not be visible when it occurs. Scombroid poisoning's symptoms include flushing and sweating, a burning, peppery taste in the mouth, dizziness, nausea, and headache. Sometimes these symptoms are followed by a facial rash, hives, edema, diarrhea, and abdominal cramps. Histamine is odorless and tasteless and is not destroyed by cooking. In addition, poor temperature control, especially improper thawing conditions can allow bacteria present on the fish to begin the decomposition process with production of histamine.

The following control measures should be practiced to guard against seafood-specific illness:

- Make sure that fresh, wholesome fish is acquired from a reputable seafood supplier.
- Refuse fish that has been thawed and refrozen. Signs that fish have been refrozen include dried or dehydrated appearance; excessive frost or ice in the package; or white, cottony blotches (freezer burn).
- Check temperatures. Fresh fish must be between 32° and 34°F (0° and 11.1°C) upon arrival.
- Thaw frozen fish quickly at refrigeration temperatures below 45°F (7.2°C).

(Chapter 6 has a complete discussion on how to purchase safe fish.)

Plant Toxins

Certain plants, which are themselves poisonous, can be classed as biological hazards. These plants have on occasion been used accidentally as food. Cooking these poisonous plants does not usually destroy the toxins that produce illness.

Plants that have been implicated in outbreaks of foodborne illness include fava beans, rhubarb leaves, jimson weed, and water hemlock. Honey from bees that have gathered nectar from the mountain laurel, milk from cows that have eaten snakeroot, and jelly made from apricot kernels have also poisoned human beings.

Of the hundreds of poisonous plants, one merits special mention—the mushroom. Poisonous and nonpoisonous mushrooms often look so much alike that the untrained eye cannot tell the difference. To avoid the problem, avoid using any mushroom not secured from an approved and reliable source. *There is no sure method, including cooking with a silver coin, that the amateur can use to detect poisonous varieties of mushrooms.*

THE CHEMICAL HAZARD

Public concern over chemicals in our food supply has grown in response to a well-publicized controversy concerning the use of pesticides on food. A *chemical hazard* is the danger posed by chemical substances contaminating food all along the food supply chain. Four kinds of chemical hazards are of special concern to the foodservice manager besides pesticides: (1) contamination of food with foodservice chemicals, such as detergents and sanitizers; (2) use of excessive quantities of additives, preservatives, and spices; (3) acidic action of foods with metal-lined containers; and (4) contamination of food with toxic metals.

Pesticides

Of the chemical agents causing foodborne illness, those receiving the most attention have historically been the pesticides. However, the danger of pesticide contamination has been minimized by the increased controls and legislative requirements of agricultural pesticides. Most pesticide illnesses occur because of massive exposure of agricultural workers to the compounds in their concentrated form.

The FDA asserts that the danger of pesticides has been exaggerated. It reports that "less than 1 percent of fresh foods recently tested for pesticide residue had residues that exceeded federal tolerance levels." Thus the foodservice operator should put the greatest emphasis on preventing accidental contamination in the operation.

The best control measures are to purchase food only from reputable, legally approved sources, and to wash all fresh fruits and vegetables regardless of their source. The foodservice operator can *and must* prevent contamination of food by chemical agents within the foodservice establishment. Pesticides should only be used in the facility by a trained professional, such as a pest control operator (PCO). If they are stored in the operation, they must be labeled properly, kept in their original containers, and stored separately from food, food-contact surfaces and other chemicals. Pesticides have caused poisoning after being accidentally mixed into flour, sprayed onto oatmeal, and mistaken for baking soda.

Foodservice Chemicals

Some detergents, polishes, caustics, cleaning and drying agents, and other products of this type commonly found in food services, are poisonous to humans and should never come into contact with food. Great care must be exercised in following label directions for the use and storage of these chemicals, under safe conditions and away from food. Poisonings have occurred when drain cleaner, paint remover, oven cleaner, and silver polish accidentally contaminated food. (For rules concerning storage of chemicals refer to Chapter 10.)

Additives and Preservatives

The threat of illness from additives is a subject of debate among scientists and legislators. Although the effects of long-term use of additives are not yet known, it cannot be denied that excessive amounts of certain food preservatives and additives have caused illness.

Another foodborne illness in question results from the use of too much monosodium glutamate (MSG), a food additive that serves as a flavor enhancer. Symptoms in susceptible consumers are flushing of the face, dizziness, headache, dry and burning throat, and nausea. The reaction to MSG does not respond to treatment with antihistamines. MSG is used in many prepackaged goods. Foodservice managers should avoid adding it to their recipes, except in following recipe directions as to exact amounts.

Agents used to preserve the flavor, safety, and consistency of foods have been linked to food contamination.

Sulfiting agents are preservatives that are used to maintain the freshness and color of vegetables, fruits, frozen potatoes, and certain wines. Excessive use of them has been linked to a number of lethal allergic reactions among sensitive individuals, particularly asthmatics. The reactions include nausea, diarrhea, asthma attacks, and in some cases, loss of consciousness.

Foodservice operators should read labels on all processed foods to determine whether sulfiting agents are present. These agents include sulfur dioxide, sodium and potassium bisulfite, sodium and potassium metabisulfite, or sodium sulfite. Sulfites are allowed to be present in some processed foods, such as maraschino cherries and raisins, because there is no acceptable substitute. For this reason, it is essential to be truthful in menu descriptions. Do not claim that there are *no* sulfites in the foods you serve unless you are absolutely sure none are present; inform consumers when specific menu items contain processed ingredients.

Never add sulfites to foods. The FDA prohibits the addition of sulfites to foods in a foodservice operation. In addition, liability may exist when certain food items are combined. Keep items that can legally contain sulfites away from those that cannot, especially on salad or dessert bars. Purchase sulfite-free ingredients for service with raw fresh fruits and vegetables.

Nitrites are preservatives used by the meat industry as a curing agent to prevent growth of certain harmful bacteria, and as a flavor enhancer. Scientists have established a link between cancer and nitrites. The risk may increase by the over-browning or burning of foods treated with nitrites. The meat industry is aware of the scientific findings that link cancer to the nitrosamines that are formed when meat is over-browned or burned, and has responded by decreasing the levels of nitrites in meats. Some manufacturers have even added vitamin C and vitamin E to nitrite-treated foods to prevent the formation of tumor-causing nitrosamines. The only concern for the foodservice manager is that, unless the consumer insists, foods should never be burned or over-browned.

To control the possibility of illness resulting from the overzealous use of food additives and preservatives, the foodservice manager should see to it that these products are never used to cover up spoilage in foods, that only approved additives and preservatives are used, and that manufacturer's instructions are carefully followed.

Food irradiation is classified as a food additive and is regulated by the FDA. This means that the FDA must approve each food item proposed for irradiation. The process of *irradiation* in food processing eliminates micro-organisms from food materials and lengthens storage life. Some spices are irradiated to control bacteria and mold.

Foodservice managers should be familiar with chemicals used in their operations. The federal government regularly publishes a list of chemicals that are Generally Recognized As Safe (GRAS), and those that are hazardous. Local health departments may provide this information, too. Chemicals are being used to improve and maintain the quality and safety of foods. However, knowing which chemicals are potential hazards is a safety step a smart manager can take to avoid chemical contamination.

Metal-Lined Containers

When certain metals come into contact with acidic foods, chemical contamination occurs. Poisoning may result when high-acid foods are stored or prepared in copper or brass containers; in galvanized (zinc-coated) containers, such as garbage cans; or in containers of gray enamelware, which may be plated with antimony or cadmium. Imported enamelware coated with lead glaze and tin milk cans used to store fruit juices have been linked to contamination. Enamelware should not be used in foodservice operations since it may chip. Chipping exposes the underlying metal, which can cause problems with acidic food items.

Poisoning can also result when high-acid foods are prepared with improper metal utensils or when paint-type brushes are used to baste foods. Foods implicated in metal poisonings are sauerkraut, tomatoes, fruit gelatins, lemonade, and fruit punches. To prevent such accidental poisonings, use containers only for the purposes for which they are intended and purchase appropriate foodservice brushes.

Toxic Metals

Some metals, like iron, are necessary components of the human diet in at least trace quantities. As is true with almost any substance, however, they become toxic in excessive amounts. For example, copper water lines accidentally exposed to carbonated beverages in dispensing machines have caused chemical poisoning incidents.

Copper poisoning is characterized by the rapid and violent onset of symptoms, particularly vomiting. Foodservice managers should check or have a professional vendor check soft drink post-mix systems to be sure that carbonated water does not backflow into the copper water lines and dissolve the copper. The copper that is dissolved is then drawn into the system and can be dispensed in the beverage.

Uncovered meats can become poisoned through contact with refrigerator shelves containing cadmium. Metal refrigerator racks should never be used as make-shift grills in the kitchen for this same reason. Also, no lead or lead-based product should *ever* be used in foodservice areas.

Management must make sure that hazardous metals are not used in foodservice preparation, storage, and service equipment.

THE PHYSICAL HAZARD

A *physical hazard* is the danger posed by the presence of particles or items that are not supposed to be a part of a food product. Physical contaminants, such as chips of glass from broken light bulbs or glasses, and metal fragments from enamelware dishes and tableware, are obvious dangers. A worn can opener, for example, can shower metal curls on the food in the can being opened. Other metal objects, such as magnets, packing staples, tacks, and pins, can accidentally fall into food. The accidental swallowing of unfrilled and frilled toothpicks has occurred. The practice of scooping up ice with a glass is definitely hazardous, since glass chips can become a part of beverage servings. Good facilities planning and the training of personnel in safe operating procedures can reduce these physical hazards. (Reducing physical hazards in facilities will be discussed in Chapter 13.)

Dangers caused by physical contaminants may also result from tampering incidents, particularly with soft-packaged food items. Items arriving at the foodservice facility should not be accepted if there is evidence of tampering. In addition, the operation should use portion control items to discourage tampering.

If items such as glass, nails, or other objects are found in foods and appear to have been intentionally placed there, the manager should alert the supplier. It may just be an isolated incident. However, this tampering obliges the manager to do whatever is needed to resolve the problem, including contacting the public health department and the police. The foodservice facility's own storage procedures should be checked to ensure that they are not part of the problem.

SUMMARY

Foodborne illness poses a potentially serious threat to public health. The foodservice manager plays an important role in preventing biological, chemical, and physical contamination of food.

The most serious risk associated with food is the biological hazard. Pathogenic bacteria, viruses, and parasites are responsible for the major foodborne illnesses. *Salmonella, Shigella,* and *Listeria* are bacteria that cause foodborne infections. Foodborne intoxications are caused by *Staphylococcus aureus, Bacillus cereus,* and *Clostridium botulinum.* Clostridium perfringens is a toxin-mediated infection. Emerging pathogens of concern are *Campylobacter jejuni, E. coli* 0157:H7, and Norwalk virus. Another virus with serious consequences is hepatitis A.

Trichinella spiralis, found in pork and meat from wild game, and *Anisakis,* found in finfish, are parasitic roundworms that cause disease.

The basic control measures to be taken against biological hazards are as follows: use food only from approved sources; store, cook and cool foods properly to prevent bacterial growth; prevent cross-contamination; and follow good personal hygiene and hand-washing practices. It should always be remembered that *people*—in one way or another—are the primary cause of food contamination.

Chemical hazards arise from the improper use of foodservice chemicals, additives, preservatives, and toxic metals. Chemicals and metal products should be used only for their intended purposes. The foodservice manager must also guard against physical hazards, like glass chips, metal fragments, and toothpicks. Foodservice managers should properly inspect incoming food items for physical contaminants or signs of tampering.

A CASE IN POINT

On February 2, 1975, the passengers and one crew member aboard a chartered aircraft flying from Tokyo to Copenhagen, with an interim stop in Anchorage, developed an illness characterized by diarrhea, vomiting, abdominal cramps, and nausea.

Breakfast was served $5^1/_2$ hours after the plane left Anchorage, and the 196 passengers began getting ill approximately 2 hours after this meal was served. Breakfast consisted of an omelette, ham slices, yogurt, bread, butter, and cheese.

Except for the one crew member who had eaten the ham, none of the plane's crew became ill. They had been served a steak dinner instead of breakfast since it was suppertime for them.

Breakfast was prepared in Anchorage by a catering company. Three cooks were involved in the preparation of the ham and omelettes. The ham had been prepared the night before and refrigerated. The next day, as the omelettes were being prepared, one of the cooks placed the ham slices on top. The food was stored at room temperature during the 6 hours required for the preparation of all of the omelettes.

Following preparation, this food was placed for $14^1/_2$ hours in a holding room where the temperature was measured at 50°F (10°C). Beginning about 7:30 A.M., the breakfast food was loaded onto the plane. The food was again stored at room temperature in the galley ovens until it was heated just prior to serving.

This large foodborne outbreak resulted from ham that had been handled by a cook who had an inflamed cut on his finger from which *Staphylococcus aureus* was later cultured.

What else went wrong that contributed to the outbreak?

STUDY QUESTIONS

1. Why is it important to develop multiple barriers to bacterial growth?
2. Give three examples each of biological, chemical, and physical hazards connected with contamination and foodborne illness.
3. What is the difference between a foodborne *infection* and *intoxication?*
4. What is cross-contamination? Give two examples.
5. Define the terms *reservoir, carrier,* and *host.*
6. What are some factors in controlling *Salmonella* food contamination?
7. Describe how *Staphylococcus aureus* contamination might occur.
8. What foods are usually implicated in the transmission of listeriosis?
9. What is the reservoir for *Shigella* and *Clostridium perfringens* bacteria?
10. What rules should always be followed by the foodservice manager to prevent occurrences of botulism?
11. What are emerging pathogens?
12. Name two foodborne viruses and tell how they can be prevented.
13. What are the two most common illnesses known to be transmitted by toxins in marine fish?
14. What are the reservoirs for *Trichinella spiralis* and *Anisakis?* How can each of these parasites be killed?
15. How can metals cause poisoning in a foodservice operation?

ANSWER TO A CASE IN POINT

The ham was held at room temperature for more than a sufficient amount of time to allow the growth of the staphylococcal bacteria and the production of toxins. A total of 20½ hours at temperature ranges dangerous to food, between 50° and 70°F (10° and 21.1°C), made this ham the source of one of the largest foodborne illness outbreaks ever recorded.

Staphylococcal toxin is not easily destroyed by ordinary cooking temperatures, so the best control measure is the prevention of entry of the bacteria into the food in the first place. The risk of storing food at room temperatures for long periods of time and of allowing infected foodhandlers to prepare food is once more emphasized.

MORE ON THE SUBJECT

NATIONAL RESTAURANT ASSOCIATION. *Foodborne Illnesses.* Washington, DC: National Restaurant Association. rev. 1990, #PB-390. A booklet that folds out into a wall chart and details common foodborne diseases and their causes.

AMERICAN EGG BOARD. *A Scientist Speaks About the Microbiology of Eggs.* rev. July, 1990. #E-0010; and *A Scientist Speaks About Eggs and Salmonellae.* rev. July, 1989. #E-0007. Park Ridge, IL: The American Egg Board. These two pamphlets give detailed descriptions of eggs and their biology, as well as how to avoid biological contamination in eggs and egg products.

CENTERS FOR DISEASE CONTROL. *Foodborne Disease Surveillance.* Washington, DC: CDC, U.S. Department of Health and Human Services, Public Health Service. These summaries, published annually, are the authoritative surveys of foodborne illness and infection in the United States. They contain statistical information on incidence of disease, cite places in which food was mishandled in such a way as to cause a disease outbreak and identify the kinds of food implicated in the outbreaks. Details on every reported outbreak and analysis of all statistical information are also provided. To receive this information, write to the Massachusetts Medical Society, C.S.P.O. Box 9120, Waltham, MA 02254.

RICHARDSON, TREVA M. and WADE R. NICODEMUS. *Sanitation for Foodservice Workers.* New York: CBI/Van Nostrand Reinhold, 1982. 275 pages. This book concentrates on the major foodborne diseases and their bacteriology. It also provides information on sanitary dishwashing, effective pest control, and the dimensions of liability.

BENENSON, ABRAM S. *Control of Communicable Diseases in Man.* 15th ed. Washington, DC: American Public Health Association, 1990. 485 pages. This is an easy-to-use, paperback encyclopedia that lists foodborne and waterborne diseases in alphabetical order and by an index. Each entry includes identification, pathenogenic agent, reservoir, how it is transmitted, and how it can be controlled.

MARRIOTT, NORMAN G. *Principles of Food Sanitation.* 2nd. ed. New York: AVI, 1989. 387 pages. Chapter 2 of this book, "The Relationship of Micro-organisms to Sanitation," gives descriptions of foodborne illnesses and emerging pathogens of concern. Chapter 3 covers how foods become contaminated. The book also has charts and a glossary.

4

The Safe Foodhandler

Good personal hygiene is a critical protective measure against foodborne illness. In addition, customers frequently judge a foodservice operation by observing the personnel serving them. By establishing a systematic approach to hiring, training, and supervising employees, the responsible manager will help protect the safety of the food served to customers, as well as enhance the quality of the dining experience.

It is ironic that people are the cause *and* the victims in foodborne illness incidents. At every step in the flow of foods through the operation from receiving through final service, employees can contaminate food. When they contact their environment with their hands, perspiration, or breath, they spread bacteria and other micro-organisms. Every unguarded cough or sneeze transmits a wave of invisible micro-organisms capable of causing disease. Human excrement is also a significant factor in the spread of pathogens that can potentially contaminate food. As we have seen in earlier chapters, poor personal hygiene is one of the most frequent factors that contributes to foodborne outbreaks.

The foodservice manager who wants to provide safe and wholesome food is confronted with a paradox; somehow he or she must build a sanitary barrier between the product and the people who prepare, serve, and consume it. To do this requires a trained staff of foodhandlers who possess the knowledge, skills, and attitudes necessary to operate the safe food system. The manager must:

- Determine the specific sanitation and food safety requirements for every job description.

- Hire employees who meet the job requirements.
- Orient new and present personnel in personal hygiene and safe foodhandling practices.
- Establish employee commitment to the safe food system.
- Conduct continuous supervision of sanitary practices.
- Enforce all rules equally.
- Revise all employee and sanitation rules as the job needs, the laws, and the science of food safety changes.

THE DANGER IN AND AROUND US

An apparently healthy individual may harbor sizable numbers of micro-organisms. *Staphylococci* are found on the hair and skin and in a person's mouth, throat, and nose. The lower intestinal tract is a common habitat for *Shigella* and Clostridium perfringens. According to one estimate, up to 50 percent of healthy foodhandlers are carriers of disease agents transmitted by food.

With some diseases, such as hepatitis A, an individual is at the most infectious stage before the symptoms appear. For other diseases, once the individual becomes visibly ill, the risk of contamination is dramatically increased. A sore throat, a nagging cough, sinus pains, and other symptoms of the common cold are signs of infection with potentially dangerous consequences for the foodservice operation. The same can be said of symptoms of gastro-intestinal disorders, such as diarrhea or an upset stomach.

Even when the illness passes, some of the organisms that caused it may remain in an individual, making that person a carrier and a potential source of re-contamination. *Salmonella* may remain in a person's system for months after recovery, and hepatitis A has been found in the intestinal tract up to five years after the disappearance of disease symptoms.

Respiratory-tract infections are especially difficult to control because they can be spread so easily to large groups. An uncontrolled sneeze expels numerous droplets, each of which may contain bacteria or viruses.

The temperature of a person's skin is ideal for bacterial growth, and skin secretions provide nutrients for this growth. *Staphylococci* abound in and around boils, pimples, carbuncles, inflamed cuts, burns, and infected eyes and ears. Because of these factors, foodhandlers' hands are the most potentially dangerous serving equipment in the operation. Simple acts that in another setting would at worst be considered rude behavior—picking the nose, rubbing the ear, scratching the scalp, touching a pimple or open sore, or running fingers through the hair—are potential transportation for micro-organisms to foods.

BUILDING AN EFFECTIVE PERSONAL HYGIENE SYSTEM

Establishing good personal hygiene practices requires you to:

- Set up and enforce standards, policies, and procedures for personal cleanliness among employees and management.

- Provide facilities and equipment that encourage personal cleanliness and sanitary practices.
- Supervise practices to ensure only healthy employees are allowed to work with food.

Hiring the New Employee

The foodservice manager is obligated to protect customers and employees from individuals who have health problems or personal habits that can affect food safety. Professional ethics, good business sense, and public health laws all require protective behavior. In most communities, the health code prohibits persons who have communicable diseases that can be transmitted by food or who are carriers of such diseases from engaging in work activity that may result in the contamination of food or food-contact surfaces. However, it is important to note that the healthy employee with poor personal habits is also very likely to cause food contamination.

Since a job interview is your first in-depth contact with a prospective employee, you should be prepared with the following:

1. A complete job description. By having the specific duties for a job outlined, it will be easier to find a match with a qualified applicant.
2. A well-designed application form. In addition to routine questions on work experience, training, and other relevant information, the application should also provide information you need to check references and former employers. It is important to follow through with the references and especially contacting the former employer to verify the information the job applicant has provided on the application.
3. Questions relevant to the job requirements. For example, you might ask an experienced server, "What was the most difficult situation you encountered with a customer and how did you handle it?"
4. Enough time to discuss the job description, the applicant's qualifications, and answer questions. Make sure there are no interruptions during the interview. You may forget important information if you answer a phone or go to check on a shipment.

The truly safe foodhandler is a product of continuing observation on the part of both trained employees and management. Every workday the manager must conscientiously observe workers and be on the lookout for disease symptoms and unsafe personal habits. Above all, it is not only the responsibility of management to stress the regular and constant practice of good personal hygiene, but to be an example for the employees to follow.

Personal Hygiene Standards and Policies

After hiring an applicant, an explanation of personal hygiene policies should be a part of the new employee's orientation. The *employee rules* that are established should be easily understood and uniformly enforced. Employees should learn the rules either from the manager personally or from a highly dependable supervisor. Policies should be written in simple language, verbally explained to all staff, and made available for all employees to read in employee manuals. Post signs in employee restrooms, over handwashing stations, and on bulletin boards to reinforce the rules.

The employee rules and their wording may vary, depending on the size and complexity of the foodservice operation, but they should cover these areas: personal cleanliness; proper working uniform; and prohibited actions and habits. After an employee reads and understands these rules, he or she should agree *in writing* to follow these standards of conduct.

Personal Cleanliness

Emphasis on personal hygiene is a fundamental responsibility of the foodservice manager. Although it can be an awkward and embarrassing task for the manager to talk to mature people about personal cleanliness, people with poor hygienic practices and appearances are a poor advertisement for the establishment and will offend customers. Rules for personal cleanliness follow:

Hand Washing. The most critical aspect of personal cleanliness is frequent and thorough hand washing. Body odor offends, and lack of bathing accelerates bacterial growth, but most often it is dirty hands and fingernails that create a hazardous condition by contaminating the food product. Hand washing *must* follow any act that offers a possibility that the hands have picked up contaminants. One of the most notorious outbreaks of foodborne illness on record involved a single foodhandler who scratched a facial infection and then handled a large amount of sliced meats.

The following activities should always be followed by thorough hand washing:

- Using the restroom
- Using a handkerchief or tissue
- Handling raw food—particularly meat and poultry
- Touching areas of the body, such as ears, mouth, nose, or hair, or scratching anywhere on the body
- Touching infected or otherwise unsanitary areas of the body
- Touching unclean equipment and work surfaces, soiled clothing, or wash rags
- Smoking or using chewing tobacco
- Clearing away and scraping used dishes and utensils; performing scullery operations
- Eating food or drinking beverages.

Every one of these everyday actions and numerous others contaminate the foodhandlers' hands.

It may seem elementary, but clean hands are so critically important that all employees must be instructed in proper hand-washing procedures. A handwashing reminder can be added at certain points in preparation and operations manuals. Proper hand washing is more complicated than just running water and soap over the hands. All employees need to use the following techniques when washing their hands (see Exhibit 4.1):

Exhibit 4.1 Proper handwashing techniques

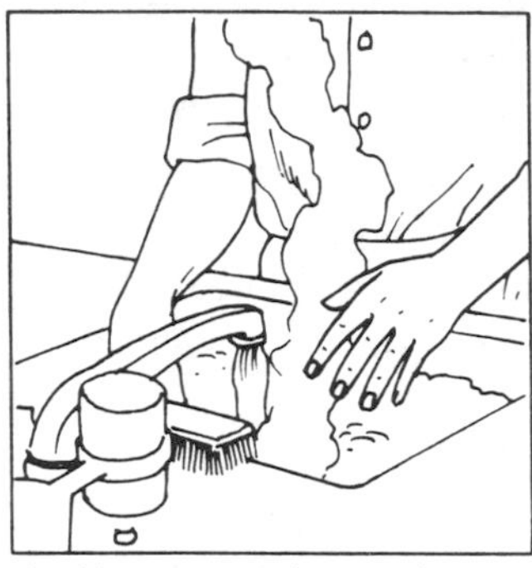

1. Use water as hot as the hands can comfortably stand.

2. Moisten hands, soap thoroughly, and lather to elbow.

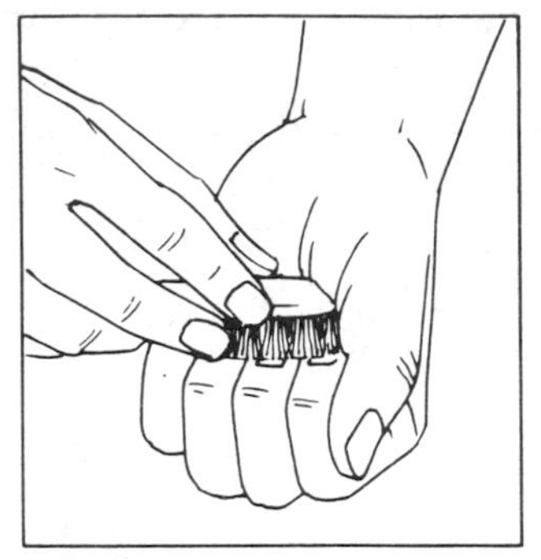

3. Scrub thoroughly, using brush for nails.

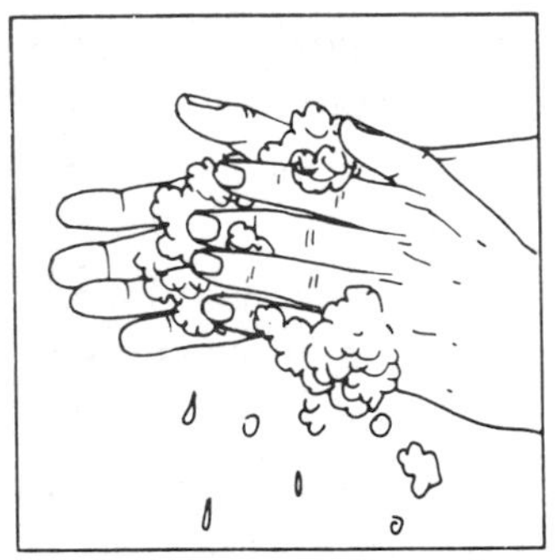

4. Rub hands together, using friction for 20 seconds.

5. Rinse thoroughly under running water.

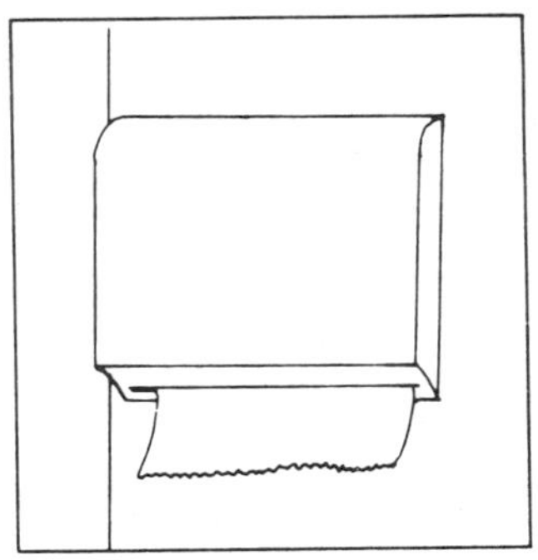

6. Dry hands, using single-service towels or hot-air dryer.

1. Turn the water on and let it run to a temperature as hot as the hands can comfortably stand.
2. Moisten the hands under the water and apply soap to them (antibacterial soap is recommended, and dispenser soap is preferred over bar soap), lathering well beyond the wrists and up the arms to the elbows, if short-sleeves are worn, to remove soil and dirt.
3. Pay particular attention to the areas between the fingers and around the nails. Use a brush for cleaning under the nails. Rinse the brush clean and store it in a sanitizing solution between and after every use.
4. Rub one hand against the other in a rotating motion using friction for 20 seconds.
5. Rinse thoroughly under the running water, allowing the water to flow from the elbows down to the fingertips. This action will rinse away contaminants. Turn water faucet off with a sanitary, single-service towel, or your elbow.
6. Dry hands thoroughly with a hot air dryer or with a new sanitary, single-service towel.

7. Do not touch anything that recontaminates the hands before returning to work. Use a sanitary paper towel to open the door to the restroom. Just because you carefully wash your hands does not mean the person leaving the restroom before you did. Drying hands on aprons, or using a handkerchief will undo the process.
8. Repeat this procedure as often as necessary to keep hands clean at all times.

It is important that even though you have followed all proper handwashing guidelines, you should avoid *as much as possible* direct hand contact with ready-to-eat foods and foods that have been cooked. Do not put hand lotion on freshly washed hands, because lotion provides moisture that encourages bacteria on the skin to multiply.

Hand Sanitizers. Recently, the use of *hand sanitizers,* a liquid used to lower the number of micro-organisms on the surface of the skin, has become more common in the foodservice industry. Hand sanitizers should not be used as a substitute for hand washing. Instead, hand sanitizers must be used along with hand washing. Hands must be washed before the use of any sanitizer. In addition, hand sanitizers must be stored and dispensed under supplier-recommended conditions that must not result in the contamination of food, utensils, equipment, or food-contact surfaces.

Gloves. Proper use of various types of plastic gloves can provide an additional sanitary barrier between the foodhandler and the food in some settings. Frequently, gloves are regarded by foodhandlers as more sanitary than bare hands. Because of this false sense of security, the foodhandler might not change the gloves after handling an item or product that might be contaminated, resulting in the cross-contamination of other food products. For example, after cutting raw chicken, the foodhandler must not handle a cooked product with the same gloves. It is important for the foodhandler to realize that gloves are just as susceptible to contamination as bare hands and must be viewed as an extension of the hands. Gloves should not be used to avoid hand washing. After any action with gloves that would require hand washing, foodhandlers must throw away the gloves, wash their hands, and then put on new gloves. In addition, plastic or rubber gloves are not appropriate for use near heat.

Fingernails. Fingernails should be trimmed and clean. Ragged nails harbor bacteria and are very difficult to keep sanitary. Long fingernails, false fingernails, and acrylic nails can break off into food and have the same disadvantages as ragged nails because food can get caught in them. Fingernails should not be polished because polish can chip and contaminate a food product. Hangnails should be clipped, treated, and covered with a clean bandage.

In maintaining a staff policy on hand care, as with other hygienic routines, it is not enough to merely set up clear rules. A manager needs to explain all rules and set an example that will help enforce the rules. Rather than warning employees to keep their fingernails trimmed (or else), the manager should tell them: "Fingernails that are long or ragged are difficult to keep clean, and they carry a tremendous number of bacteria."

Cuts and Abrasions. Unprotected wounds are sources of disease-causing bacteria and expose the employee to infection. Wounds and open sores in general must be antiseptically bandaged. Bandages on the hand must be covered with waterproof, disposable, plastic gloves that will protect the food as well as the foodhandler. In addition, care must be exercised so the bandages do not fall off in food when gloves are removed. In some cases until the injury heals, the employee may need to move to another job station, where food or clean utensils are not handled.

Hair Washing. Keep hair clean. Oily, dirty hair can carry and hold huge numbers of disease-causing bacteria. Since hair and dandruff can fall into food, good grooming habits and hair restraints are necessary to reduce the potential for contaminating food and food-contact surfaces.

Bathing. The manager should require that employees bathe daily or more often depending on the type of job they do. The written rules should state the acceptable standards in this regard and the rules should be monitored consistently.

Proper Working Uniform

The clothing of foodservice employees plays an important role in the prevention of food contamination, making it essential that employees observe strict standards of personal attire. Employee rules should cover the items of work clothing, hair-coverings, and jewelry.

Work Clothes. Soiled work clothes are unacceptable for two reasons. First, dirty clothing carries disease-causing microorganisms. One contact with soiled clothing is enough to start the contamination cycle: from the clothing, to the hands, to the food. The foodservice manager must insist on clean uniforms and encourage changes as often as they are necessary. Second, dirty clothes give a bad impression to patrons, who may wonder if the food being served is as unsanitary as the clothes of the person serving it. Employees must also refrain from wiping their hands on their clothing, using clothing as hand protection to move hot food vessels, or from wearing clothing that needs constant adjustment, such as an apron with shoulder straps or shoes with loose or untied laces. Uniforms should be simple to avoid contact with food and equipment, for example, billowy sleeves should be avoided.

Ideally, work clothes should be put on in the foodservice establishment and not worn while commuting to work. An area away from food preparation should be set up for changing clothes with lockers provided to safely store the clothing and belongings of the employees. If a foodservice establishment is too small to have a changing room, then employees should be trained in procedures for wearing their work clothes to the establishment. They should be instructed that when they are wearing their work clothes, no stops should be made on the way to work. Street clothes should not be worn in food preparation or serving areas.

Employees' shoes should not have platforms, high heels, or absorbent soles. Work shoes must also have a closed toe so the foot is completely covered. This design reduces the possibility of slips and falls. A shoe cleaning device, such as a rough-surfaced door mat or a special brush for that purpose, should be provided, so shoes can be cleaned before entering the building.

Hair Restraints. To comply with federal, state, and local health codes, employees, crew, and managers are required to wear hairnets, headbands, barrettes, hats, or caps to keep hair from contaminating food. Hair restraints serve a double purpose in that they also discourage employees from running their fingers through their hair, scratching their scalps, or otherwise contaminating their hands by touching their hair. For those employees who have beards, a beard restraint should be worn for the same reasons as a regular hair restraint.

Jewelry. Items of jewelry, such as rings, bracelets, and watches, collect soil and are difficult to keep clean. Jewelry can also become caught in machinery or catch on sharp or hot objects. It should not be worn by foodhandlers while on duty.

Prohibited Habits and Actions

Unfortunately, even if employees are aware of safe foodhandling practices and the reasons behind them, and follow these practices, they may not always stop to think of sanitation when they seek more convenient ways of doing their job. The manager must monitor employee practices and include in the rules a specific list of the personal habits that are not allowed.

The following practices are typical of the habits that must be eliminated:

- Failing to wash hands after handling raw food or after using the toilet
- Using wiping cloths to remove perspiration from the face
- Stacking plates of food in order to carry more at one time
- Washing hands in sinks used to prepare foods
- Spitting on the floor or into sinks
- Unguarded coughing or sneezing in a food-preparation area
- Picking up bread, rolls, butter pats, or ice with bare hands instead of with tongs
- Handling place settings or food after wiping tables or bussing soiled dishes
- Touching the food-contact surface of glassware and tableware with bare hands (for example, touching the tines of a fork instead of the handle or picking up a water glass with fingers inside the glass).

When forming your own list, remember one simple and easy rule: *Hands that have touched contamination must never touch food without prior hand washing and avoid hand contact as much as possible with ready-to-eat foods and cooked food, even after hands have been washed properly.* The employee rules should paint a picture of a model foodhandler who:

- Is in good health
- Has clean personal habits
- Handles food safely
- Appreciates the need for sanitary practices.

In addition, smoking, chewing gum or tobacco, and tasting food require special rules.

Tobacco Use and Gum-Chewing

Employees must not smoke or use tobacco in any form while preparing or serving food, or while in areas used for equipment and utensil washing, or food preparation. Employees should only be permitted to smoke or chew tobacco in designated areas where the use of tobacco will not result in food contamination, such as in employee break areas.

Smoking can endanger the health of both the employee and the customer. It is impossible to smoke without exposing the fingers to droplets of saliva. Small and unnoticed, these droplets can contain thousands of bacteria, which can contaminate anything the fingers touch. Exhaled smoke sends saliva droplets and other contaminants into the air. Contamination can also operate in reverse: micro-organisms may pass from a soiled object to the hands, to the cigarette, to the lips and mouth. It is absolutely essential that foodhandlers who have been smoking wash their hands thoroughly before returning to work.

Gum-chewing is also a potential food-contaminating action. Blowing bubbles, stretching the gum with the fingers, and parking gum under counters or on the floor are among gum-chewing habits that must be forbidden around foodhandling. Gum is also a potential physical hazard. On occasion it has been found in food.

Eating or Tasting Food

Eating and drinking in the food preparation area must be prohibited. Food tasting, however, is necessary in many kitchens. The key here is to avoid any possible contamination of the food being prepared. The commonly pictured practice of the chef tasting from the stockpot using the stirring spoon *must be stopped.* A recommended procedure is to ladle a small amount of the food into a small dish, tasting from that dish with a separate spoon, then removing the dish and spoon to be washed.

Management's Responsibility

Setting up the rules necessary for safe foodhandling is a beginning, not an end. Management must also make it possible for sanitation rules to be observed. In order to do this, management must provide adequate facilities and assign jobs in a way that ensures the practice of good sanitation rules.

Adequate Facilities

Even if you have strict sanitation rules and a complete training program in place, employees will still find it difficult to practice good personal hygiene if proper facilities are unavailable. Small establishments should have adequate washing facilities and an area to hang coats and store personal belongings safely. Larger establishments should also include a changing area.

A dressing room, where employees can change into work clothes and clean up, with lockers for safekeeping clothing and personal objects would be the best way to instill good personal hygiene. Because wearing jewelry is prohibited, a secure place needs to be provided for individuals to leave these items during working hours. Facilities should be conveniently located, clean, well-lighted, and uncluttered, otherwise, they can become reservoirs for potential contaminants and employees will avoid using them.

A break room or lounge area separate from the foodhandling areas is also desirable and is required under ordinances in some locations. This area should be in a place where employees can smoke and eat without endangering food, equipment, or utensils. Facilities that do not have break areas should instruct their employees to eat or smoke in a designated section of the dining area.

Employee restrooms, preferably separate from those for customers, must be provided and equipped with self-closing doors. Remind employees never to wear their aprons into the restroom. If employees need to shower at the beginning or end of the day, adequate shower facilities and supplies should be provided for them.

Hand-washing stations for employees must be located in the restrooms and in other convenient locations throughout the kitchen or food-preparation area. If the lavatory is too hard to reach or blocked by boxes or trash cans, employees may wash their hands in a sink in which food or utensils are cleaned or not wash their hands at all. Ideally, the hand-washing station needs to have faucets that are foot-, knee-, or elbow-operated, or that have automatic sensors to avoid possible recontamination of the hands by touching faucet handles. Self-closing, hand-operated faucets may also be acceptable (see Exhibit 4.2).

The hand-washing station must be supplied with hot and cold water through a mixing faucet, at a temperature between 110° and 120°F (43.3° and 48.9°C). If the water is uncomfortably hot or cold, employees will avoid necessary hand washing.

Soap dispensing equipment should be provided in sufficient quantity at, or near, the wash stations. Sensor- or remote-operated liquid soap dispensers are strongly recommended for use in a food service. Many local health departments do not accept hand-operated soap dispensers and bars of soap because they believe these soaps and dispensers can be a means of transferring micro-organisms. The soap selected should not be too harsh on the employees' hands because it might discourage frequent hand washing.

For hand drying, management must supply disposable paper towels or forced-air blowers. There are too many possibilities for contamination with retractable cloth dispensers. Some authorities consider air blowers the most sanitary drying method, but employees may be inclined to consider them too slow and might resort to the unsanitary practice of wiping their hands on their clothes and aprons. If disposable paper towels are available, waste receptacles that do not require touching the unit and that can be easily cleaned must be provided.

Exhibit 4.2 A proper handwashing station

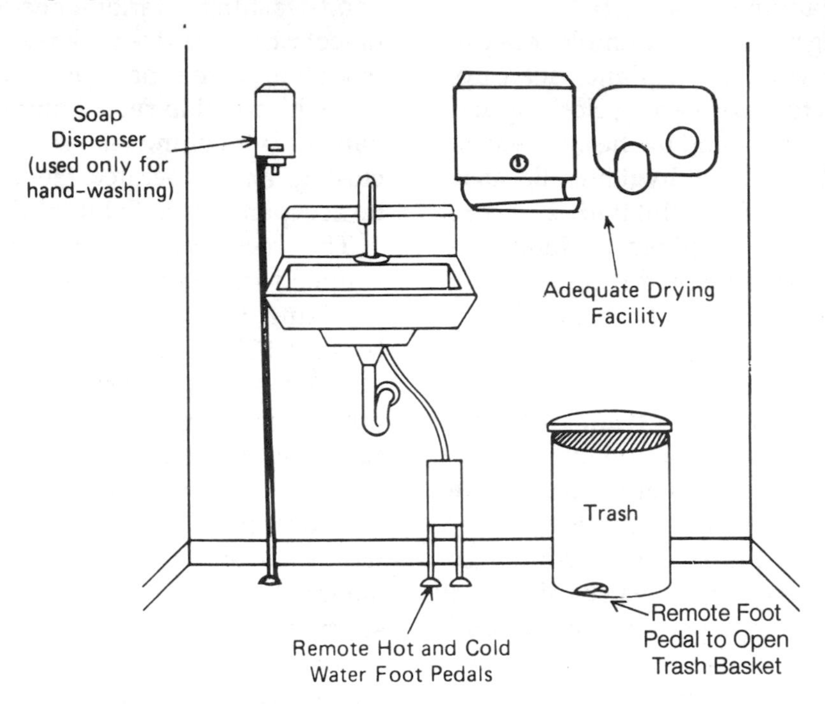

Hand-drying facilities may also be available at food-preparation and utensil sinks to be used by employees performing tasks at these sinks to discourage using aprons or wiping cloths to dry hands. Employees should wash hands at a regular handwashing station after food preparation or any type of scullery operations. Check handwash facilities periodically to make sure that hot water, soap, and towels are available and that equipment or boxes are not blocking access to the sinks.

Job Assignments

When foodservice managers develop job descriptions and assign job responsibilities, they need to look at raw/cooked meat preparation, dirty/clean dish handling, and clearing/setting tables with the risk for cross-contamination in mind. By planning tasks to prevent cross-contamination, the manager re-emphasizes sanitary practices and minimizes the amount of time needed for supervision. A continuous assignment to one given area may help employees follow staff sanitation rules. For example, an employee who is expected to cook and wrap food, clear off tables, and then return to the food-preparation area, will have a harder time avoiding contaminating foods than an employee who is assigned to one specific function or area.

Training the Safe Foodhandler

Assuming that proper procedures, policies, and facilities have been established, the next step the manager must take is to introduce a training program. Employees will need to acquire the necessary knowledge, skills, and attitude to follow the rules. By developing an ongoing program of instruction, motivation, and supervision, a manager will reinforce the importance of following good sanitation rules. A one-time lecture or crash course will not do the job. (Chapter 16 of this text will be devoted entirely to training methods for instructing foodhandlers in sanitation.)

Supervision for Sanitation

No system for safe foodhandling practices can succeed without constant monitoring. *Supervision* is the audit of the training system, making sure that the rules are being followed.

Supervising Employees

In the heat of action during peak serving periods, even the most careful employee may make mistakes. One person may sweep a lock of hair back with a bare hand; another may disjoint raw poultry and then, without washing his or her hands, slice baked ham. Still another may neglect to trade a heavily soiled uniform for a clean one. To stop these errors, a supervisor must see that employees do not use unsafe practices that can threaten food safety. Good supervision reinforced with a continuous training program will help all employees—new and experienced—follow the rules.

In a discreet manner, the manager should observe employees daily for infected cuts, burns, boils, respiratory symptoms, and other evidence of infection that can be transmitted through food. Employees should be encouraged to report their illnesses and injuries to their supervisor. If trust is established with employees through open communication, the employees are more likely to come forward with this information.

Monitoring safe food service means developing good employee supervision routines by incorporating these mental questions:

1. Are employees following the clean-hands policy?
2. Are they practicing good personal hygiene?
3. Are they wearing clean uniforms?
4. Are they free of body odors?
5. Are they wearing hats, hairnets, or other hair and beard restraints?
6. Are they touching a sore or scratching their head or face?
7. Are they smoking, eating, or chewing gum or tobacco in food-preparation or serving areas?
8. Are their fingernails short and clean?
9. Are they spitting in sinks, on the floor, or in a disposal area?
10. Are they coughing or sneezing into their hands?
11. Are they wearing rings, dangling bracelets, or other jewelry while handling food?
12. Are they using wiping cloths to remove perspiration from their faces? Do they tuck a towel into their apron or back pocket for hand wiping?

Excluding Ill or Infected Employees

The decision to exclude ill or infected employees from work can be a difficult one. Situations where the employee is needed or where no sick pay will be provided make the decision even harder.

Under the law in many localities, the foodservice manager who knows or suspects that an employee has certain contagious diseases or is a carrier of certain diseases that can be transmitted by food must notify health authorities immediately. Depending on the disease, the employee may continue to work after medical treatment. (The employee may, for example, be required to wear plastic gloves because of a skin infection, or be assigned to another job that does not involve direct contact with food.)

More serious ailments may require treatment that excludes the employee from the establishment for a considerable period of time. In extreme cases, an individual may be restricted from foodservice employment indefinitely. In these cases, it is best to obtain a physician's statement.

Employees with AIDS

Because there has never been a scientifically documented case of Acquired Immune Deficiency Syndrome (AIDS) transmission through food, it is believed that AIDS poses no threat of foodborne illness to the consumer (see Exhibit 4.3). AIDS cannot be spread through casual contact to customers or co-workers. Perhaps the only real threat it poses to the foodservice operator is in the fear and misconceptions it can create in the public, both patrons and employees. However, a person with AIDS is more susceptible to acquiring a foodborne illness through foodhandling.

Management should know the legal issues concerning an employee with AIDS or an employee who has tested HIV-positive. The Americans with Disabilities Act (ADA) provides comprehensive civil rights protection to individuals with AIDS and bans discrimination against people with disabilities in jobs and public accommodations. Employers are prohibited from firing people or transferring them out of foodhandling duties simply because they have AIDS or test HIV-positive.

Management also has a strict responsibility to maintain the confidentiality of any employees who report that they have AIDS or any other illness. Employers have been sued for compromising that confidentiality.

Self-Supervision for Managers and Supervisors

While the manager and the supervisor are observing the employees, the manager and supervisor should also be supervising themselves. *One of the strongest tools of management is to set a good example.* The cook who sees the manager walking through a food preparation area while smoking a cigarette probably thinks, "If the boss can do it, why can't I?" The manager or supervisor who does not follow the rules will not lead effectively. When managers and supervisors remember that their primary function is to serve safe food to their customers using safe methods, they will take the time to learn and to effectively train their employees.

A manager confronted with a supervisor who does not set a good example may

Exhibit 4.3 AIDS is not a foodborne illness

Source: American Red Cross

want to try a subtle technique to change that attitude. The supervisor can be asked to make a sanitary survey of another department or operation. Unsanitary practices observed by the supervisor in an inspection will suggest possible defects in his or her own department, and will hit home the need for rules, training, and supervision.

SUMMARY

Micro-organisms found on the skin, in cuts and burns, on hair, in the respiratory system, in the intestinal tract, or in other areas are transmitted from the body to the food through poor personal habits, such as unclean hands, as well as by uncovered coughing and sneezing. Stressing good personal hygiene is essential.

The manager is responsible for building a sanitary barrier, which involves many steps to protect customers from contaminated food. The manager must carefully interview job applicants. Each job description should be developed with the necessary sanitation requirements outlined. Clearly *written* rules that state safe foodhandling practices and prohibit unsafe personal habits must be established. It is the manager's responsibility to provide adequate facilities and equipment for the employees to follow the established employee rules. The manager must also provide continuous training of employees, using supervision to correct deviations from the rules. Finally, the managers and supervisors have to set good personal examples by always following the employee rules themselves.

A CASE IN POINT

Shigellosis was the illness transmitted to a number of people who attended a "Hawaiian Luau Night" at a social club.

The event had been catered by Chic Catering. A wide variety of chilled meat salads and fresh fruit and vegetables were served at the party. After the local health department received medical histories from ill party attendees, a sanitarian visited Chic Catering to question the caterer and staff. Keith, the assistant cook and server at the event, reported having had diarrhea on the day of the luau. When asked, the caterer, Jim, said that Keith had made a number of trips to the bathroom, but that it hadn't interfered with his work. Jim said that some of the salads might have been at a room temperature of 78°F (25.6°C) for several hours because they ran out of ice to chill them. The hand-washing facilities available to the catering staff had been inadequate as well.

Before leaving the catering facility, the sanitarian observed that the handwashing station in the kitchen did not have a supply of paper towels and was blocked off by several boxes.

What controls could have prevented this outbreak?

What does Jim need to do to correct the sanitation breakdown?

STUDY QUESTIONS

1. Why is unguarded sneezing hazardous in a foodservice operation?
2. Why is it important to have a well-designed application form?
3. Why should the manager write down and post employee rules for personal cleanliness and safe foodhandling?
4. Describe the correct handwashing procedures for foodservice employees.
5. Why should a manager explain the reasons behind employee sanitation rules?
6. A foodhandler accidentally cuts a finger but wishes to continue work as a salad preparer. What should the manager do?
7. Where in a foodservice operation may employees be permitted to smoke?
8. Why is smoking around food a hazardous activity?
9. What two purposes does wearing hair restraints serve?
10. What should the employee rules state about the tasting of food?
11. Where should employee hand-washing stations be located?
12. What is one method for supervising employees in sanitary food practices?

ANSWER TO A CASE IN POINT

The controls that could have prevented this outbreak are personal hygiene and supervision by management. For food service, ill employees who continue working can be more hazardous than in other industries. Foodservice managers must develop and uniformly enforce a policy stating that employees who are ill will not be given foodhandling duties, but will either be reassigned other tasks or sent home until they are well or no longer contagious. Jim should have developed open communication with his employee so that Keith would know to stay home when he is sick.

Because Keith did not wash his hands after he used the restroom, he contaminated the meat salads and fresh fruit and vegetables with *Shigella* bacteria. Since Keith was preparing and serving chilled, ready-to-eat foods, hand washing was critical.

Hand-washing facilities must be well-equipped and accessible to allow employees to practice the necessary steps for hand washing as often as necessary. Paper towels or air dryers must always be provided. Jim needs to make sure that his employees are able to practice proper hand washing both at job locations and in the kitchen at Chic Catering.

Jim, as the manager, needs to train Keith and all his other employees in the proper personal hygiene habits needed for safe food preparation and handling, as well as telling them the practices that are prohibited. He particularly needs to tell them the importance of hand washing and show them exactly how they need to wash their hands. After training them, Jim needs to monitor and supervise their personal hygiene habits and practices to ensure that they are following the employee rules.

MORE ON THE SUBJECT

LONGRÉE, KARLA, and GERTRUDE G. BAKER. *Sanitary Techniques in Food Service.* 2nd ed. New York: Macmillan, 1982. 271 pages. This text provides practical guidance for the foodhandler in sanitation. Designed as a teaching aid for vocational training, the book contains many charts and tables depicting safe operating procedures. It also includes new material on foods prepared away from the serving premises, new equipment, changes in the food supply, and changes in microbiological problems.

THE EDUCATIONAL FOUNDATION OF THE NATIONAL RESTAURANT ASSOCIATION. Chicago: The Educational Foundation, 1991. The video "Introduction to Food Safety: Employee Health and Hygiene," is the first in the four-part series for employee training. Also, *Serving Safe Food: A Guide for Foodservice Employees* covers personal hygiene habits to follow, and those that should be prohibited during working hours.

THE NATIONAL RESTAURANT ASSOCIATION. *Basic Facts About AIDS for Foodservice Employees.* Washington, DC: National Restaurant Association, 1987. PB#010. An 8-page pamphlet that informs foodhandlers about this disease.

PART II

The Flow of Food Through the Operation

5

Establishing the Foodservice Safety System

If you add the benefits you receive from a well-designed and run food safety program to the costs in dollars and human suffering you save by preventing a foodborne outbreak, it makes sense to implement the most effective food safety program possible. To design a program that maintains a high level of food safety, you start by focusing on the foods that have been implicated most often in disease outbreaks.

As we have seen in earlier chapters, any food can transmit disease and should be handled properly. But those foods that have been involved most often, and which are classified as potentially hazardous, require special attention. This *potential* for causing illness exists because the nature of food makes it favorable to the growth of micro-organisms which can make people sick.

Also, studies of foodborne disease outbreaks have identified several common errors in handling certain potentially hazardous foods. These errors provide the conditions micro-organisms need to get into food, grow in food, or survive the food preparation process. A well-designed food safety program will monitor all food preparation activities where errors in handling can happen and make a special effort to eliminate those errors.

Foodservice operations that handle only a few potentially hazardous foods and use simple preparation procedures, such as a hot dog stand, will generally have fewer chances for disease outbreaks. They may need a simpler food safety system than an operation with a variety of items on a menu, such as a 24-hour coffee shop. Along the same lines, operations, such as nursing homes and long-term care centers, have to be especially careful with the foods they serve. The procedures these operations develop have to be more detailed because the people they serve are more susceptible to foodborne illness, and the symptoms are more severe for these people.

A well-designed food safety system covers the entire operation. The system will concentrate on potentially hazardous foods and how these foods move through the operation. By focusing on the factors that are critical to food safety, operators can set their priorities concerning time and use of resources. The food safety system we will examine in this chapter will enable you to:

- Identify foods in an operation most likely to be implicated in a foodborne illness outbreak.
- Follow a potentially hazardous food through the flow of food.
- Determine which conditions lead to foodborne disease and eliminating or changing them by implementing appropriate procedures.

THE HACCP/S.A.F.E. SYSTEM

One food safety and self-inspection system that provides foodservice operators with the tools they need to develop a cost-effective food safety program is *Hazard Analysis Critical Control Point,* more commonly known as HACCP. The National Restaurant Association has developed a similar system named Sanitary Assessment of the Food Environment (S.A.F.E.). Both systems highlight potentially hazardous foods and how they are handled in the foodservice environment.

The HACCP system was first developed by the Pillsbury Company in 1971 for the National Aeronautics and Space Administration (NASA) to make sure the food served to the astronauts in outerspace was absolutely safe. The symptoms of a foodborne illness could prove disastrous for anyone in outerspace. That HACCP system had to ensure zero defects during foodhandling by monitoring the whole preparation process. It had to identify and correct errors before they happened, rather than using the traditional method of test sampling the finished product to identify foods with high levels of contamination.

The HACCP system proved so valuable, successful, and straightforward in food processing that modified HACCP systems are being adapted for use in foodservice operations. The HACCP/S.A.F.E. systems are receiving increased use and acceptance in the foodservice industry.

HACCP IN FOOD SERVICE

The path that food travels in a foodservice operation is known as the *flow of food.* It begins with the manager's or operator's first decision to include an item on the menu. After this decision, the flow of food continues with the development of the recipe, delivery of products, storage, preparation, holding or display, service, cooling and storing leftovers for use the next day, and reheating.

At each step in the flow of foods, various risks for contamination exist. The chance that a condition or set of conditions in foodservice will lead to a hazard is called *risk.* A *hazard* is unacceptable contamination (of a biological, chemical, or physical nature), unacceptable microbial growth, or unacceptable survival of micro-organisms of a concern to food safety, or persistence of toxins. (Chapter 3 described all of these hazards in detail.)

As risks are determined, foodservice operators need to identify critical control points in the flow of foods. A *critical control point* is an operation (practice, preparation step, procedure), by which a preventive or control measure can be applied that would eliminate, prevent, or minimize (a) hazard(s). The HACCP process begins by identifying critical control points based on the risk that an action will cause a hazard and by assessing the severity of that hazard.

The foodservice operator can make changes in procedures as needed to prevent or minimize hazards and set up procedures to monitor high-risk activities which may require corrective actions. Both monitoring and corrective actions are to ensure that a hazard will not happen or will be corrected immediately.

For example, one of the earliest control points occurs during delivery. Let's say a manager at the Midtown Restaurant decides to run beef stew as a special. At the receiving dock, a shipment of beef chuck roasts arrives from the restaurant's usually reliable supplier. Steve, a foodservice employee trained to check the food shipments, discovers that the wrapping covering the beef roasts is torn and the surface of the meat is slightly sticky. His operation's requirements for acceptable beef roasts are that meat should have no discolorations, be firm and elastic to the touch, and be at a temperature of 45°F (7.2°C) or below.

Steve knows that temperature abuse of a beef roast, a potentially hazardous and high-risk food, will allow the naturally present bacterial contamination on the raw product to grow. The result could be a highly contaminated product. He checks the temperature of the meat with a properly calibrated and sanitized, bimetallic stemmed thermometer. The temperature of the beef roast is above 45°F (7.2°C). It does not meet the temperature requirement, so Steve informs his manager of the problem shipment. Judy, the manager, returns with Steve and backs up his finding. Judy takes corrective action and rejects the order based on the temperature and the other conditions of the beef roast.

The flow of food is best illustrated by designing a flowchart (see Exhibit 5.1). All potentially hazardous foods on your menu should have a flowchart. This is not as difficult as it sounds because similar potentially hazardous foods follow similar processes, using essentially the same basic flowchart, for example, chicken soup and beef noodle soup are similar, while split pea soup and bean soup are similar. From the receiving of the food all the way through the preparation process to the final serving of the food, a flowchart organizes your process.

Exhibit 5.1 A HACCP flowchart for beef stew

Critical Control	Hazard	Standards	Corrective Action if Standard not Met
		Receiving	
Receiving beef	Contamination and spoilage	Accept beef at 45°F (7.2°C) or lower; verify with thermometer	Reject delivery
		Packaging intact	Reject delivery
		No off odor or stickiness, etc.	Reject delivery
Receiving vegetables	Contamination and spoilage	Packaging intact	Reject delivery
		No cross-contamination from other foods on the truck	Reject delivery
		No signs of insect or rodent activity	Reject delivery
		Storage	
Storing raw beef	Cross-contamination of other foods	Store on lower shelf	Move to lower shelf away from other foods
	Bacterial growth and spoilage	Label, date and use FIFO rotation	Use first; discard if maximum time is exceeded or suspected
		Beef temperature must remain below 45°F (7.2°C)	Discard if time and temperature abused
Storing vegetables	Cross-contamination from raw potentially hazardous foods	Label, date, and use FIFO rotation	Discard product held past rotation date
		Keep above raw potentially hazardous foods	Discard contaminated, damaged, or spoiled products
		Preparation	
Trimming and cubing beef	Contamination, cross-contamination, and bacteria increase	Wash hands	Wash hands
		Clean and sanitize utensils	Wash, rinse, and sanitize utensils and cutting board
		Pull and cube one roast at a time, then refrigerate	Return excess amount to refrigerator
Washing and cutting vegetables	Contamination and cross-contamination	Wash hands	Wash hands
		Use clean and sanitized cutting board, knives, utensils	Wash, rinse, and sanitize utensils and cutting board
		Wash vegetables in clean and sanitized vegetable sink	Clean and sanitize vegetable sink before washing vegetables
		Cooking	
Cooking stew	Bacterial survival	Cook **all** ingredients to minimum internal temperature of 165°F (73.9°C)	Continue cooking to 165°F (73.9°C)

(Continued)

Exhibit 5.1 *(Continued)*

Critical Control	Hazard	Standards	Corrective Action if Standard not Met
		Cooking (continued)	
		Verify final temperature with a thermometer	Continue cooking to 165°F (73.9°C)
	Physical contamination during cooking	Keep covered, stir often	Cover
	Contamination by herbs and spices	Add spices early in cooking procedure Measure all spices, flavor enhancers and additives, and read labels carefully	Continue cooking at least 1/2 hour after spices are added
	Contamination of utensils	Use clean and sanitized utensils	Wash, rinse, and sanitize all utensils before use
	Contamination from cook's hands or mouth	Use proper tasting procedures	Discard product
		Holding and Service	
Hot holding and serving	Contamination, bacterial growth	Use clean and sanitary equipment to transfer and hold product	Wash, rinse, and sanitize equipment before transferring food product to it
		Hold stew above 140°F (60°C) in preheated holding unit, stir to maintain even temperature	Return to stove and reheat to 165°F (73.9°C)
		Keep covered	Cover
		Clean and sanitize serving equipment and utensils	Wash, rinse, and sanitize serving utensils and equipment
		Cooling	
Cooling for storage	Bacterial survival and growth	Cool rapidly in ice water bath and/or shallow pans (<4"deep)	Move to shallow pans
		Cool rapidly from 140°F (60°C) to 45°F (7.2°C) in four hours or less	Discard, or reheat to 165°F (73.9°C) and re-cool one time only
		Verify final temperature with a thermometer; record temperatures and times before product reaches 45°F (7.2°C) or less	If temperature is not reached in less than four hours, discard; or reheat product to 165°F (73.9°C) and re-cool one time only

(Continued)

Exhibit 5.1 *(Continued)*

Critical Control	Hazard	Standards	Corrective Action if Standard not Met
		Cooling (continued)	
	Cross-contamination	Place on top shelf	Move to top shelf
		Cover immediately after cooling	Cover
		Use clean and sanitized pans	Wash, rinse, and sanitize pans before filling them with product
		Do not stack pans	Separate pans by shelves
	Bacterial growth in time or after prolonged storage time	Label with date and time	Label with date and time or discard
		Reheating	
Reheat for service	Survival of bacterial contaminants	Heat rapidly on stove top or in oven to 165°F (73.9°C)	Reheat to 165°F (73.9°C) within two hours
		Maintain temperature at 140°F (60°C) or above; verify temperature with a thermometer	Transfer to preheated hot holding unit to maintain 140°F (60°C) or above
		Do not mix new product into old product	Discard product
		Do not reheat or serve leftovers more than once	Discard product if any remains after being reheated

Is it necessary to design a flowchart for nonpotentially hazardous foods as well? Crackers, for example, need to be checked at delivery to make sure that the package is unopened, there are no tears (a possible indication of rodent contamination), and that the crackers are fresh. After the foodservice employee has stored the crackers, the only monitoring that needs to be done is to check the freshness date, to keep the packages dry and unopened, and to watch for signs of insect or rodent infestation. The opportunities for mishandling the crackers are relatively few and temperature abuse of the crackers will have little effect on their safety because they are a poor growth medium for bacteria. There is no need to design a flowchart for crackers.

In the case of the beef stew and other potentially hazardous foods, there are many chances in the food preparation and serving flow that could cause the food to become contaminated or could increase the possibility of microbial growth. These are the foods that the HACCP system targets for flowcharting. Emphasizing high-risk foods and handling procedures will maximize efficiency of time and supervision for a foodservice manager.

HOW TO SET UP A HACCP SYSTEM

HACCP can fit any operational step in the growing, harvesting, processing, preparation, and serving of foods. The National Advisory Committee for Microbiological Criteria for Foods has identified seven major principles involved in operating a HACCP system. These include:

1. Assessing hazards at each step in the flow of food and developing the procedures to lower the risk for each step
2. Identifying the critical control points
3. Setting up control procedures and standards for critical control points
4. Monitoring critical control points
5. Taking corrective actions if there is a break from procedures set up at a critical control point
6. Developing a record-keeping system that documents the HACCP plan
7. Verifying that the HACCP system is working.

Stated simply, the HACCP system for food service involves identifying and controlling points from receiving to serving at which:

- Food can become contaminated
- Contaminants can increase
- Contaminants can survive.

It is at these points that the foodservice manager can control possible hazards. The first step in implementing a HACCP system is to identify foods at risk, and operations at which a hazard might occur. Hazards can exist with raw food products that have been purchased and received, or may result through errors made in storage, preparation, display, or service. *It is important to keep hot food hot, cold food cold, and all food clean, or free from cross-contamination.*

Some hazards are more dangerous or severe than others; and some are more likely to cause problems during the course of food preparation than others. That is why we pay particular attention to potentially hazardous foods. These foods are high risk because they provide favorable conditions for bacterial growth; therefore, the likelihood, or risk, that bacteria will grow is high.

Step 1: Assessing Hazards

Select foods to be studied. Examine your menu and recipes, conduct a detailed review of your menu, and identify those foods, such as meat, seafood, poultry, and egg and dairy dishes, that are potentially hazardous (see Exhibits 5.2 and 5.3).

Take a close look at the size and type of food establishment that you run. Your suppliers, equipment, and personnel will give you an idea of what you should have on your menu. For example, if you manage a small, family restaurant, you may find it difficult to safely monitor all the steps in serving chicken cordon bleu made from scratch, but you could serve baked turkey breast. If a recipe is too difficult for you to manage with your available foodservice employees and your facilities, perhaps you should consider dropping that item from the menu or purchasing it in a prepared form from a reliable supplier that has a verified HACCP system. As mentioned in Chapter 1, you

Exhibit 5.2 Examples of potentially hazardous foods

Poultry	Meat
Fish	Meat Products
Shellfish	Milk
Egg Products	Milk Products
Raw Eggs	Puddings
Gravies	Sauces (other than sauces of a high-acid nature)
Soups (such as creamed)	Foods High in Protein
Custards	Cream-filled Goods
Potato and Protein Salads	Low-acid Foods
Cooked or heated vegetables and plant products, such as:	
Potatoes (baked or mashed)	
Tofu	
Beans (cooked)	
Winter squash (cooked)	
Rice (cooked)	

can reduce the risk of an outbreak by *avoiding* menu items that pose the risk.

In general, you can reduce the risk of a foodborne outbreak by reducing the number of steps involved in the preparation and serving of potentially hazardous foods. It is also important to note that the nonpotentially hazardous foods must also be handled carefully. For example, most fruits will not provide a medium for rapid bacterial growth, but they do provide a growth medium for molds and yeasts. Fruits and vegetables need to be washed thoroughly and protected from new contamination by employees during foodhandling procedures.

Now that you have surveyed the potential risk for the foods, evaluate each of

Exhibit 5.3 Typical hospital menu

(Potentially hazardous foods are boldface)

Roast Turkey, Cranberry Jelly

Pot Roast of Beef, Brown Gravy

Poached Salmon Filet, Lemon Butter

Baked Potato **Cornbread Dressing**

Buttered Broccoli **Steamed Carrots**

Sliced Tomato Salad, Vinaigrette Dressing

Bran Muffin Parkerhouse Roll

Butter

Sherbet **Coconut Cream Pie** Baked Apple

Source: *Management by Menu* by Lendal H. Kotschevar. Copyright © 1987 by The Educational Foundation of the National Restaurant Association

these procedures for what might go wrong in terms of contamination. After you determine the hazards, assess their severity. *Severity* is the seriousness of the consequences of the results of the hazard. Rank these hazards according to severity as well as their probability of happening. For example, during the cooking process, the severity of the hazard caused by undercooking is very high and requires careful monitoring.

Step 2: Identifying Critical Control Points

After the hazards have been determined and assessed, it is necessary to identify critical control points during preparation where hazards can be controlled or prevented. In general, good personal hygiene and avoidance of cross-contamination are important in preventing contamination at each step. Cooking and cooling are critical control points because thorough cooking kills vegetative bacteria, and rapid cooling prevents bacterial growth. Create a flowchart of the steps required during preparation of the potentially hazardous foods you wish to monitor. Procedures to prevent, reduce, and eliminate recontamination hazards must also be identified at each step on your flowchart (refer to Exhibit 5.1). This flowchart will also be used in Chapters 6–8, which will explain each step in detail.

Step 3: Setting up Procedures for Critical Control Points

Establish requirements that must be met at each critical control point. It is important that these requirements be observable and measurable, using such factors as times, temperatures, and sensory measures. The type of procedures and requirements to be developed could include:

- Washing hands
- Washing, rinsing, and sanitizing food-contact surfaces and utensils
- Thorough cooking food to appropriate temperatures above 140°F (60°C)
- Holding cooked foods at temperatures above 140°F (60°C)
- Rapid cooling and keeping food below 45°F (7.2°C).

For example, recipe directions for a baked chicken breast should include:

Exhibit 5.4 Sample of a recipe with requirements for food safety included in italics

Turkey Salad CCP—Critical Control Point

Yield: 110 *Each Portion: 1 cup*

Sanitation Instructions: Wash, rinse, and sanitize all equipment and utensils before and after use. Wash hands before handling food, after handling raw foods, and after any interruption that may contaminate hands.

Ingredients	Weights/Measures	Procedures
Turkey parts (legs and thighs)	70 lbs.	1. Wash turkey thoroughly under cold potable running water being careful to avoid splashing any of the water.

(Continued)

Exhibit 5.4 *(Continued)*

Water Salt Bay leaves Peppercorns	10 gal. 7 oz.	2/3 cup 9 leaves 10–12	CCP	Place turkey parts in steam kettle or stock pot; add water to cover, salt, bay leaves and peppercorns. Bring to boil; reduce heat; simmer (about one and a half hours.). Add water as needed to keep turkey parts covered. *Continue cooking until the turkey parts are 165 °F (73.9 °C).*
			CCP	2. Remove turkey parts. Place in shallow four inch pans in a single layer. *Refrigerate on the top shelf uncovered until temperature of the turkey is 70 °F (21.1 °C), then cover. Refrigerate covered at least two hours or overnight. Check temperature to ensure it has dropped to 45 °F (7.2 °C) or lower within four hours.* Label and date pan.
			CCP	*One pan at a time, remove turkey parts from the refrigerator;* remove meat from bones. Cut into 1/2–1 inch pieces on a clean and sanitized cutting board.
			CCP	*Return cut-up turkey to refrigerator. Store covered on the top shelf until used* in Step 5.
Celery	12 lbs.	2-1/4 gallon	CCP	3. Wash the celery. *Blanch in boiling water for 1 min. Refrigerate covered on top shelf.*
Preparation				
Sweet peppers	1 lb. 8 oz.	1 qt.		4. Wash the sweet peppers. Seed and pith, and cut in 1/4 inch pieces. Place in a bowl large enough to hold all ingredients.
Onions	8 oz.	1-1/2 cups		Peel the outer layer from the onion. Wash and cut into 1/4 inch pieces. Add to sweet peppers. Chop celery. When chopping celery, sweet peppers, and onions use a clean and sanitized cutting board.
				5. Mix these ingredients thoroughly with a slatted spoon. Add turkey parts.
Lemon juice (chilled) Commercial mayonnaise Salt Red Pepper Sauce	 3 lb. 4 oz. 4 oz.	1 cup 6 1/2 cups 6 tbsp. 6 tbsp.	CCP	6. Blend these liquid ingredients and salt and pepper together. Add to turkey parts and vegetables. Mix lightly and thoroughly with a large slotted spoon. *Cover, label with date, and refrigerate until ready to serve.*

1) washing hands; 2) washing, rinsing, and sanitizing the cutting board and knife that are used for slicing the chicken breast; and 3) the actual product temperature of 165°F (73.9°C). As many requirements as possible should be written into all recipes along with other procedures (see Exhibit 5.4 for a detailed sample recipe). This also means that the appropriate facilities and equipment must be available and employees must be trained to follow the recipe directions.

Step 4: Monitoring Critical Control Points

To monitor your critical control points, use your flowchart to follow potentially hazardous foods through the entire process to compare your operation's performance against the requirements you have set. Make corrections as needed.

For example, you should have targeted your refrigerated storage area as a point where errors could occur. A delivery of fresh turkey breasts has just arrived at your facility. After checking that the product temperature is 45°F (7.2°C) or lower, the turkey breasts are placed in the walk-in refrigerator. To ensure that the raw turkey breasts remain below 45°F (7.2°C) in storage, you check the hanging thermometer inside the walk-in unit to see that the air temperature reads 40°F (4.4°C) or lower. To make sure that no cross-contamination occurs, check the area to see that the raw turkey breasts are not stored above any previously cooked foods or any foods that will not be cooked, for example, fresh cole slaw. Correct placement of the raw turkey in the walk-in unit prevents the possibility of contaminated liquid leaking from the turkey breasts onto any foods that are served raw or with no further cooking.

Another step to target is cooking. Your requirement is to cook turkey breasts to an internal temperature of 165°F (73.9°C). The monitoring is done by using a correctly calibrated, clean and sanitized thermometer to measure the product temperature until it reaches 165°F (73.9°C). By performing these simple steps of observing and measuring, you have *monitored* your critical control points.

Step 5: Taking Corrective Action

A careful examination of your controls in action could point out an area of deficiency; if it does, a back-up procedure should be enacted. For example, you discover that your chef has been preparing turkey breasts all at one time, exceeding your requirement for the time the meat is exposed to temperatures above and below the legal requirements of 45° to 140°F (7.2° to 60°C). By monitoring the temperature, you discovered a high-risk situation in your preparation procedures. It becomes clear to you that you need to correct your chef's procedures. In this particular instance, instruct your chef to prepare turkey breasts in small batches, and to return each batch to the cooler after preparation and whenever preparation is interrupted. Therefore, the time that the turkey breasts are exposed to the temperature danger zone is kept to a minimum. These corrective actions should be added to your flowchart.

Step 6: Setting up a Record-Keeping System

Any system that works well for your particular operation may be used for this step. Possible systems include supervisor spot-checks of temperatures, flowcharts added to operation manuals, and written logs. These logs could be completed at a shift change or at the beginning or end of a meal period. Exhibit 5.5 shows a sample log. (See Chapter 15 for a discussion of a HACCP-based regulatory log.)

Step 7: Verifying That the System Is Working

To check how well your controls and procedures are working, retrace the flow of foods, this time with each control in place. Study logs of time/temperature errors and corrective actions taken in order to evaluate your program and to detect *dry lab,* the entering of data without taking actual measurements. Breaks from time/temperature standards or the need for frequent corrective actions indicate a need for a change in your system. Observe the entire operation with an analytical eye. You may find that some controls are not possible or that better methods exist for meeting controls. This may mean completely changing a procedure, or buying a prepared product. The HACCP system provides for continual change and improvement. It can be altered as menu items, equipment, and personnel change.

THE ROLE OF TRAINING

If the key to food safety success is the HACCP program, then the door is the foodservice employee. Once the HACCP system has been designed and the foodservice operator has provided the system and equipment needed to prepare food safely, its success is up to the managers and their employees. The manager and every employee must be trained in basic principles of food safety and the use of the operation's HACCP system.

Employee training (which will be covered in more detail in Chapter 16) involves providing the knowledge and skill needed to do the job right. Employees must be taught how improper foodhandling makes people sick and what can be done to prevent the foodhandling errors that occur most often. Employees must learn how to measure temperatures of different types of potentially hazardous foods as the foods move through the entire food flow. They need to know which heating procedures kill bacteria and which cooling procedures prevent bacterial growth. In addition, the importance of personal hygiene and cleaning and sanitizing procedures must be stressed. Most importantly, managers must support employees in developing the attitude that the procedures of the operation's HACCP system are the responsibility of each and every employee working in the operation.

OTHER HACCP AND NON-HACCP CONSIDERATIONS

While HACCP systems place a strong emphasis on the flow of food through the operation and the prevention of microbiological hazards, it is also important to inspect those systems that support the HACCP system or help satisfy the principles of food safety and sanitation.

Exhibit 5.5 Sample log for HACCP record keeping.

Hazard Analysis Critical Control Point Worksheet

mo. day yr.
Date: ___ ___ ___ TIME: Start ___:___ A.M./P.M. End ___:___ A.M./P.M.
Product ______
Ingredients ______
Sources ______

Time	Temp.	Procedure/Observation	Comment/Interpretation

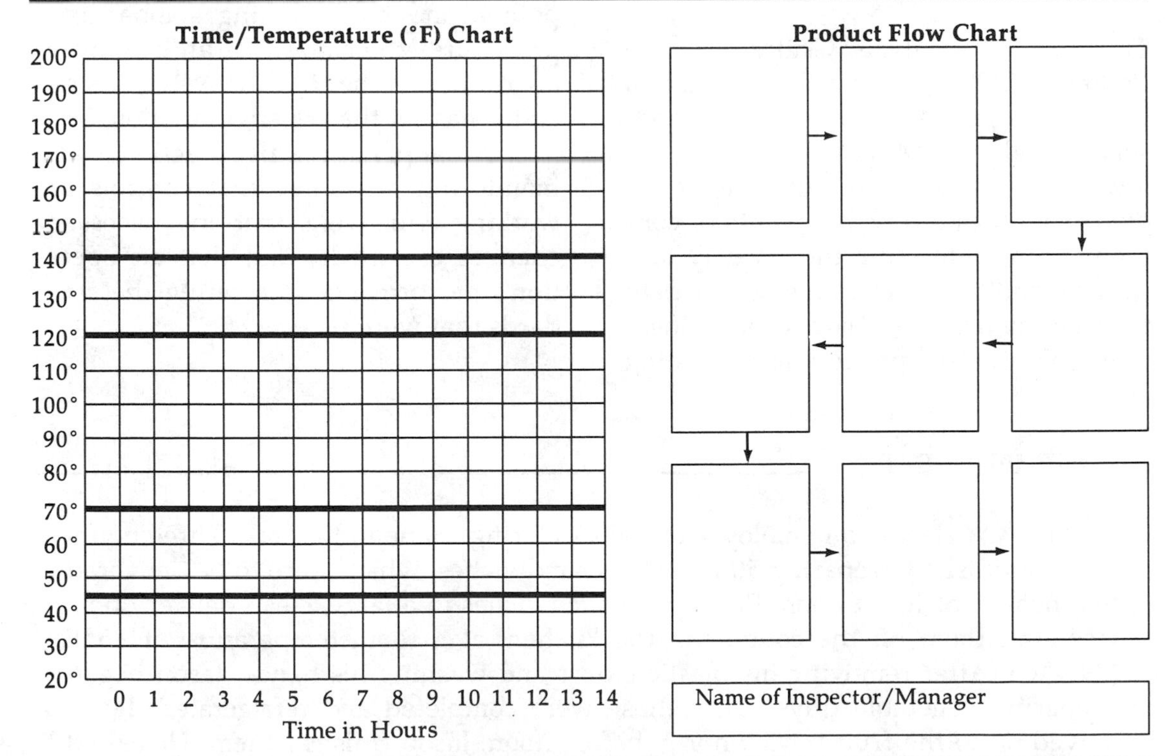

Source: Adapted from New York State Department of Health

The selection, operation, and maintenance of equipment can be essential to the success of the HACCP system. Inspect heating and cooling units and other equipment for proper operation to provide an extra level of support within the HACCP system.

The proper labeling and storage of detergents and sanitizers may not be included in food flowcharts, but proper storage and use must be monitored to prevent chemical contamination that could lead to foodborne illness. In addition, an effective pest control program controls the presence of vermin that can contaminate food. (Chapter 12 will cover this in detail.)

SUMMARY

The Hazard Analysis Critical Control Point (HACCP) system of self-regulation is becoming more common in the foodservice industry. A HACCP system allows an establishment to evaluate its operation, locate possible points of contamination, determine the severity of a hazard, and take preventive measures to protect against a foodborne illness outbreak. Self-inspection and training help ensure that the correct steps are being followed, and that safety and quality are maintained.

HACCP is a scientifically based management system for food safety. It stresses the process of foodhandling, rather than focusing on the facilities and aesthetics. This does not eliminate the importance of facilities, instead, it changes the focus to the maintenance of a complete food safety management program.

To implement the system, an operator must be able to set priorities for the existing hazards according to their severity and risk. Controls must then be set up at each step of food preparation. By examining the menu and recipes, by creating a flowchart indicating critical control points, and by designing standardized procedures to ensure that all employees are trained in the HACCP system, an operator has set the system in motion.

Follow-up, corrective action, and monitoring guarantee that controls are working effectively, that any necessary changes are made, and that an operation's performance is meeting the standards that were set.

A CASE IN POINT

At 7:00 A.M., Jason, an employee at Cal's Catering, came in to work. Three hours later he finished preparing 50 roast beef sandwiches, which were to be delivered to a nearby office at noon. Placing the sandwiches in a large, clean delivery box, Jason set them on the counter in the kitchen, at a room temperature of 85°F (29.4°C). After removing his plastic gloves and washing his hands, Jason began preparing vegetable trays. Once those were completed and refrigerated, Jason moved on to the fruit trays. Shortly before noon, Jason finished them. He called the delivery-van driver to take the foods to the office party.

The food was greatly enjoyed, and the party was a rousing success—until early the next morning, when the office workers called in ill.

What food safety factor was violated?

How might Cal's Catering adjust its procedures to make sure that this error is not repeated?

STUDY QUESTIONS

1. What is HACCP? What is S.A.F.E.?
2. What are the seven steps to follow in setting up a HACCP-based system?
3. When a menu is reviewed, what foods should be identified for flowcharting?
4. What is a hazard? Give an example.
5. What is a critical control point? Give an example.
6. How can a menu and recipe be used as controls?
7. How can controls and steps be evaluated?
8. What should a flowchart include? How can this information be used?
9. What is the purpose of HACCP training?

ANSWER TO A CASE IN POINT

Roast beef, a potentially hazardous food, must be kept out of temperatures that encourage the growth of bacteria as much as possible to ensure its safety. Jason kept the beef unrefrigerated for at least five hours as he prepared sandwiches and the other food. Although he *did* refrigerate the vegetable trays that he subsequently prepared, he failed to refrigerate the sandwiches, which remained in the kitchen at a temperature of 85°F (29.4°C).

To make sure that this error is not repeated, management must specify the maximum time that beef can remain unrefrigerated. Procedures that require the refrigeration of sandwiches need to be written into the recipe or operations manual. Management must also inform foodhandlers of the time limit, and monitor operations to see that employees maintain the standards.

MORE ON THE SUBJECT

NATIONAL RESTAURANT ASSOCIATION. *Make a S.A.F.E. Choice: A New Approach to Restaurant Self-Inspection.* Washington, DC: The National Restaurant Association, 1987. #PB002. 24 pages. This booklet covers the application of the principles of S.A.F.E., and gives the overview and scope of self-inspection programs. It also discusses potentially hazardous foods, major foodborne illness sources, and crisis management. It includes many charts and graphs.

BRYAN, FRANK L. *HACCP—Hazard Analysis Critical Control Point—Manual.* Washington, DC: Food Marketing Institute, 1989. This manual gives examples of monitoring systems for many products prepared in foodservice operations. It also contains sample forms that aid in monitoring the system.

BRYAN, FRANK L. *Microorganisms in Food 4: Application of the Hazard Analysis Critical Control Point System to Ensure the Microbiological Safety and Quality of Foods.* Oxford, England: Blackwell Scientific Publications, Ltd., 1988. This book describes the principles of food safety and of the HACCP systems. The application of HACCP in every phase of food production and service is presented with many examples used throughout the book.

BRYAN, F. L., C. A. BARTLESON, C. O. COOK, P. FISCHER, J. J. GUZEWICH, B. J. HUMM, R. C. SWANSON, and E. C. D. TODD. *Procedures to Implement the Hazard Analysis Critical Control Point System.* Ames, IA: International Association of Milk, Food and Environmental Sanitarians, 1991. This how-to book describes the procedures to set up a HACCP system. Forms and tables for creating flowcharts, determining hazards, and assessing critical control points are included to guide the development of a HACCP system.

NATIONAL ADVISORY COMMITTEE FOR MICROBIOLOGICAL CRITERIA FOR FOODS. "HACCP Principles for Food Production." Washington, DC: U.S. Department of Agriculture, Food Safety and Inspection Service, 1989. This paper outlines and discusses the seven principles involved in operating a HACCP food safety program.

6

Purchasing and Receiving Safe Food

It is impossible to prepare a wholesome meal with contaminated or spoiled ingredients. As we discussed in Chapter 5, food supplies must be in excellent condition when they arrive in the receiving area. They must be purchased from approved sources only and examined for signs of spoilage and contamination before being used. This is one of the first control points in the food preparation process.

Governmental regulating agencies help reduce the distribution of unsafe products, but the final responsibility for product safety in the foodservice establishment rests with the foodservice manager. State and federal governmental regulatory agencies can neither inspect *all* food products nor check the products at *every* point in the distribution network. The additional processing steps and longer lines of supply have increased the risk of hazards that can cause contamination. For example, a frozen food product, which is in good condition when it leaves the manufacturer's plant, may be damaged by thawing in the railroad car or truck that carries it to the wholesaler, and again, in the delivery truck that carries it to the foodservice operator.

In this chapter, we will consider:

- General rules for inspection of food as it arrives at a foodservice establishment
- Governmental programs to help ensure a safe food supply
- Specific signs of spoilage in food products.

SOURCES OF SAFE FOOD

To a great extent, the foodservice manager must rely on assurances from suppliers that the food coming into the establishment is of sanitary quality. It would be unrealistic for the manager to control or inspect every step of the production and processing chain of the countless food supplies delivered to the average restaurant, commissary, or kitchen. However, the foodservice manager does have the final authority to accept or reject the supplier that appears at the back door of the establishment.

If possible the foodservice operator should consider checking suppliers from time to time by asking to inspect their facilities. For example, many multi-unit foodservice companies have quality assurance personnel who visit the processing plants and supplier operations from which they make purchases.

A foodservice operator inspecting a supplier should ask these questions: Do you have a verified HACCP system covering the foods under consideration for purchase? If applicable, are delivery trucks adequately refrigerated? Are your employees trained in sanitary practices? Are you willing to make deliveries when my employees are not rushed so that food can be handled quickly and properly? Is the product quality consistent? Are the products packaged safely?

Remember, quality control in foodservice industry starts with the supplier. Make sure that your operation uses *only* reliable suppliers with verified HACCP systems. Check health department reports on suppliers. With foodservice supply houses constantly coming into or leaving the marketplace and with continuing innovations, such as prepared frozen entrees, mechanically tenderized meats, microwave ovens, modified atmosphere packaging, and onsite packaging, a foodservice operation's inventory of suppliers must be constantly updated and upgraded.

USING A THERMOMETER

Before you learn the details about various inspection methods used for receiving certain foods, you need to know the importance of the thermometer. The thermometer can be the single most important item in the protection of food.

Choosing the Right Thermometer

A number of thermometers are used in the foodservice facility. Some are used to measure temperatures of equipment and storage areas. Others are used to measure the temperatures of food. Built-in or hang-type thermometers are usually required in refrigerator and freezer units (see Exhibit 6.1). Thermometers are also built into hot-holding equipment and machine dishwashers. Since temperature control is an integral part of food

Exhibit 6.1 Refrigerator-freezer thermometer

Photograph courtesy of Tel-Tru Manufacturing Company, Rochester, N.Y.

sanitation, choosing the appropriate food thermometer and using it properly are of critical importance.

The most versatile type of thermometer for measuring food temperature is the bi-metallic stemmed thermometer (see Exhibit 6.2). These thermometers must be numerically scaled, easily readable, and accurate to ±2°F (±1°C). The metal stem should be at least five inches long with the lower two inches being the sensing area for immersion into foods. Make sure that you buy the kind of bi-metallic stemmed thermometer that has a calibration nut so the device can be adjusted to maintain maximum accuracy.

Thermometers with a scale ranging from 0° to 220°F (−17.7° to 104.4°C) can be used to measure the temperatures of incoming shipments of frozen and refrigerated food temperatures in refrigerators, freezers, and hot-holding equipment; and the temperatures of sanitizing solutions. Due to the risk of physical contaminants to the food and damage to the thermometer, a bi-metallic stemmed thermometer with a plastic lens cover should not be left in food during cooking in an oven, microwave, or on a stovetop range. Mercury-filled or other glass thermometers should *never* be used to measure food temperatures because they can break. In addition, only National Sanitation Foundation (NSF) foodservice approved thermometers should be used.

Other Food Thermometers

The digital thermometer, which presents a rapid, easy-to-read temperature display, is available in a variety of sizes and styles, from bayonet and pocket thermometers to those that are panel-mounted (see Exhibit 6.3). Many are equipped with interchangeable temperature probes, which can be used for surface, immersion, penetration, and air measurement. These thermometers can be used to monitor temperatures in all sorts of foods and food-storage areas.

The Time Temperature Indicator (TTI) is frequently used to monitor temperatures during the transportation or storage of sous vide, modified atmosphere packaged or cook-chill foods. This thermometer has liquid crystals that give a special label on the food package an irreversible color change when the item reaches an unsafe temperature.

Other thermometers are designed specifically to measure certain food items. These include candy, meat, and deep-fry thermometers.

Exhibit 6.2 Thermometer for checking internal temperatures of food. Be sure the dimple on the stem of thermometer is completely immersed or surrounded by the food product for an accurate reading

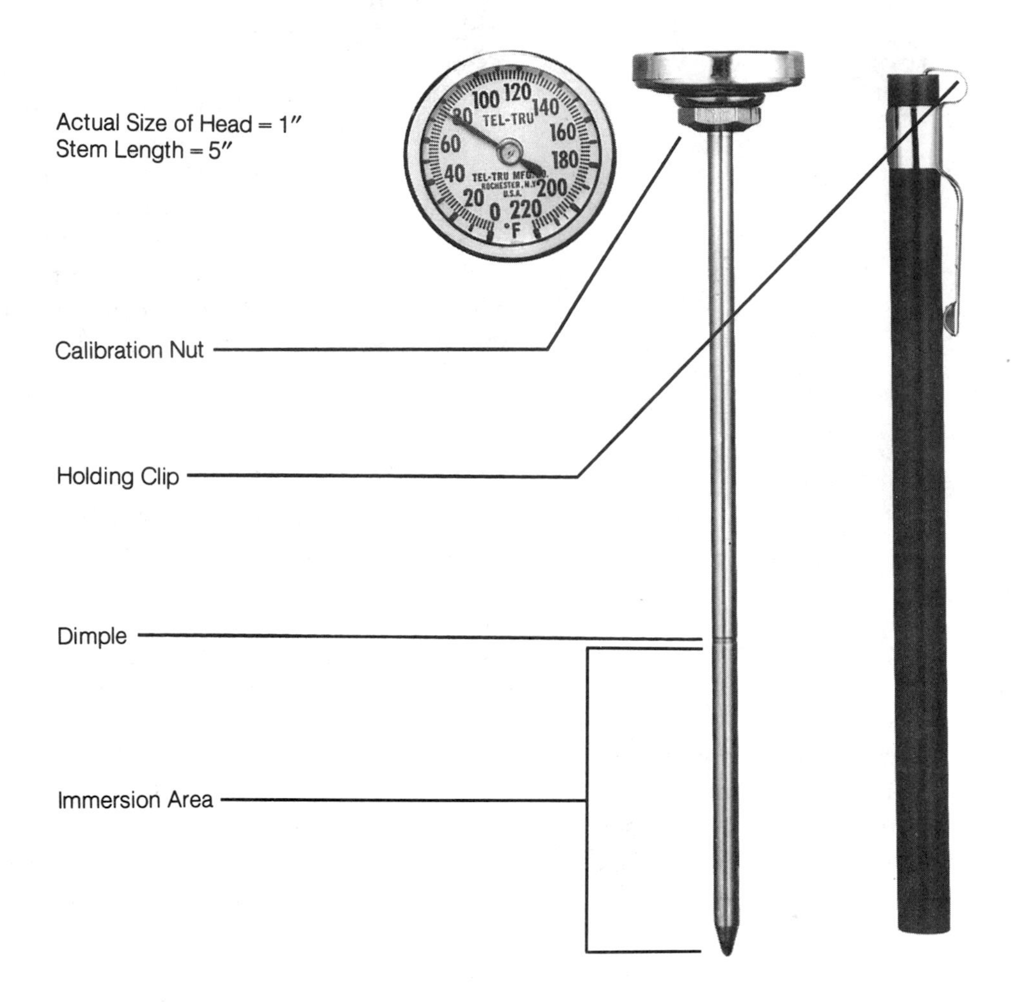

Photograph courtesy of Tel-Tru Manufacturing Company, Rochester, N.Y.

Using Food Thermometers

After selecting an appropriate thermometer, the manager must be sure that each device is accurate, is used properly, and that it does not become a contaminant itself. The following rules should be observed.

- Make sure that all thermometers and their cases remain clean. Wash, rinse, sanitize, and air dry thermometers

Exhibit 6.3 Digital thermometer with interchangeable sensing probe

Photograph courtesy of Tel-Tru Manufacturing Company, Rochester, N.Y.

before and after each use to eliminate any contamination. A sanitizing solution appropriate to food-contact surfaces can be used to sanitize the thermometer.

- Take the temperature in the geometric center of the food. This is usually the center of the thickest part, although composition of the food being tested may alter the heat distribution. After inserting the sensor area into the food product, wait for the needle to stop moving or the numbers on the digital readout to significantly slow. When the needle has been still for 15 seconds, record the reading.
- Recalibrate or adjust the accuracy of the thermometer periodically, and especially after an extreme temperature change, or if the thermometer has

been dropped. Calibrating can be done by using either the *ice point method* (for cold foods) or the *boiling point method* (for hot foods). To use the ice point method, insert the sensing area of the thermometer into a 50/50 ice and water slush until the indicator stabilizes. Then adjust the calibration nut so that the indicator reads 32°F (0°C). To use the boiling point method, insert the sensing area into boiling water until the indicator stabilizes. Then adjust the calibration nut so that the indicator reads 212°F (100°C). The boiling point differs daily with atmospheric pressure changes and depending on the altitude, so adjustments must be made to suit individual altitude levels. The boiling point lowers about 1°F (0.6°C) for each 550 feet above sea level.

- Hanging or sitting thermometers can become damaged in a refrigerator and give you a false high or low reading. Use your calibrated bi-metallic stemmed thermometer or other approved thermometer to verify the accuracy of each hanging thermometer by placing them side by side and taking a comparison reading. An inaccurate hanging thermometer cannot usually be calibrated. It must be replaced.
- Some digital thermometers can be calibrated. Change the batteries based on the amount of use or if an indicator light on certain models lights.

RECEIVING AND INSPECTING FOOD

Receiving food is a control point and for certain foods it can be a critical control point, for example, raw shellfish and pasteurized milk. Setting up procedures for inspection and standards for acceptance

Exhibit 6.4 A HACCP flowchart for safe receiving procedures for beef stew ingredients

Critical Control	Hazard	Standards	Corrective Action if Standard not Met
		Receiving	
Receiving beef	Contamination and spoilage	Accept beef at 45°F (7.2°C) or lower; verify with thermometer	Reject delivery
		Packaging intact	Reject delivery
		No off odor or stickiness, etc.	Reject delivery
Receiving vegetables	Contamination and spoilage	Packaging intact	Reject delivery
		No cross-contamination from other foods on the truck	Reject delivery
		No signs of insect or rodent activity	Reject delivery

and then following them closely is vital to preventing foodborne illness. Add the receiving sections from your flowcharts for potentially hazardous foods to the procedures manual (see Exhibit 6.4). At this receiving stage in the flow of food, rejecting unacceptable products serves as your control: if a food item is not up to your standards, do not accept it. A back-up menu plan is a good idea in case you do have to reject a delivery.

The following rules should help the foodservice operator check the quality of foods entering the establishment:

- Inspect foods immediately upon delivery. Not only will this ease returns or credits from the supplier but it will also make certain that perishable foods are not kept out of proper storage conditions for more than the briefest time. A close inspection using spot checks is necessary since the possibility always exists that any food (even government-inspected and -approved supplies) may have been mishandled in transit.
- Try to schedule deliveries during off-peak hours. This arrangement will make it easier for employees to examine foods properly. Also, stagger the delivery times so not all shipments arrive at the same time. This will also help ensure that enough employees and time are available to inspect everything that comes in and to properly store items as soon as possible.
- Mark all items for storage with a date of arrival, or the *use by* date to ensure proper rotation of stock. (This procedure is discussed in Chapter 7.)
- Plan ahead for the arrival of shipments. Make sure that sufficient refrigerator/freezer space is available. In some well-equipped operations, there is a refrigerator and a freezer in the receiving area for quick and temporary storage. Larger operations may provide facilities for washing certain items of fresh produce to prevent insects and excessive soil from being brought into the operation. Washed produce needs to be drained or dried before storage.
- Make certain that employees are trained properly in inspection functions. Receiving employees should be able to judge quality; check temperatures of refrigerated and frozen foods, eggs and dairy products, fresh meat, fish, and poultry products; know code dates; identify thawed and refrozen foods; detect damage; spot insect infestations; and so on.
- Keep the receiving area well lighted and clean to discourage pests.

Checking Special Package Temperatures

When you inspect deliveries, you should observe and record the temperature, as well as notice the appearance of specific food items. The following procedures describe how to check the temperatures of packaged food items.

Eggs

Check the temperature of a shipment of whole shell eggs by breaking open one or two eggs into a glass and checking the temperature with a sanitized thermometer. Make sure the entire sensing area of the thermometer is submerged. Eggs should be received at a temperature of 45°F (7.2°C), or below.

Milk

If milk is delivered at a temperature above 45°F (7.2°C), do not accept the shipment. Check the temperature by opening one carton of milk and taking the milk's temperature with a sanitized thermometer. If milk is packaged in a plastic bulk dispenser, check the temperature by folding the soft plastic around the thermometer. Be careful not to puncture the plastic with the thermometer.

Modified Atmosphere Packaged Foods (MAP)

Raw, cooked, or cured meats; and smoked or raw fish are sometimes packaged in clear plastic or shrinkwrapped packaging that is called *modified atmosphere packaging* (MAP). Check the temperature of these products by holding the thermometer tightly between two packages, being careful not to puncture or break the wrapping. Closely examine the temperature indicator on the package to see if the product was temperature abused at any point in transit. If the color indicators do not match, reject the shipment.

If packages are refrigerated, make sure that the package is at the recommended temperature for the item; if they are frozen, they should be 0°F (−17.7°C) or below. Time and proper refrigeration of MAP foods, especially certain products that require minimal or no extra preparation, is the only barrier to microbial growth. Also check the date recommended by the manufacturer for shelf life to be sure it has not expired.

Refrigerated Entrees

These new-generation foods are packaged partially or fully cooked. They only require reheating. Check the temperature in the same manner in which you would check MAP foods. Make sure that these entrees are received at a temperature of 45°F (7.2°C) or below, or at manufacturer's specifications. Ready-to-eat, chilled entrees and food products that require no cooking or heating must also be received at 45°F (7.2°C) or below.

Packaged Frozen Foods

For frozen foods, temperatures can be checked by opening the box or cutting the sidewall of a case of the product being careful not to cut the packages inside. Insert the probe of the thermometer into the case, placing it between the packages in the case, again being careful not to pierce interior packages. Make sure that the temperature is not above 0°F (−17.7°C). Once the temperature has been determined, reseal the case with tape, write the date, and initial it, in order to explain the condition of the case to other employees.

The remaining sections of this chapter give some pointers on the inspection of specific food items.

Meat

When meat arrives at a foodservice facility, it should be inspected on the basis of several factors: temperature, color, odor, texture, and packaging. It is required that the foodservice operator purchase only meat that has been inspected by the USDA or the state department of agriculture. Such meat is stamped with a circle containing the abbreviations for "inspected and passed" and the number identifying the processing plant. The stamp will not appear on every cut of meat, but it should be present on every inspected carcass and also on cases or bags.

The foodservice operator can determine from the supplier whether cuts of meat are from inspected carcasses. It is generally advisable to ask suppliers for written proof of government inspection of meats.

The *inspection service* of the USDA should not be confused with its *grading service.* Grading is a voluntary service paid for by the packers. *Grades* refer to the quality and palatability of the meat, not its sanitary or wholesome condition. "USDA Prime" is not necessarily any more sanitary or wholesome than "USDA Choice." The USDA grade information is printed in a shield-shaped symbol. (See Exhibit 6.5.)

Exhibit 6.5 Inspection and grade stamps for meats: *top,* wholesomeness; *bottom,* quality

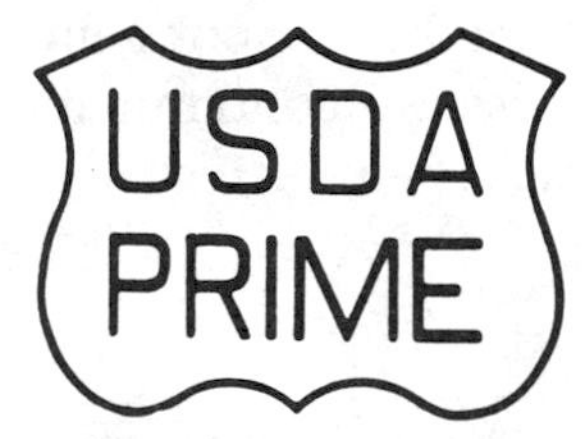

Source: U.S. Department of Agriculture

Inspection is *mandatory;* it is done to ensure the safety and wholesomeness of food items. Wholesomeness does not mean freedom from disease-causing micro-organisms. What the USDA means by wholesomeness is that the carcass and viscera of the animal were examined and no signs of illness were indicated, and conditions met sanitation standards. Wild game, game birds, and other exotic meats, such as alligator and snake, may not be approved by the regulatory agency that inspects your facility, unless it has been slaughtered and processed in an approved facility. Grading, as mentioned above, is *voluntary,* and is done to identify the relative quality of food items.

With all meats, discoloration is a cause for concern. Brown, green, or purple blotches are all signs of microbial attack. Black, white, and green spots may indicate molds or could be freezer burn.

Unfrozen meat should be firm and elastic to the touch. No meat should feel slimy, sticky, or dry. Generally, the appearance of slime, caused by microbial growth on meats stored under conditions of high temperature and humidity, is the very first indication of deterioration in meat. A sour smell is grounds for rejection, except in the case of deliberately aged beef.

Fresh meat temperature should be checked before it is put in storage. Upon arrival, chilled meat should have a maximum internal temperature of 45°F (7.2°C) or below. Frozen meat should be below a temperature of 0°F (−17.7°C) and show *no* signs of thawing. (See Chapter 7 for information on storing meat.)

When the meat arrives, the packages should be inspected for broken cartons, dirty meat wrappers, and tears, which are all indicators of poor sanitary practices and possible contamination. These products must be rejected.

Beef

Beef should be a bright, cherry red, although deliberately aged beef may be darker and packaging may influence the color. A purplish tinge may simply mean that the beef has not been exposed to the air for a long enough period of time, though it can also be a sign of deterioration. Do not accept fresh beef that is turning brown or greenish. Beef usually spoils first at or near the surface of the cut. Check for evidence of *freezer burn,* the loss of water from the surface of a frozen food. Freezer burn results in a lightened area on the surface of beef, which creates an undesirable appearance and texture. Be particularly careful inspecting ground beef, since it is the most subject to spoilage.

Lamb

Fresh lamb is light red in color if it has been properly exposed to air. Fresh lamb should be rejected if the color is brown, or if there is a whitish surface covering the lean meat. Check for freezer burn on frozen cuts.

Pork

The fat portions of pork should be white and firm, the lean portions light pink. Deterioration is usually evident by darkening of lean meat and discoloration and rancidity of the rind. Like beef, pork spoilage is first evident near the surface of the cut. In a few cases, pork has been reported to spoil in the interior near the bone. Test for this condition by pushing a cleaned and sanitized knife into the flesh and smelling the knife tip. A sour odor is cause for rejection.

Other Meat Products

Sausage should be free of slime and mold, which indicate decomposition. External mold is common on dry sausages, such as hard salami, but it can be washed off and is considered harmless if it is confined to the casing. Its temperature should be recorded using the method for modified atmosphere packaged products discussed earlier in this chapter. Meat that is packaged using a process called *wet aging* needs to be checked to see if the packaging is torn or too loose.

Poultry

All poultry (chickens, ducks, turkeys, and so on) to be sold for human consumption is inspected by either the federal or state government. The USDA, through its Food Safety and Inspection

Service (FSIS) inspects poultry for wholesomeness. Federal legislation passed in 1968 also provides for involvement by the states in poultry inspection.

Although the carcasses have been inspected, processed poultry products may or may not be inspected—this is a voluntary program that the processors request. It is wise to ask suppliers of processed poultry if their operation has been inspected by federal or state inspectors.

There is a grading system for fresh and frozen poultry quality, performed by the USDA for the individuals or firms who desire on a voluntary basis to pay for this service. The shield with the grade stamp may appear on any kind of chilled, frozen, or ready-to-cook poultry or poultry parts, either on a section of the carcass or the packaging. Grading may also be done in cooperation with a state, in which case the official grade shield may include the words "Federal-State Graded." Exhibit 6.6 depicts the USDA inspection and grade stamps for poultry.

Indications of spoiled or inferior poultry are readily observable. Soft, flabby flesh usually means an inferior product. A purplish or greenish overall cast, or a greenish discoloration around the neck and neck opening may mean staleness, improper bleeding after killing, or improper handling. Other signs of spoilage include an abnormal odor, stickiness under the wings and around the joints, and darkened wing tips.

Fresh poultry should be graded A, surrounded by crushed ice, and delivered at a temperature below 45°F (7.2°C). Temperatures below 28°F (−2.2°C) may significantly extend shelf life. It is important

Exhibit 6.6 Inspection and grade stamps for poultry: *top*, wholesomeness; *bottom*, quality

Source: U.S. Department of Agriculture

to remember to handle all poultry products carefully because even though it is inspected, poultry is commonly contaminated with *Salmonella.*

Eggs

Purchase only the amount of eggs needed for one or two weeks. These eggs should be grade AA or A, with clean, uncracked shells. The USDA shield on egg cartons indicates that federal regulations are enforced to maintain quality and reduce contamination (see Exhibit 6.7). The shells should not be cracked, checked, or dirty. (For a complete discussion on egg handling see Chapter 3.)

Use only vendors who deliver eggs in refrigerated trucks. Eggs should be stored in their original containers and refrigerated immediately upon delivery.

Exhibit 6.7 Inspection and grade stamps for eggs: *top*, wholesomeness; *bottom*, quality

Source: U.S. Department of Agriculture

Until they are needed, keep eggs at a refrigerated temperature of no more than 45°F (7.2°C).

To test the freshness of a shipment of eggs, break one open. If the white clings to the yolk, and the yolk is firm, high, and does not break easily, the egg is acceptable (see Exhibit 6.8). In a very fresh egg of top quality, the white will stand up well on its own. Sniff the egg. There should not be any odor to a fresh egg.

Liquid, frozen, and dehydrated eggs are being used increasingly by foodservice establishments. These products are pasteurized as required by law. They must be refrigerated or frozen, as appropriate, at delivery and at all storage points. They must also be checked for use-by dates.

Fish

Few food items are less able to withstand improper handling than fresh and frozen fish. Fish is highly susceptible to deterioration. Fresh fish should be packed in crushed or flaked, self-draining ice. Upon arrival, fish should be between 32° and 45°F (0° and 7.2°C).

Fresh fish of acceptable quality can be distinguished from old fish by marked differences in appearance. Fresh fish have bright red, moist gills, the eyes are bulging and clear. The flesh and belly areas are firm and elastic. Fresh fish do *not* have a noticeably strong, fishy odor. The flesh should not pull away from the bones easily, and the scales should stick to the flesh. The skin should be vibrant and bright.

An unacceptable fish presents a total contrast. The gill slits are gray, or gray-green, and dry. The gills are darker and dull red, brownish, or gray. The eyes are cloudy, red-bordered, and sunken. The flesh is soft and yielding. If finger pressure is applied, the impression will remain. If fish has an ammonia odor, the deterioration is advanced. Fish may also contain parasites, tumors, abscesses, and cysts.

Fish fillets intended for raw or lightly cooked consumption should be frozen. Acceptable frozen fish must show no indications that it has been allowed to thaw. Fish that has been thawed and refrozen before it reaches the foodservice establishment may have a sour odor and an off-color. The paper in which it was wrapped will likely be moist, slimy, and discolored. The bottom of the shipping carton may

Exhibit 6.8 Egg quality by grade

	Grade AA	Grade A	Grade B
Break out appearance	Covers a small area.	Covers a moderate area.	Covers a wide area.
Albumen appearance	White is thick and stands high; chalaza prominent.	White is reasonably thick, stands fairly high; chalaza prominent.	Small amount of thick white; chalaza small or absent. Appears weak and watery
Yolk appearance	Yolk is firm, round and high.	Yolk is firm and stands fairly high.	Yolk is somewhat flattened and enlarged.
Shell appearance	Approximates usual shape; generally clean,* unbroken; ridges/rough spots that do not affect the shell strength permitted.		Abnormal shape; some slight stained areas permitted; unbroken; pronounced ridges/thin spots permitted.

*An egg may be considered clean if it has only very small specks, stains, or cage marks.

Source: U.S. Department of Agriculture

have excessive ice formations or liquid, indicating that refreezing has taken place; and the container itself may be deformed by internal pressure. Brown coloring at the edges of a fillet is also a sign of refreezing. These signs that indicate the fish was temperature abused and whatever bacteria were present might have grown.

Shellfish

Shellfish can be placed in two basic categories. Crustacea, such as lobster, crabs, and shrimp, is one. The other is molluscan bi-valve shellfish, such as oysters, clams, and mussels. Shellfish can be shipped in various forms: live, fresh, frozen, whole, in-shell, or shucked. When shellfish are shipped live, they must be alive upon delivery and packed in nonreturnable containers. The FDA oversees the interstate shipping of shellfish, but the foodservice manager should verify the safeness of molluscan shellfish and crustacea suppliers.

The shell of a live lobster should be hard and heavy. Live lobsters and crabs should show signs of movement. The tails of live lobsters should curl under when the animals are picked up. The shells of clams and oysters will be closed if they are alive. Partly open shells might indicate that the clams, mussels, and oysters are dead. To determine if they are, tap the shells; if the shells close, they are alive.

The edible portions of frozen lobsters should have firm flesh. If shellfish give off a strong odor, they should be rejected. Frozen lobsters are acceptable for use as long as there is no evidence of shrinkage or a change in the normal contours of the unshelled lobsters. All molluscan shellfish must come from reliable, certified sources.

As a safety measure, the FDA requires that foodservice operators keep shellstock identification tags on file in the establishment for 90 days after the receipt of clams, mussels, and oysters. Upon delivery, the foodservice operator writes the date on the tag (see Exhibit 6.9).

Fresh Produce

Fresh produce must be handled with extreme care because of its perishability. Pinching, squeezing, or unnecessary handling upon delivery will bruise fruits and vegetables, leading to premature spoilage. Upon delivery, produce cartons should be checked for signs of insect infestation and for insect eggs and cockroach egg cases.

For fruits, taste is the best test of quality. Many foodservice operators rely on appearance as an indication of quality, but this standard may not, in all respects, be dependable. Blemishes can be present even though the flavor and quality are unimpaired.

Fruits and vegetables show spoilage in a variety of ways. The foodservice operator should be able to identify not only produce that is obviously spoiled, but produce that will spoil under storage in a very short time. Appendixes A and B list the signs of freshness and spoilage in common fruits and vegetables.

Dairy Products

Unpasteurized milk and dairy products are a potential source of micro-organisms that cause foodborne illnesses, including salmonellosis, campylobacteriosis, and listeriosis. Therefore, all cartons and bulk containers that enter a foodservice establishment must carry the pasteurization label. Cream, dried milk, cottage cheese, Neufchâtel cheese, and cream cheese must all be made from pasteurized milk.

All market milk is Grade-A quality. Local, state, and federal milk-control programs combine to assure the consumer that the milk they buy is unadulterated, taken from safe cows, and processed using sanitary methods.

Generally, milk intended for use as a beverage must be packaged in individual containers, not larger than one pint, or may be served from a bulk dispenser unit. Milk and milk products may be poured from commercially filled containers of less than one-half gallon capacity for use in cooking. Dried milk or milk in bulk containers of 5 to 10 gallons may be used for cooking and serving purposes, as local regulations permit.

Fresh milk has a sweetish taste. Sour, bitter, or moldy-tasting milk must be rejected. In many states, the health departments require that milk be marked with a date. Milk that is delivered after the expiration date marked on the

Exhibit 6.9 Facsimile of shell-stock identification tag

THIS PACKAGE CONSISTS OF OYSTERS

____ Gals.

____ Bu.

THIS TAG IS REQUIRED TO BE KEPT ON CONTAINER UNTIL EMPTY AND THEREAFTER KEPT ON FILE FOR 90 DAYS.

Packed by

Address: Bon Secour, Ala.

Distributed By ____________

Address ____________

Date Reshipped

SHELLFISH DREDGED FROM

BAY GARDENE

LOCAL AREA OR BED NO. LA 51

DATE

01/10/84

TO BE RETAINED BY RECEIVER FOR 90 DAYS.

TO:

OYSTERS

ALABAMA STATE BOARD OF HEALTH

BUREAU OF SANITATION

DIVISION OF INSPECTION MONTGOMERY, ALA.

No. 00576

SHIPPER'S NAME AND ADDRESS

FROM:

Bon Secour, AL 36511

Certificate No. Ala 49

BELOW TO BE FILLED IN BY RECEIVER

DATE REC'D

LOT. NO.	LOT CONSISTS OF

Source: Alabama State Board of Health

container must be rejected, as well as milk with a temperature above 45°F (7.2°C) on delivery. Milk with off flavors—possibly from odor absorption or from abnormal conditions in the cows—may still be wholesome, but nonetheless should be rejected. It will not taste good to customers.

Butter should have a sweet, fresh flavor, uniform color, and firm texture. It should be free of mold, specks, and other foreign substances and should be received in clean, unbroken containers. Butter that is rancid or that has absorbed foreign odors should not be accepted.

In the United States, the composition of cheese is regulated by a government standard of identity specifying the ingredients that may be used, the maximum moisture content, the minimum fat content, and the requirements for pasteurization or holding of the milk to remove harmful bacteria. Cheese should be checked to see that each type has its characteristic flavor and texture, as well as a uniform color. If the cheese has a rind, the rind should be clean and unbroken. Cheese that is dried out or has mold that is not natural to the product should be rejected.

Frozen Foods

Frozen foods should be inspected for signs of thawing and refreezing as well as for other signs of deterioration. Obvious signs of thawing and refreezing include fluid or frozen liquids at the bottom of the food carton, or the presence of large ice crystals in the product itself. The temperature of frozen foods, including meat, should not be above 0°F (−17.7°C). Ice cream, however, may be delivered and stored at a temperature of 6° to 10°F (−14.4° to −12.2°C).

Canned Foods

The foodborne illness botulism is so dangerous that all deliveries of canned goods must be inspected carefully. Check the can exteriors following the steps outlined in Exhibit 6.10. External indications of damage and possible contamination are illustrated in Exhibit 6.11 and described in the following list:

1. *Swelled top or bottom.* One or both ends of a can may bulge outward as a result of gas produced by bacterial or chemical action inside the can. Whether the ends spring back when touched by a finger or not, discard the can. Sometimes both ends of a can will appear flat, though one will bulge outward when the other is pressed—reject.
2. *Leakage.* Any can with a leak should be rejected.
3. *Flawed seals.* Improper sealing at the top or side of the can is cause for rejection.
4. *Rust.* Rusted cans should be rejected. They're either too old to use or the contents may be contaminated. Rust may conceal pinholes through the can, which lead to contamination.
5. *Dents.* Dents along the side or top seams of a can, or dents that make it impossible to open the can with a manual can opener, are cause for

Exhibit 6.10 Inspection of can exteriors: (*A*) hold can in both hands and observe can surfaces for obvious defects; (*B*) if can has a paper label, observe label for stains that may be evidence of leakage, rust, etc.; (*C*) put slight pressure on one end and observe any movement of either end; (*D*) then turn can over and repeat the procedure for the other end

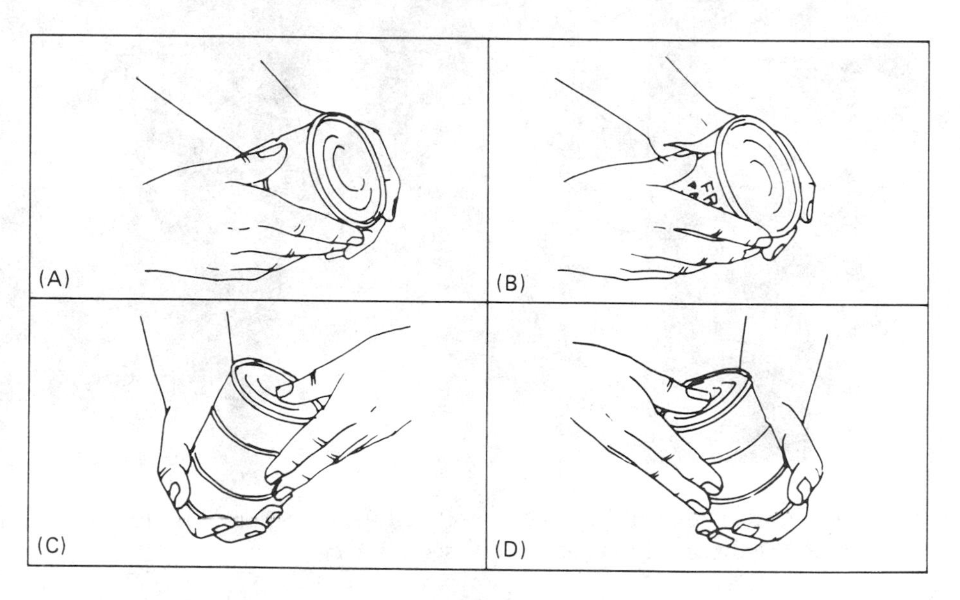

Source: "Guidelines for Evaluation and Disposition of Damaged Food Containers." Bulletin 38-L, 2d ed. Copyright © 1979 by the National Food Processors Association. Reprinted by permission.

rejection. The seams have probably been broken. Check with the local health departments on regulations concerning dented cans. Some jurisdictions do not allow *any* dents.

After examining the exterior of the cans, spot checking should be done on the contents. A can that seems undamaged on the outside may still contain contaminated food. Any canned goods that appear abnormal in odor, color, or texture; that are foamy; or that have a milky-colored liquid, which is not natural to the product, must be rejected. *Never taste-test goods that have these characteristics.* People have died from botulism poisoning contracted from tasting *and spitting out* contaminated food.

Dry Foods

Cartons of dried fruits and vegetables, cereals, sugars, and flour should be dry and undamaged. Punctures, tears, or slashes in the package may indicate insect or rodent entry. If the outside of a container is damp or moldy, the condition may extend to the contents, raising the possibility of advanced growth of micro-organisms. Most dry foods are poor media for bacteria, but a touch of

Exhibit 6.11 Serious defects in cans that are cause for rejection

Bulged ends (except beer or other carbonated beverage cans), including bulged lids caused by dents.

Pull-top containers with fractures of the lid score lines.

Badly rusted with deep pits that are nearly perforated.

Crushed to the point that they cannot be stacked on shelves or opened with can openers.

moisture can radically change this situation. Dry foods *must be kept dry.*

Dry foods that are not of the color or odor standard for the product also need to be rejected. Spoiled baked goods, for example, may have a slimy appearance. The best way to spot tiny insects or insect eggs in cereal or flour is to sprinkle a bit of the product on brown paper. Look for webbing in the product or sometimes small holes in the packaging.

Aseptic and Ultra-Pasteurized Packaged Foods

A food product that is packaged *aseptically* has been heat treated and is free of pathogenic micro-organisms. Some milk products, such as individual half-and-half creamers, as well as some fruit juices, and other products are being aseptically processed or ultra-pasteurized. A Grade A dairy product that is labeled *ultra-pasteurized* may not always be in an aseptic package. Therefore, the temperature of the milk or dairy products must be below 45°F (7.2°C) for delivery, storage, preparation, service, and display.

Grade A milk and dairy products that are labeled UHT and whose package instructs, "keep refrigerated after opening," have been aseptically packaged. These products can be stored safely without refrigeration. Some liquid coffee creamers are being processed by this method now. Make sure that upon delivery aseptic and ultra-pasteurized packaged foods are carefully checked for safe temperatures.

Modified-Atmosphere Packaged Foods (MAP)

Modified atmosphere packaging (MAP), such as vacuum packaging and sous vide (both of which will be discussed in greater detail in Chapter 7), involve processes that usually result in an oxygen-free atmosphere in the package. Such conditions can encourage the growth of *Clostridium botulinum* and other anaerobic pathogens.

The FDA does not permit a foodservice operation to prepare sous vide on-site. Sous vide indicates raw or partially cooked food placed in a pouch with the air removed. The pouch is sealed, then cooked and refrigerated until needed. To make sure that you obtain safe food items, purchase MAP products only from approved suppliers. Suppliers must give evidence that they understand the equipment, the procedures, and the concepts that are essential for safe MAP products.

When inspecting such items, accept only packages that are received intact. Fish and fish products, which have a history of botulism, present a great risk when MAP is used. In addition, reject any packages that contain bubbles, which can indicate the presence of gas or toxins from pathogenic micro-organisms. Follow the manufacturers' instructions for the temperatures needed during transit and delivery.

REJECTING SHIPMENTS

As you establish control points and critical control points at each step in the flow of food, you will see that if food is not safe and wholesome when it is delivered, it can pose serious dangers as it advances through the other stages in the flow of food. You monitor this control point by rejecting food that does not meet your standards.

Employees who receive food need to be instructed to notify a manager if a delivery is questionable. To reject a delivery from a supplier be sure to:

- Identify what is wrong with the delivery
- Reject the order tactfully, but firmly
- Obtain an adjustment or credit.

When you must reject a shipment, follow your established company policies for doing so. Become familiar with the procedures that should be followed and with any written documents that must be completed and processed. You should record the reasons why the order was rejected and report it to the supervisor if necessary.

SPOILAGE WITHIN THE FOODSERVICE FACILITY

The same guidelines that apply to preventing spoiled and contaminated foods from entering the establishment apply to the spoilage of food *within* the establishment. (The standards for storing food and food products will be covered in Chapter 7.)

When it comes to evaluating specific items of food, it is often difficult to tell the difference between spoilage and contamination. *Spoilage* is damage to the edible quality of food, which is observed by appearance, odor, or texture changes. *Contamination* is the unintended presence of harmful substances in food or water, which may not be visible or detected through taste or odors. For the foodservice manager, the best rule to follow is *when in doubt, throw it out.*

SUMMARY

Purchasing safe food supplies may be a critical control point in the flow of foods. Although various state and federal governmental agencies regulate the production and processing of foods, including meats, finfish, shellfish, poultry, canned goods, eggs, and dairy products, it is still the responsibility of the foodservice manager to check the sanitary quality of foods used in the establishment.

A complete inspection needs to be made of all incoming supplies. Plans must be made so that deliveries can be handled promptly and correctly.

The thermometer is an extremely important instrument in the prevention of foodborne illness and contamination. Management must make sure that thermometers are available and used properly. Thermometers must be clean and sanitized before and after every use.

Meats, poultry, and finfish should be checked for color, texture, and temperature on delivery. Live molluscan shellfish and crustacea must be delivered alive or properly packed fresh or frozen. Produce must be fresh and wholesome. Milk, eggs, and other dairy products must be checked for temperature and freshness. Frozen goods need to be checked for temperature and signs of thawing and refreezing. Canned goods must be inspected for signs of damage, and dry goods must have dry and undamaged packaging. Modified atmosphere packaged, aseptic, and ultra-pasteurized products all need to be checked for intact packaging and for the temperature recommended by the manufacturer.

Employees responsible for inspecting incoming supplies should be able to identify damaged, contaminated, and spoiled products, and how to properly reject a shipment according to company policies. On the receiving dock and within the establishment, products of suspect quality must be rejected or discarded.

A CASE IN POINT

On Monday, a number of food deliveries at the White Oaks Nursing Home arrived during the lunch hour. The food products were all different—cases of frozen salisbury steaks; canned vegetables; frozen shrimp; fresh tomatoes; fresh chicken for that night's dinner; and a case of potatoes.

Betty, the new assistant manager, thought that the best thing to do since she was extremely busy was to put everything away and check it later. She asked the storeroom attendant to put the frozen salisbury steaks and shrimp in the freezer and the fresh chicken in the refrigerator. The attendant placed the fresh tomatoes and potatoes into dry storage, while Betty placed the canned vegetables there as well.

"Whew," said Betty. "I'm glad that's over!" Then she went back to the front of the house.

What has been done incorrectly?

What could be the possible result?

STUDY QUESTIONS

1. Give examples of questions that should be asked when a foodservice manager checks out the sanitary methods used by a supplier.
2. List at least three general rules for safely receiving food products.
3. Name two types of food thermometers that measure internal temperatures.
4. How do you calibrate a bi-metallic stemmed thermometer?
5. Describe the different methods for checking the temperature of frozen foods, shell eggs, and bulk milk.
6. Describe the delivery requirements for MAP foods.
7. Describe three signs of spoiled poultry.
8. Describe the appearance of fresh and old fish.
9. Give a reason to reject a shipment of milk.

10. Name five external signs that indicate a canned food item should be rejected.
11. What is the inspection procedure for dry food items?
12. Describe the procedure for rejecting a food delivery.

ANSWER TO A CASE IN POINT

Usually, *later* never comes, so we have to assume that the food items at White Oaks Nursing Home went unchecked until each product was ready to be used. Even then, the cook or the server could be in such a hurry that the inspection may be cursory or nonexistent. Therefore, a control point has been bypassed. Consider the following:

1. Betty, or her manager, should have arranged with each supplier to bring the food deliveries at a time when business was slower so that Betty and other employees would have had time to inspect each shipment. She should also try to stagger the deliveries so that everything does not arrive at once.
2. Betty, or the attendant helping her, should have marked the date of arrival on each case of products received before they went into storage.
3. The canned goods were not inspected for damage, nor were the fresh vegetables checked for signs of spoilage, before being put away.
4. The frozen foods may or may not have been delivered at the proper temperature. The shrimp should have been checked to be sure no thawing and refreezing had taken place.
5. The poultry and any other potentially hazardous foods should have been checked for correct temperatures and appearance.

Remember, a good purchasing/receiving program is a check of the quality and the sanitation of food products. All in all, the failure to inspect the delivered food could prove disastrous for the operation—and for the residents of the nursing home. The quality and safety of the items received have become uncertain.

In the next chapter, Angie, another employee at the White Oaks Nursing Home, will prepare the questionable chicken, unaware that it had been improperly received.

MORE ON THE SUBJECT

NATIONAL FISHERIES INSTITUTE. *SeaCare: A Quality Program for Food Service Operators.* Arlington, VA: National Fisheries Institute, 1990. A training package that includes a 12-page manual and a 20-minute video tape that both cover purchasing, receiving, storing, handling, preparing, and serving all kinds of seafood.

UNITED FRESH FRUIT AND VEGETABLE ASSOCIATION. *Fresh Fruit Selection and Care,* and *Fresh Vegetable Selection and Care.* Alexandria, VA: United Fresh Fruit and Vegetable Association, 1990. #PL1414 and #PL1418. Detailed brochures describe the criteria for purchasing fresh produce.

ASSOCIATION OF FOOD AND DRUG OFFICIALS. *A Pocket Guide to Can Defects.* York, PA: Association of Food and Drug Officials. rev. 1987. 25 pages. This booklet gives definitions of each class of damage—critical, major, and minor—and examples of what constitutes each category. There are numerous photos to illustrate each example.

FROZEN FOOD COORDINATION COMMITTEE. *Code of Recommended Practices for the Handling and Merchandising of Frozen Foods.* Washington, DC: Frozen Food Coordination Committee, 1982. This publication supplies a set of specific recommendations for the handling of frozen foods.

7

Keeping Food Safe in Storage

In a HACCP-based system, proper storage is another line of defense from microbial growth and contamination. While storage is necessary, the quality of most food does not improve over time, and incorrect storage practices have the potential to cause serious and costly problems. There is a direct relationship between cost control, food safety, and the need to maintain good storage practices. For example, insects can infest dry stores and contaminate food if sanitation standards are not maintained. The replacement of these damaged products is costly.

This chapter will discuss methods for preventing the contamination and spoilage of foods in storage. We will cover the following main topics:

- Fundamental principles of storage in foodservice facilities through flowcharting.
- Elementary rules for the use of refrigerators, freezers, and dry-storage facilities.
- Appropriate storage procedures for most common foods.

STORAGE PRINCIPLES

Despite the wide variety of products found in a foodservice facility, a few general principles can be successfully applied to most storage situations. The rules that follow cover storage of all types of foods:

1. Follow the rule of *First In, First Out* (FIFO). Sticking to this principle means that goods should be used in the order in which they are delivered. For example, do not put today's frozen beef roasts into the freezer in front of those delivered last week. Create a system, such as dating goods on receipt and placing new deliveries *behind* those already in storage to guarantee that FIFO is followed.
2. Keep potentially hazardous foods out of the temperature danger zone, which is 45° to 140°F (7.2° to 60°C).
3. Store food only in areas designed for storage. There is no excuse for storing food products near chemicals, in toilet areas, in boiler or furnace rooms, under stairways, or in vestibules. Such procedures are dangerous because of the risk of contamination and are prohibited by local health codes.
4. Keep all goods in clean, undamaged wrappers or packages. A dirty wrapper can attract pests or contaminate food as it is being opened. Packages that are torn or damaged are a sign of potential contamination or pest damage. Packaging should not be reused. Unless special conditions apply, wrap products in material that is moisture-proof and airtight. If goods are removed from the original packaging, they should be placed in clean and sanitized food-grade containers with tight-fitting lids.
5. Keep storage areas clean and dry. This rule applies to dry storage, refrigerators, freezers, and heated cabinets.
6. Keep vehicles for transporting food within the establishment clean. It is senseless, for example, to wrap meat properly, refrigerate it at the optimum temperature, store it in a clean refrigerator, and then carry it to the kitchen on a cart used to transport garbage.

The sections of the flowcharts for potentially hazardous foods that covers storage should be added to your procedures manual (see Exhibit 7.1). For the foodservice employees who will be in charge of storage, the instructions should also be in a handy simplified form, such as a wall chart.

TYPES OF STORAGE

Storage involves four major areas: (1) refrigerated storage for short-term holding of perishable and potentially hazardous food items; (2) freezer storage for longer-term storage; (3) deep chilling in specially designed units for short periods of time; and (4) dry storage for somewhat longer holding of less perishable items. Each storage area has its own particular sanitation requirements.

Refrigeration

The failure to properly cool hot food leftovers, and store cold food items out of the temperature danger zone, at or below 45°F (7.2°C) is often a major factor in foodborne illness outbreaks. To prevent procedural mistakes, you should anticipate problems and establish controls to ensure proper refrigerated storage.

By maintaining potentially hazardous foods at low internal temperatures of

Exhibit 7.1 A HACCP flowchart for safe storage procedures for beef stew ingredients

Critical Control	Hazard	Standards	Corrective Action if Standard not Met
		Storage	
Storing raw beef	Cross-contamination of other foods	Store on lower shelf	Move to lower shelf away from other foods
	Bacterial growth and spoilage	Label, date and use FIFO rotation	Use first; discard if maximum time is exceeded or suspected
		Beef temperature must remain below 45°F (7.2°C)	Discard if time and temperature abused
Storing vegetables	Cross-contamination from raw potentially hazardous foods	Label, date, and use FIFO rotation	Discard product held past rotation date
		Keep above raw potentially hazardous foods	Discard contaminated, damaged, or spoiled products

45°F (7.2°C) or less, refrigeration slows bacterial growth. Storing raw food products below cooked foods or foods that will receive no cooking and covering foods helps prevent cross-contamination of refrigerated foods. However, the time factor is also important. No food can last forever—even in refrigerated storage. Some deterioration eventually will occur with bacteria. Refrigerators should not be used for long-term storage.

Operating Practices

The most efficient refrigerator can become a hazard or fail to hold foods at the proper temperature if employees do not follow standard operating practices.

Refrigerators should not be overloaded. This practice not only makes cleaning difficult but it prevents air circulation needed to maintain proper temperatures. Opening and closing the door of a refrigerated unit too often allows warm kitchen air to flood the interior, which can affect temperature and food safety. In addition, the time that the door *stays* open should be minimized to help maintain unit temperatures. The use of cooler curtains is recommended for walk-ins. Adding plastic insulating strips may help maintain the temperature in walk-ins.

The colder a food item is kept, the safer it is. It is important to note that keeping food cold also protects its quality. For example, quality can be better protected by holding fresh, raw fish in crushed, self-draining ice, at 32°F (0°C). In general, the lower the product temperature is kept, the longer the shelf life of the product. *Shelf life* is the time a product can be stored without serious change in quality.

Different types of refrigeration units serve various purposes. It is essential that each type is used only for its intended purpose. A storage refrigerator, for example, is intended to be used only to maintain the temperature of cold foods and is not capable of cooling large amounts of hot foods. A blast chill unit or a tumble-chill unit is made to receive and to pull down the temperature of hot food. (A complete discussion of types of refrigerators and their sanitation features is in Chapter 9.)

Foods, whether raw or prepared, that have been removed from their original package for storage in a refrigerator should be placed in clean, nonabsorbent, and covered containers. Food cartons should not be stored in such a way as to interfere with the circulation of cold air. For the same reason, shelving should be the open, slatted type. Lining the shelves—with aluminum foil, for example—may make them look neater, but it drastically reduces refrigeration efficiency. Do not store food in a can once the can is opened, as the can acts as an insulator and keeps the cold air from reaching the food product.

Ideally, separate refrigerators should be maintained for each major food category. For example, the optimal storage temperature for fish, meats, and poultry would cause some fruits and vegetables to freeze. Where this flexibility does not exist, meats, fish, and dairy products should be stored in the coldest part of the unit, away from the door. Dairy items in particular should be tightly covered to prevent odor absorption. It is much safer to store ready-to-serve food in separate units from raw foods. However, where it is not feasible to store ready-to-serve food separately, prepared foods are required to be stored *above,* not below, raw foods to prevent cross-contamination.

Refrigerator Temperatures

A maximum refrigerator air temperature of 40°F (4.4°C) or lower must be maintained to keep the food product below 45°F (7.2°C). The unit should be regularly checked with a reliable thermometer. Ideally, foods should be kept at the coldest temperature possible to maintain safety and quality. When separate refrigerators are used for different foods, the optimum range for air temperatures and humidity factors for quality control should be as follows:

- Meat and Poultry: 32° to 40°F (0° to 4.4°C); 75 to 85 percent relative humidity
- Fish: 30° to 34°F (−1.1° to 1.1°C); 75 to 85 percent relative humidity
- Live Shellfish: 35° to 45°F (1.7° to 7.2°C); 75 to 85 percent relative humidity
- Eggs: 38° to 40°F (3.3° to 4.4°C); 75 to 85 percent relative humidity
- Dairy Products: 38° to 40°F (3.3° to 4.4°C); 75 to 85 percent relative humidity
- Most Fruits and Vegetables: 40° to 45°F (4.4° to 7.2°C); 85 to 95 percent relative humidity

Temperatures should be monitored and checked regularly in all refrigerators. There should be a hanging thermometer for each temperature zone of the refrigerator. One should be in the warmest part of the unit, near the door, and one in the coolest part, near the back, to make sure that the proper minimum temperatures are maintained. Check the temperature frequently by reading each thermometer. The accuracy of any thermometer should be checked frequently as well. With good air circulation, the temperatures should be consistent. It is important that food product temperatures be checked frequently with a metal-stemmed thermometer.

Many commercial foodservice refrigerators are equipped with externally mounted or built-in thermometers. They allow the foodservice manager and employees to monitor the temperature at any time without opening the door.

Deep Chilling

Deep or super chilling can be used to safely hold foods for short periods of time without damaging quality. *Deep chilling* differs from regular freezing and refrigeration. It has been found that bacterial growth can be sufficiently decreased by storing certain foods at temperatures between 26° and 32°F (−4° and 0°C). These temperatures are lower than those of regular refrigerators and higher than the temperatures for freezers. Foods, such as poultry, meat, seafood, and a few other protein items, may be held at these temperatures without forming ice crystals or otherwise losing quality. This method has been shown to increase the shelf life of food products. Further, it usually does so without the compromise of quality that hard freezing may cause in foods.

Specially designed units may be used for this purpose, or refrigerators kept at deep-chill temperature ranges may be used if they have the compressor capacity. The product temperature must be reduced quickly and maintained closely in order for high quality to be preserved.

Freezer Storage

Freezers are not substitutes for good advance planning of menus and food purchases. Storage freezers should *not* be used for freezing chilled foods. Storage freezers are designed to receive and keep frozen foods at 0°F (−17.8°C) or below. Slow freezing can damage the quality of perishable items, and lengthy freezer storage means increased opportunities for contamination and spoilage.

Freezing will not improve the culinary quality of foods; with most products, the reverse may be true. Among items that may deteriorate in freezer storage over time are hamburger, fatty fish (mackerel, salmon, bluefish), turkey, pork, creamed foods, custards, gravies, sauces, and puddings.

Guidelines

The foodservice manager should make certain that both the freezer unit and frozen foods are handled with care. The following recommended guidelines should be put into practice:

1. Freezers must be maintained at an air temperature of 0°F (−17.8°C) or lower. Many experts recommend that the freezer temperature be kept between −10° and 0°F (−23° and −17.8°C). Even slight variations above 0°F (−17.8°C) can damage food quality, especially for meat and fish. It has been found, for example, that food deteriorates several times faster at 15°F (−9.4°C) than at zero. Defrost damage is cumulative. Each incident of thawing adds to the damage and allows the growth of

some bacteria, such as *Listeria.* Neither can be corrected by refreezing. Partial thawing of food may happen in a self-defrosting freezer that goes through a warm-up phase one or more times a day.

2. Frozen food products should be placed in frozen storage facilities of approved design immediately after delivery and inspection, and removed from storage only in quantities that can be used immediately.

   The rule on immediate storage of certain frozen foods does not strictly apply if the delivered items are going to be prepared and served the same day. In every case, the products should be kept frozen or refrigerated, and not held at room temperature under any circumstances.

3. Only frozen or pre-chilled foods should be put into the freezer unit. Warm food products will raise the temperature in the unit.

4. Frozen food inventories should be rotated on a First In, First Out (FIFO) basis. Labeling each product with its description, date of entry, and noting the manufacturer's expiration date can make this process fairly straightforward.

5. Reach-in freezer units should be defrosted as frequently as necessary to maintain efficiency. The operator's manual for the freezer should give some indication for proper frequency. To maintain food safety and quality, frozen food items should be used up first or moved to another freezer during the defrosting of a freezer unit.

6. An easily visible thermometer or other device to record temperature should be present in each freezer unit. As with refrigerators, it may be useful to have more than one thermometer in order to check for hot spots and temperature variations.

7. Foods should be placed in a freezer in a way that provides spaces for cold air to move around and through the food.

Operating Practices

Foods intended for frozen-food storage should be wrapped or packaged in moisture-proof material or containers. This practice minimizes loss of flavor, discoloration, dehydration, and odor absorption.

Whenever circumstances permit, frozen-food products should be stored in the original cartons in which they were shipped. If the carton is broken or the original cartons take up too much space, food items should be repackaged in airtight containers for storage.

Employees should be trained to open freezer doors only when necessary and to remove as many items as possible that they need at one time. Freezer cold-curtains can be used to guard against heat gain.

Dry Storage

The amount of storage space required for dry food items, such as rice and canned goods, varies with the type of operation, the menu, the number of customers served, the purchasing policies, the delivery frequency, and so on. Whatever the dimensions of the space allotted, the area should be well-ventilated, well-lighted, clean, and protected from pests and excessive moisture and heat.

Environmental Control

Temperatures of 60° to 70°F (15.6° to 21.1°C) are adequate for most goods placed in dry storage. A temperature of 50°F (10°C), however, is ideal and increases the shelf life of virtually all dry products. A relative humidity of 50 to 60 percent is satisfactory for the storage of most goods. Adequate ventilation preserves dry products and hinders infestation by certain insects. A thermometer, and possibly a hygrometer, should be prominently displayed in all storage locations. A *hygrometer* is an instrument used to measure the humidity. The accuracy of these devices should be verified from time to time and the temperature and humidity checked often.

Operating Practices

The rules to follow in relation to dry-storage areas include:

- Practice the FIFO rotation method.
- Keep all containers of food tightly covered.
- Clean up all spills immediately.
- Do not place any food items on the floor.
- Do not store trash or garbage cans in food storage areas.

STORAGE OF SPECIFIC ITEMS

Up to this point we have considered the general rules for refrigerator, freezer, and dry storage. Now let's take a look at some of the correct storage procedures for specific items of food. Appendix C lists the recommended storage times, cabinet air temperatures, and procedures for food stored on a short-term basis. Not all refrigerators are capable of producing the 32°–36°F (0°–2.2°C) ideal holding temperature. However, refrigerators and freezers should be capable of the following operating air temperatures:

- Refrigerators, short-term: 40°F (4.4°C) or below
- Storage freezers: 0°F (−17.8°C) or below.

Appendix D suggests maximum freezer storage times, and Appendix E recommends maximum storage periods for foods kept in dry storage. Refer to these charts for details on the food products discussed in the following sections of this chapter. Keep in mind, however, that not all experts agree on storage periods—the times given in these charts should be taken as general guidelines only. In addition, the times suggested are *maximum* storage times. To maximize safety and quality, the manager should see to it that supplies are usually stored for shorter periods.

Meat

Meats should be placed in storage immediately after delivery to the foodservice operation. For meat placed under refrigeration, 32° to 40°F (0° to 4.4°C) is the recommended unit temperature range.

Quarters and sides of beef, which are used in larger operations, may be hung in refrigerated storage without a covering, provided that nothing is stored under them and that the hook on which they are hung is cleaned and sanitized. Raw cuts of meat should be wrapped loosely to allow for air circulation, except for ground beef, which turns brown when it is exposed to air. In addition, these operations will have a separate walk-in, which maintains colder temperatures for this meat.

Frozen meats must be held at a temperature of 0°F (−17.7°C) or below. They should be wrapped and sealed in moisture-proof paper or containers before being placed in the freezer. Faulty wrapping or long storage may result in freezer burn.

Unless processed meats are delivered in the frozen state, they should not be frozen. Products such as ham, bacon, and luncheon meats deteriorate quickly when frozen.

Poultry and Eggs

In general, poultry is more perishable than meat. Whole birds should be wrapped loosely, refrigerated at a unit temperature of 32° to 40°F (0° to 4.4°C). Refrigerated poultry should be used within three days of receipt. Giblets and cooked, cut-up pieces of poultry should be kept no more than one or two days in the refrigerator.

Fresh eggs should be stored in their original containers in a refrigerator. Until they are needed, keep eggs at a refrigerator air temperature of no more than 45°F (7.2°C). Washing eggs may reduce their quality, wholesomeness, and safety and should not be done in the foodservice establishment. Washing is unnecessary because eggs are washed and sanitized during processing.

Dry eggs in their reconstituted form are considered potentially hazardous products and should be treated with the time and temperature controls that apply to liquid eggs. Dried egg products should be refrigerated or kept in a cool, dry place away from light. When packages are opened, the products should be stored in the refrigerator. Keep frozen products frozen and thaw in the refrigerator as needed. Liquid egg products should be refrigerated before and after the package is opened.

Seafood

If refrigerated at a product temperature of 32°F (0°C) and covered with self-draining ice, fresh whole fish may be stored up to three days after delivery. Only crushed or flaked ice should be used in icing fish. Cubes or large pieces of ice can damage the fish, therefore encouraging spoilage.

When stored, fish should always be wrapped in order to prevent possible damage from waterlogging. Fillets, steaks, and other retail cuts of fish should be kept in air-tight, moisture-proof packaging and stored embedded in self-draining, flaked or crushed ice. If kept at 32°F (0°C) but not iced, the fish should be used within 48 hours. Live shellfish should be stored in their original container at 40°F (4.4°C).

Dairy Products

Most dairy foods readily absorb strong odors, including flavors from other foods kept in the area. For this reason, dairy products should be kept tightly covered and stored away from sources of strong odors, such as fish, peaches, onions, or cabbage. An air temperature below 40°F (4.4°C) is necessary. Milk, cottage cheese, and cream should not be used after the date marked *sell by* or *good until* on the carton or delivery container.

Dairy products should not be held at room temperature, except when they are being served. Dairy products being used in cooking should not be at room temperature for more than two hours. Milk that has been held at room temperature should never be poured back into a refrigerated carton.

Fruits

Most fruits keep best in the refrigerator, though apples, avocados, bananas, and pears ripen best at room temperature. Berries, cherries, and plums should *not* be washed before refrigeration, since moisture increases the likelihood that mold will grow. Wash these fruits before preparation and/or serving. Contrary to common opinion, citrus fruits are best stored at a cool room temperature. The USDA recommends that citrus fruits be stored at product temperatures of 60° to 70°F (15.6° to 21.1°C). For fruits that *are* refrigerated, a humidity of 85 to 95 percent is ideal.

Vegetables

Most fresh vegetables are best kept under refrigeration. The air temperature range should be about 40° to 45°F (4.4° to 7.2°C), with the same relative humidity as fruits, 85 to 95 percent.

Vegetables that quickly spoil or lose flavor include lima beans, cauliflower, and cucumbers. Vegetables need some exposure to air, but too much air will cause them to lose necessary moisture. Onions should not be stored with potatoes or moist vegetables that may absorb odors.

The best way to store potatoes is at an air temperature of 45° to 50°F (7.2° to 10°C) in ventilated containers in a dry, dark place. Potatoes will, however, keep for two weeks after delivery at an air temperature of 70°F (21.1°C). Sweet potatoes, mature onions, hard-rind squashes, eggplants, and rutabagas should not be refrigerated, but held at air temperatures of about 60°F (15.6°C).

Modified Atmosphere Packaged Foods

Modified atmosphere packaging (MAP), such as vacuum-packaging, can help extend the shelf life and quality of food products under *proper storage* conditions. However, improperly used it could cause foodborne illness. These special processes reinforce the need for proper storage.

Vacuum Packaging

Vacuum packaging will not stop the growth of bacteria in food. In fact, it may mask potentially lethal consequences of anaerobic growth by *Clostridium botulinum.*

When used in a retail operation, vacuum-packaged foods must be maintained at temperatures recommended by the manufacturer at all times. Vacuum-packaged fish must be strictly maintained at a product temperature of 38°F (3.3°C) or below. Shelf life must not exceed the date indicated on the processor's packaging. In addition, consumer packages must be labeled prominently and conspicuously with temperature requirements and clear instructions as to shelf life.

Sous Vide

Sous vide is the French term for *under vacuum.* Sous vide cooking entails vacuum-packaging raw, fresh foods in impermeable plastic pouches and sealing them with special equipment. The sealing results in air being forced from the pouches, placing them under vacuum. The foods are then fully or partially cooked, rapidly cooled, and stored under refrigeration until further preparation is needed. At that time, the food in the pouches is reheated for service.

The FDA states that only licensed food processing operations are allowed to use the sous-vide process on-site. They allow a foodservice operation to purchase sous vide products only from a reputable, reliable supplier that has a verified HACCP system. The storage temperature of sous vide products is product-specific, but generally 32° to 38°F (0° to 3.3°C). Since sous-vide cooking presents certain conditions for survival and growth of some micro-organisms (see Chapter 2), it is extremely important to follow manufacturers' storage and preparation guidelines.

New-Generation Refrigerated Foods

New-generation refrigerated foods, such as fresh pasta, meat, and vegetable salads, may not show the typical characteristics of spoilage. But factors such as improper temperature control, exceeding the recommended shelf life, and partial processing, can facilitate the growth of hazardous micro-organisms.

When storing new-generation refrigerated foods, the upper-limit product temperature of 45°F (7.2°C) at which traditionally packaged foods can be stored does not apply. Pathogenic bacteria that grow at refrigeration temperatures, such as *C. botulinum* and *Listeria monocytogenes,* are potential dangers. These foods should be stored at a refrigerator temperature of 40°F (4.4°C) or lower.

Canned Goods

Storage of canned goods should meet all the general requirements for dry storage. It is important to remember that canned goods are subject to spoilage and deterioration in quality over long periods of time.

The optimum storage temperature for canned goods is 50° to 70°F (10.0° to 21.1°C) with a relative humidity of 50 to 60 percent. Higher temperatures are likely to accelerate bacterial growth and food deterioration. Too much moisture may cause rusting of the cans.

The storage times listed in Appendix E are based on ideal conditions. Note that acidic canned foods such as tomatoes and berries keep for a shorter period of time than nonacidic foods. Acidic canned goods also can cause pinholes in containers.

Canned goods should always be wiped with a clean cloth before opening to prevent external soil and debris from getting into the contents.

Baking Supplies and Grain Products

Salt and sugar stored under proper humidity and temperature are the only foodstuffs that can be stored almost indefinitely. Other dry goods deteriorate over time, with cereals and noodles being especially subject to loss of quality. Many cereal and grain products are favorite targets of vermin and can easily become moldy or musty.

SUMMARY

Improper storage of food is a potential source of contamination and spoilage within the foodservice establishment. Although different kinds of food have different storage requirements, all food should be kept in areas that are clean and used only for the storage of food. In addition, all food stocks should be rotated on the First In, First Out (FIFO) principle.

Refrigerators are used for short-term storage of potentially hazardous and some perishable foods. There are two types: walk-ins and reach-ins. To maintain ideal food temperatures two separate refrigerators should be used—one for meat and poultry, fish, dairy products and eggs and one for fruits and vegetables. A close watch should be kept on the temperatures of all refrigerated storage areas and food products. Raw food products should never be stored above ready-to-serve products.

Freezers are useful for long-term storage of many food items. The foodservice manager should know which foods deteriorate quickly when frozen. Normal foodservice freezers should not be used to freeze volumes of food, only to hold already frozen foods.

Temperature, ventilation, absence of insects and rodents, and dryness are all important factors in evaluating a dry storage area.

The recommendations for maximum storage periods in refrigerators and freezers should be followed carefully. Although dry goods generally have a longer shelf life than goods that are refrigerated or frozen, it should be remembered that they, too, eventually lose quality, even if kept under proper conditions.

A CASE IN POINT

At 2:00 P.M. on Monday, the foodservice employees at the White Oaks Nursing Home were busy cleaning up from the lunch meal and beginning to prepare dinner. Pete, a kitchen assistant, took the still-hot vegetable soup left over from lunch and put the stockpot in the refrigerator to cool. Angie, a cook, began deboning the raw chicken breasts (see "A Case in Point" in Chapter 6) for the evening meal. She then placed the chicken breasts uncovered in the same refrigerator where Pete placed the soup, being careful to place them on the top shelf away from the soup. Next, Angie iced a carrot cake she had baked in the morning. When she was finished, she also put the carrot cake in the refrigerator, directly beneath the chicken breasts.

What storage errors were made?

What item, already questionable, is now at further risk?

STUDY QUESTIONS

1. What are the two major hazards that can occur in food storage?
2. What is meant by the First In, First Out (FIFO) principle?
3. List five rules that apply to refrigerated-, frozen-, and dry-food storage.
4. Why should shelves in a refrigerator be slatted?
5. Name several foods that cannot be frozen.
6. What is deep chilling? What makes it different from cooling and storing food?
7. What is the correct air temperature that should be maintained in a frozen-food storage unit?
8. What is the ideal temperature for dry storage?
9. Raw meat should be properly wrapped for storage in a refrigerator. What is the exception to the wrapping rule?
10. Describe the conditions for the sanitary storage of fresh fish.
11. Why should berries, cherries, and plums *not* be washed prior to refrigeration?

ANSWER TO CASE IN POINT

Since a refrigerator is designed to receive cold foods and maintain their temperature, Pete should have cooled down the vegetable soup before putting it in the refrigerator. His action may have raised the temperature of the unit, which provided an unsafe environment for the raw chicken breasts, which we know from the previous case in Chapter 6 had already been exposed to improper temperatures.

Next, the ready-to-eat carrot cake was placed directly beneath the raw chicken breasts. The chance that fluid from the raw chicken could splash onto the carrot cake when the chicken breasts are removed for cooking is indeed likely since neither the chicken nor the cake were covered. The chicken should have either been stored on the lowest shelf beneath the cake or, as in some operations, in an entirely different refrigerator designated for holding only raw food items.

The temperature of a refrigeration unit can rise from 39°F to 55°F (3.9° to 12.8°C) whenever hot leftovers from supper are placed in the unit. The temperature-rise problem can be overcome by dividing large quantities of food into shallow pans and by rapidly pre-cooling the food in special low-temperature, quick-chilling refrigerators. Quick-chilling refrigerator systems are designed to hold shallow pans and smaller units of foods for fast chilling before removal to regular refrigeration units. They should be used as interim cooling systems and *not* as cold storage refrigerators.

The type and capacity of the refrigerator are also of critical importance. Most refrigerators are cold holding units designed to receive foods at 45°F (7.2°C) or below and keep them cold. They do not have the capacity to receive large amounts of hot foods and cool them down. Special refrigerators with pull down capability may be necessary for operations that cool large amounts of hot foods on a regular basis.

MORE ON THE SUBJECT

UNITED STATES DEPARTMENT OF AGRICULTURE. *Preventing Foodborne Illness—A Guide to Safe Food Handling.* Washington, DC: USDA, Food Safety and Inspection Service. 1990. Home and Garden Bulletin #241. 24 pages. This booklet discusses how to keep food safe from contamination in handling and storage, with sections on canned foods, vacuum-packaged goods, storing cold food, and other food protection issues. It has clever illustrations, a glossary, a food safety quiz, and a list of additional resources. Write to USDA/FSIS Publications Office, Room 1165-S, Washington, DC 20250 for a free copy.

NATIONAL RESTAURANT ASSOCIATION. *Buying, Handling and Using Fresh Fruits* and *Buying, Handling and Using Fresh Vegetables.* Washington, DC: National Restaurant Association, 1984. 20 pages. Both of these booklets include how to store over 30 varieties each of fruits and vegetables to eliminate spoilage and reduce costs.

LEY, SANDRA J. *Foodservice Refrigeration.* New York: CBI, 1980. 354 pages. This book offers a fairly complete discussion of all aspects of the refrigeration of foods. Chapter 5 is most useful in the selection of refrigerators.

AMERICAN EGG BOARD. *The Egg and You.* rev. January, 1991. F-0490. 16 pages; and *The Egg Handling and Care Guide.* 1990. E-0004. 8 pages. Park Ridge, IL: The American Egg Board. Both of these booklets discuss the safety steps in storing, handling, and preparing eggs and egg products. They also discuss quality and proper appearance, among other topics. The American Egg Board also has posters available on egg safety and quality.

8

Protecting Food in Preparation and Serving

Now that the food has been purchased and stored safely, it is essential that it be *prepared* and *served* safely. It is at this point in the flow of foods that the greatest risk for contamination and temperature abuse can occur.

Let's assume that the preparers and servers are in good health, with no infections that could contaminate food. We will further assume that from the beginning you, as the manager, have followed the guidelines for a safe food service, adhering to the basic rule: "Start with the menu."

A menu (see Chapter 5) must be compatible with the capabilities of the staff, the facility, and the customer's expectations. Selecting menu items, designing methods of preparation, and choosing recipes are essential in avoiding problems related to the flow of food. Most importantly, train the staff not to change the recipe in any way that would compromise the safety of the food.

Since food can be served in a variety of ways, from the traditional by servers to home-delivery, it is important to design a food safety system that addresses all the points at which safety can be compromised in the flow of food from the supplier to the customer. This chapter will detail the following points:

- Control of the time that potentially hazardous foods remain in the temperature danger zone
- Sanitary service of food
- Protection of food in central kitchens and other units.

Exhibit 8.1 The temperature danger zone for food: 45° to 140°F (7.2° to 60°C)

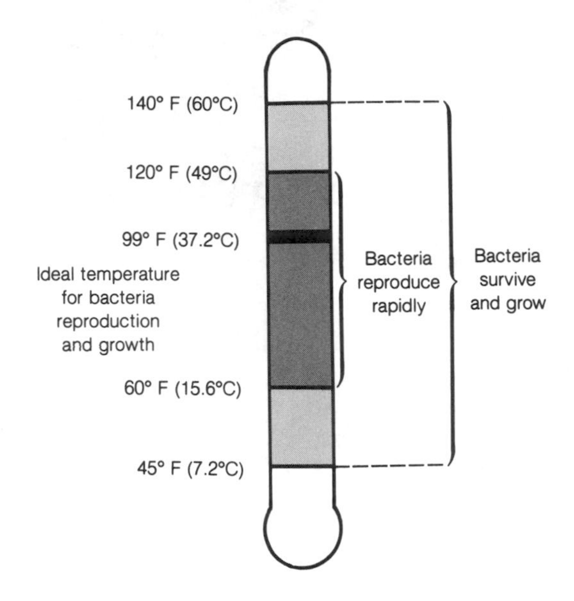

TIME-AND-TEMPERATURE PRINCIPLE

The failure to adequately control food temperature is the one factor most commonly implicated in outbreaks of foodborne illness. Since disease-causing bacteria are capable of rapidly multiplying at temperatures from 45° to 140°F (7.2° to 60°C), this is known as the *temperature danger zone* (see Exhibit 8.1).

In Chapter 2 you learned that the most important rule of food protection is the *time-and-temperature principle*: All potentially hazardous food must be rapidly cooled and kept at an internal temperature below 45°F (7.2°C) or above 140°F (60°C) during transport, storage, handling, preparation, display, and service. [Note: Some health codes specify 40° to 140°F or 145°F (4.4° to 60°C or 62.8°C) as the lower and upper limits of the temperature danger zone for potentially hazardous foods.] The total accumulated time potentially hazardous foods are exposed to the temperature danger zone must not exceed four hours.

When potentially hazardous foods are heated or cooled, pass these foods as quickly as possible and as few times as possible through the temperature danger zone. *The exposure time adds up with each stage of handling and serving.* It starts with receiving and storing, and continues through preparation, holding, serving, and reusing. Although it is vital to minimize the time food spends in the temperature danger zone, it is especially important to remember that bacteria grow more rapidly at temperatures in the middle of the danger zone, 70° to 120°F (21.1° to 48.9°C), than at the extremes. Food becomes dangerous faster at 98°F (36.7°C) than at 48°F (8.9°C).

The examples that follow will illustrate the dangers involved in violating the time-and-temperature principle and the techniques that can be used to make sure that this principle is followed.

Thawing Foods Properly

In many operations, the first step in the food preparation process is the thawing of frozen foods. The method selected to thaw frozen food and the manner in which it is carried out can spell the difference between preventing growth of micro-organisms or allowing micro-organisms to increase to high levels. Let's take a look at a common error in the thawing of frozen potentially hazardous foods.

Suppose a cook is preparing a 20-pound frozen turkey. It will take at least two days to defrost the turkey in the refrigerator, but the cook is in a hurry. The turkey is laid out on a work table in the kitchen to thaw overnight. The air temperature is about 75°F (23.9°C). The skin and outer layers are eventually exposed to temperatures favorable for the growth of bacteria once the bird is thawed.

Problems can be associated with this overnight thawing procedure:

- When bacteria have multiplied subsequent cleaning and cooking may not prevent serving a dangerous dose of disease-causing agents to the customer.
- The bacteria present on the turkey will contaminate everything in the area: the cook's hands, the work table, the cutting board, knives, and other utensils. If these utensils are used on other products without first being cleaned and sanitized, cross-contamination of other foods will occur.

Freezing food prevents most bacteria from multiplying—*it does not kill them.* Bacteria that are present on a product before it is frozen often can multiply once the product has been thawed. For this reason, thawed foods (like all potentially hazardous foods) must be kept out of the temperature danger zone.

Because improper thawing and holding of thawed potentially hazardous foods can lead to high levels of bacteria, thawing becomes a critical control point in the preparation process.

Determine if foods need to be thawed in advance of cooking. Some items like frozen vegetables, hamburger patties, and pie shells frequently are designed to be cooked from frozen temperatures all the way through to the proper end-serving temperature. The temperature danger zone can be passed through quickly and without interruption.

Other foods like frozen poultry and roasts need to be thawed before final preparation procedures can proceed. Again, the thawing and preparation methods selected must be designed to minimize time in the temperature danger zone and opportunities for cross-contamination.

In addition to cooking from the frozen state, acceptable methods of thawing frozen foods include:

1. Under refrigeration at temperatures of 45°F (7.2°C) or less. It requires adequate refrigeration space as well as enough advance planning on the part of the manager to have an adequate supply of properly thawed food available when the next preparation step is due to begin. Raw potentially hazardous foods should

also be put in pans on the lowest refrigeration shelves so they cannot drip or spill onto other foods in the refrigeration unit.

2. Under potable (drinkable) running water at a temperature of 70°F (21.1°C) or below for no more than two hours. When this method is used the water must have sufficient velocity to shake loose food particles into the overflow. Care must be exercised to prevent cross-contamination when using this method of thawing. It is very easy for water dripping off the food product to contaminate other food-contact surfaces in the preparation area. In addition, clean and sanitize all equipment and utensils after the food is thawed.
3. In a microwave oven, but only if the food will be transferred immediately to conventional cooking facilities as part of a continuous cooking process; or if the entire, uninterrupted cooking process takes place in the microwave oven. This is because the microwave thawing process may actually begin the cooking process.

For larger food items, such as a 20-pound turkey, using potable water or the microwave for thawing is not recommended.

Preparation and Cooking at the Correct Temperature

Let's say that the turkey mentioned earlier has been thawed properly and is being prepared for cooking. The chef follows proper procedures to clean and trim the turkey and takes other steps to get it ready for the oven.

Although properly thawed, we cannot assume that the turkey is free of bacteria because poultry is commonly contaminated with *Salmonella, Campylobacter,* and other disease-causing micro-organisms. Properly cooking the turkey to a safe, minimum internal temperature of 165°F (73.9°C) will kill *Salmonella* and destroy many other bacteria present on the turkey. Proper cooking is a critical control point. A lower cooking temperature allows bacteria to survive. Bacterial growth can increase if there is improper cooling or room-temperature storage.

Heat transfer from the oven to the turkey is slow. Parts of the turkey might be cooked tender without having reached internal temperatures of 165°F (73.9°C) or higher, which are necessary to kill the bacteria present.

Do not prepare stuffing *inside* poultry, because stuffing acts as insulation, further reducing heat penetrating from the oven to the interior of the bird. Rather, stuffing must be cooked separately from the bird to make sure proper temperatures are reached and micro-organisms are killed. In at least one recorded case, turkey thighs had reached an internal temperature of 200°F (93.3°C), while the stuffing inside was still at 90°F (32.2°C). Since desired minimum internal temperatures cannot always be reached, stuffing must never be cooked in the bird.

Cooked Food Temperature Requirements

Whatever cooking method (conventional oven, convection oven, etc.) is selected, the risk of bacteria surviving is best controlled by cooking the food product to an internal temperature that will ensure the safety of the food. Temperature standards for cooked food should be written into operating procedures and recipes. Minimum temperature standards have been developed for a number of cooked foods and written into federal model codes as well as state and local health regulations (see Exhibit 8.2). It is also important to remember this fundamental point: *conventional cooking procedures cannot destroy bacterial spores nor inactivate their toxins.*

Poultry, Stuffed Meat, and Stuffing

Poultry, stuffed meat, and all stuffing should be cooked to a minimum internal temperature of 165°F (73.9°C). This temperature should be reached *without* any interruption of the cooking process.

Pork

Pork and any food containing pork, such as sausage or bacon, should be cooked until all parts are heated to at least 150°F (65.6°C). An internal temperature of at least 150°F (65.6°C) is considered a margin of safety to kill any *Trichinella* larvae that may have infested the pork. An exception is that when pork and pork products are cooked in a *microwave oven,* they must be heated to an internal temperature of 170°F (76.7°C). This precaution is necessary because the post-cooking standing times are necessary to allow the internal temperature to reach all parts of the meat before serving.

Beef

Rare beef steaks should be cooked to a minimum internal temperature of 130°F (54.4°C) or as the customer requests.

Exhibit 8.2 Minimum safe internal temperatures for various hot foods

Product	**Temperature**
Pork, ham, sausage, and bacon in a microwave	170°F (76.6°C)
All foods previously served and cooled that are reheated	165°F (73.9°C) within two hours
All poultry and game birds	165°F (73.9°C)
Stuffed meats	165°F (73.9°C)
Stuffing	165°F (73.9°C)
Pork, ham, and bacon in another heating element	150°F (65.6°C)
Potentially hazardous foods	145°F (62.8°C)
Beef roasts (rare)	130°F (54.4°C) for two hours
Beef steaks (rare)	130°F (54.4°C) (or as customer requests)

Rolled roasts must be cooked to a minimum internal temperature of 140°F (60°C) for at least 12 minutes. When cooking an entire beef roast, the weight and mass of the roast as well as the type of oven used (see Exhibit 8.3) will determine the cooking time. Be certain that all parts of the meat are cooked to a minimum internal temperature of 130°F (54.4°C) for a minimum of two hours. Managers who offer rare roast beef to patrons must follow additional roasting and holding procedures. If foods can be cooked to higher internal temperatures of 165°F (73.9°C) without loss of culinary quality, an additional margin of safety may be achieved.

Using Thermometers

Since internal product temperature is a critical control point in food preparation, it is essential that managers and employees are absolutely sure that the product temperatures have been reached. Cooks and chefs should never rely solely on their own instinct, product time in the oven, appearance, or the feel of a cooked food.

Thermometers should always be used to check the temperature of products before the cooking process is ended. As discussed in Chapter 6, the thermometer used should be metal-stemmed, and numerically scaled, accurate to ±2°F (±1°C). A correctly calibrated digital thermometer may also be used. Glass thermometers must *never* be used. The internal temperatures should be checked in several places in a product, especially the thickest part, which will confirm thorough cooking.

Exhibit 8.3 Oven temperature settings for roast beef

Dry Heat	350°F (176.7°C) [preheat to at least 250°F (121.1°C)]
Convection Oven	325°F (162.8°C) [preheat to 325°F (162.8°C)]

Avoiding the Accumulation of Hazards

While cooking to inadequate temperatures greatly increases the risk of a hazardous level of bacteria survival, it is certainly not the only error that allows for the growth of micro-organisms. Errors in handling and temperature control can happen at any stage in the preparation procedure. Now consider some of the typical problems that may arise in the preparation of a beef stew.

The stew will contain beef, various chopped vegetables, bases, condiments, and spices. Let's assume the beef, which has been trimmed from larger cuts, has been carefully handled by your supplier and received and stored at the proper temperatures.

Now ready to prepare the stew, the chef, recognizing the many steps involved in the preparation, is tempted to say, "It won't hurt to bring out everything I need now because I'm going to use it before long." If the preparation process is interrupted or the amount of stew to be prepared is quite large, the "before long" may turn into unnecessary hours the beef spends within a temperature range at which micro-organisms may multiply. The potential growth time for micro-organisms, combined with soil or debris from the raw vegetables, whether they have been washed thoroughly or not, may cause a hazardous level of micro-organisms. Additional problems may occur later if the stew is allowed to be exposed to time/temperature abuse, and if the temperature at the end of cooking is inadequate.

The best way to analyze the hazards is to develop your flowchart (as described in Chapter 5) outlining the major ingredients and steps in the preparation of the stew (see Exhibit 8.4). This flowchart identifies where a procedure might introduce an opportunity for contamination or temperature abuse. It also specifies actions that will prevent these problems and states corrective actions to take if problems occur. This information would then be used to refine receiving and storage procedures, the basic preparation manual, specific recipes, and other key operational procedures.

Additional sections of the flowchart will be detailed later when we discuss cooling, hot holding and service, and reheating beef stew.

Handling Batters and Breading

Many of the foods that are breaded and battered are potentially hazardous, such as chicken, oysters, and shrimp. Batters made with eggs are also potentially hazardous and need special handling procedures.

Potentially hazardous mixtures present opportunities for contamination and time/temperature abuse. These problems should be addressed in your recipes and flowcharts. To simplify your preparation procedures, you may decide to purchase frozen items that go directly from the freezer into the oven or fryer rather than to prepare battered and breaded items from scratch. In developing preparation procedures and recipes for the preparation of battered and breaded foods, the foodservice operator must follow these procedures:

- Prepare batters with pasteurized egg products to avoid the risk of using potentially hazardous shell eggs. Even though you are using pasteurized egg products, you need to avoid contamination and temperature abuse. (A discussion of egg handling follows later in this chapter.)
- Prepare batters in small batches that will be used in short periods of time to help control the time that foods are in the temperature danger zone.
- Bread foods for later cooking in small batches and then immediately refrigerate them.
- Thoroughly cook breaded or battered items. Breadings and similar coatings can act as heat insulators and inhibit heat transfer during the cooking process. Deep frying potentially hazardous foods, particularly seafood, must be done carefully and thoroughly to make sure that disease-causing micro-organisms are killed. Make sure that fryers are not overloaded and that the temperature of the oil is allowed to recover to the required temperature before lowering the next batch. The temperature of the oil in the deep fryer is a procedure to monitor. The time the

Exhibit 8.4 A HACCP flowchart for safe preparation, handling, and cooking procedures for beef stew

Critical Control	Hazard	Standards	Corrective Action if Standard not Met
		Preparation	
Trimming and cubing beef	Contamination, cross-contamination, and bacteria increase	Wash hands	Wash hands
		Clean and sanitize utensils	Wash, rinse, and sanitize utensils and cutting board
		Pull and cube one roast at a time, then refrigerate	Return excess amount to refrigerator
Washing and cutting vegetables	Contamination and cross-contamination	Wash hands	Wash hands
		Use clean and sanitized cutting board, knives, utensils	Wash, rinse, and sanitize utensils and cutting board
		Wash vegetables in clean and sanitized vegetable sink	Clean and sanitize vegetable sink before washing vegetables
		Cooking	
Cooking stew	Bacterial survival	Cook **all** ingredients to minimum internal temperature of 165°F (73.9°C)	Continue cooking to 165°F (73.9°C)
		Verify final temperature with a thermometer	Continue cooking to 165°F (73.9°C)
	Physical contamination during cooking	Keep covered, stir often	Cover
	Contamination by herbs and spices	Add spices early in cooking procedure Measure all spices, flavor enhancers and additives, and read labels carefully	Continue cooking at least 1/2 hour after spices are added
	Contamination of utensils	Use clean and sanitized utensils	Wash, rinse, and sanitize all utensils before use
	Contamination from cook's hands or mouth	Use proper tasting procedures	Discard product

food items are in the oil is another procedure to monitor, as is the size of the batch being fried. The time, temperature, and batch size are product-specific and should be written into your recipe and flowchart.

- Discard the small amount of the left-over breading or batter at the end of the shift to avoid the risk of contaminating fresh batches. After a potentially hazardous food has been dipped into a batter or dredged in breading, the mass of breading or batter must be considered contaminated.

By writing the batch size, container size, use of refrigerated ingredients, fryer procedures, and disposal of left-over batter into recipes and procedure manuals, you have established the control points needed to serve safe battered and breaded food.

Preparing Eggs and Egg-Based Mixtures

For many years, whole, uncracked, clean shell eggs were not considered potentially hazardous foods. Eggs were often left unrefrigerated with little concern. Whole shell eggs are also important ingredients in meringues, egg nog, Caesar salad dressings, mayonnaise, Hollandaise sauce, and other items that receive little or no cooking.

However, as a result of recent outbreaks of salmonellosis traced to *Salmonella enteriditis* from clean, whole, uncracked shell eggs, the FDA has reclassified whole shell eggs as potentially hazardous and issued guidelines on how shell eggs should be handled, prepared, and served.

Basically, eggs are to be treated as any other potentially hazardous food, such as poultry, beef and meat products, sea food, and dairy products. They must be kept refrigerated at a product temperature of 45°F (7.2°C). When cooked eggs are held for hot service, they must be held at a product temperature of 140°F (60°C) or higher. In addition, the common practice of pooling large amounts of eggs for later preparation and holding is now prohibited.

In menu items containing eggs that cannot be cooked to 140°F (60°C) or higher, you should use pasteurized eggs instead of shell eggs. To reduce the risk of an egg-related foodborne outbreak, managers should evaluate all menu items, recipes, and preparation procedures that use eggs, and substitute pasteurized eggs in items that receive no cooking or are not heated to a product temperature of 140°F (60°C) or above, such as meringues and Caesar salad dressings. Failure to make such adjustments resulted in a foodborne outbreak of over 1,000 people at a convention banquet.

Managers should also evaluate and find alternatives for unacceptable practices, such as holding scrambled eggs slightly runny (less than 140°F [60°C]) for banquet service, and cooking omelettes to a soft interior. It is important that pasteurized eggs also be protected from time-and-temperature abuse and poor foodhandling practices that might contaminate them.

Preparing Protein Salads and Sandwiches

Chicken, tuna, potato, and pasta salads have been implicated in foodborne outbreaks and are by definition potentially hazardous foods. They should be prepared from items that have been properly cleaned, prepared, cooked, and cooled.

Chicken and tuna salad should be prepared from *chilled* ingredients. This could involve something as simple as moving cans of tuna and jars of unopened mayonnaise from the storeroom to the cooler the day before the tuna salad is prepared.

Maintaining these salads at safe temperatures and protecting them from cross-contamination is critical.

Sandwiches can be especially dangerous if large quantities have been prepared ahead of time and held unrefrigerated. Keep in mind that even when sandwiches are refrigerated, the insulating properties of the bread may prevent rapid cooling of the filling.

SERVING FOOD IN A SANITARY MANNER

Even though food may have been properly handled all the way from purchasing through the cooking and holding process, the service of food is another point where contamination from employees, utensils, or customers can endanger food safety. By both selecting a service method that fits the type of foodservice facility and setting up procedures to eliminate actions of employees and customers that could contaminate food, the foodservice manager minimizes the risk of a foodborne outbreak.

Kitchen and Serving Employees

When kitchen employees and servers handle dishes and utensils, the parts of the dishes and utensils that will contact the customer's mouth must never be touched. Plates must be held by the bottom or at the edge; cups by handles or the bottoms; and silverware by the handles. Storing silverware with the handles up makes the last-mentioned practice easy to observe. Employees should also use tongs to dispense rolls and bread (see Exhibit 8.5).

In some operations, soups, stews, or salads are portioned at a service station, food bar, or even at the table by a server or by the customer. This procedure increases the risk of contamination unless strict attention is given to avoiding contamination and maintaining proper temperatures. Employees should not handle with bare hands cooked food or food that will not be cooked.

Another aspect of service requires constant monitoring—the tendency of servers and others to handle clean place settings or to serve food without washing their hands after they have wiped tables or bussed soiled dishes. The potential for contamination in these practices is great, and employees must be made keenly aware of the hazards. Supervisors need to plan and schedule bussing, table setting, and restocking tasks to avoid contamination from soiled dishes. Supervisors also have to be alert for other unsanitary actions on the part of employees, such as licking fingers, and touching their hair or face. (For a complete discussion of personal hygiene refer to Chapter 4.)

Holding Hot Foods

It should be clear from the examples in the previous sections of this chapter that certain foods have to be cooked to specific minimum temperatures to kill micro-organisms that may contaminate foods during preparation and handling. What may not be readily clear is the increase in microbial growth which may occur if previously cooked food remains at temperatures favorable for growth for prolonged periods of time. This could

Exhibit 8.5 Sanitary manner of carrying utensils and serving food

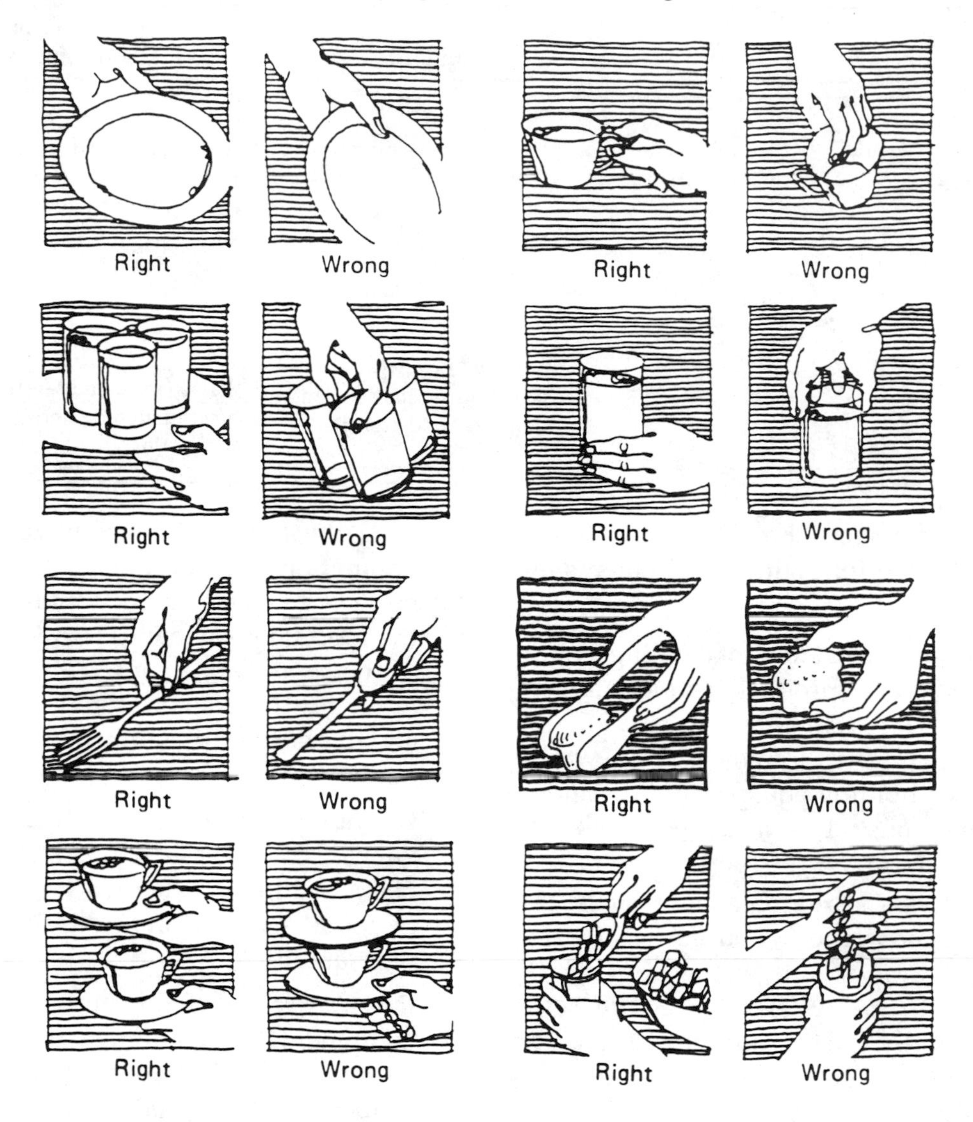

happen to food on a steam table, in a double-boiler, a bain marie, or a chafing dish on a buffet line or food bar.

It may be tempting to keep food just hot enough to serve, but not at a temperature hot enough to prevent bacterial growth or to kill bacteria. A problem arises when, for example, the chef says to hold the prime ribs at a warming temperature below 120°F (48.9°C) to keep them rare, or directs that the soup be kept just hot enough so it doesn't get too thick or skin over. Any apparent conflict between culinary and sanitary quality *must always be decided in favor of food safety* (see Exhibit 8.6).

Exhibit 8.6 A HACCP flowchart for safe holding and service procedures for beef stew

Critical Control	Hazard	Standards	Corrective Action if Standard not Met
		Holding and Service	
Hot holding and serving	Contamination, bacterial growth	Use clean and sanitary equipment to transfer and hold product	Wash, rinse, and sanitize equipment before transferring food product to it
		Hold stew above 140°F (60°C) in preheated holding unit, stir to maintain even temperature	Return to stove and reheat to 165°F (73.9°C)
		Keep covered	Cover
		Clean and sanitize serving equipment and utensils	Wash, rinse, and sanitize serving utensils and equipment

The manager should set up procedures so the following requirements are met:

- Maintain food in hot-holding equipment at a product temperature of 140°F (60°C) or above. It is not enough to rely on the thermostat of the warming or holding equipment—thermometers must be used to check the temperature of the food product.
- Stir foods at reasonable intervals. If a deep container is used, while the lower parts are being sufficiently heated, the upper surfaces are being cooled by the surrounding air and the heat may not be penetrating all parts of the food.
- Cover or protect containers to guard against contaminants falling into the food and also to retain heat.
- Select proper utensils for portioning or serving. Cups, bowls, and other utensils with short handles should be avoided. A long-handled ladle or dipper will keep the server's hand away from the food and help prevent contamination. The serving utensil should be stored properly when not in use, in the food with the handle extended out of the food. Dispensing utensils, such as tongs, may also be stored on a clean and dry surface in a properly designed holder.
- Do not prepare food further in advance than necessary. Even under the best conditions, extended hot-holding of foods will not improve their culinary quality.
- Do not reheat food in hot-holding equipment. The equipment is intended to receive foods already heated to a proper internal product temperature and to hold them at that temperature. It is usually not designed to receive cold or cool food and heat it to temperatures that are lethal for micro-organisms.
- Cooled foods must be reheated to a product temperature of 165°F (73.9°C) *before* being transferred to hot-holding equipment.

Holding breaded, fried, and baked dishes can also present problems. Holding temperatures of 140°F (60°C) or above will tend to dry out or overcook some food. The tendency to hold these items at lower than approved temperatures is not acceptable. If it is necessary to keep food below 140°F (60°C) to preserve its flavor, the food should be prepared on an as-needed basis and served immediately. Some quick-service operations that experience difficulty in holding foods above 140°F (60°C) for their flavor (for example, hamburgers) have implemented very strict holding time specifications of ten minutes or less to prevent any microbial growth during the holding process. Foodservice managers who hold beef roasts at less than 140°F (60°C) must utilize equipment designed for that purpose and follow the manufacturer's directions.

Self-Service Operations

Self-service areas can increase convenience for the guest and help hold down labor costs for the foodservice operator. Managing a salad or other food bar must be done with a solid plan that includes measuring food temperatures at least every two hours.

Food Bars

Once a novelty, food bars have become a common feature in many types of foodservice operation from dinner houses, to quick-service operations, from college cafeterias to supermarkets. Buffets, pasta bars, taco bars, seafood bars, and dessert bars allow customers to select their favorite items and the amount desired.

Customers at salad bars may use these practices:

- Picking up foods with their fingers, such as carrot sticks, olives, and pickles.
- Eating from their plates and from the food line while moving through the line.
- Dipping their fingers into salad dressings and then licking them to taste the dressings.
- Returning food items to the salad bar so as not to waste food.
- Returning for an additional helping with the same plate they used originally.
- Putting their head under the sneeze guard to reach items in the back of the food bar.

Unsupervised children present special problems because they may be unaware of food bar etiquette. Since it is difficult for an operation to monitor the conduct of its patrons, the first step in managing a food bar is to determine if your type of facility and your employees are able to maintain a sanitary operation. Several quick-service operators experimented with salad bars, then determined that they could maintain high standards of sanitation more effectively by using prepackaged salads.

A salad bar or any other buffet-style, self-service food bar requires constant supervision. Assign employees to monitor the area. This will discourage customers from unsanitary practices. Employees handing out fresh plates for return visits to the food bar also discourage customers from reusing their soiled salad or buffet plates. In addition, an operation might post politely worded signs, providing customers with information about food bar courtesy.

Proper design of self-service areas can encourage sanitary practices. Sneeze guards, or food shields, are required over display counters and salad bars. These guards should be placed in a direct line between the mouth/nose of the average height customer and the food being displayed with a maximum height of 48 inches off the floor (see Exhibit 8.7). All food items need to be identified. Salad dressings should be identified and individual long-handled scoops provided so that customers are not tempted to stick their fingers in for a taste.

Maintaining proper temperatures and efficient product rotation in the food bar is critical as with any food holding service. Potentially hazardous foods held on a buffet, food, or salad bar must be maintained at product temperatures of 140°F (60°C) or higher for hot holding, and 45°F (7.2°C) or below for cold items. Product rotation that replaces smaller amounts of food on a frequent basis and does not mix replacement food with food already on a food bar, greatly reduces the risk of contamination and microbial growth.

Exhibit 8.7 Food shield for buffet-style service. Photograph courtesy of Cambro Manufacturing Company, Huntington Beach, California.

Other Service Considerations

Dairy products, packaged foods and condiments, as well as bread, rolls, and breadsticks need to be monitored for customer safety. Plate garnishes, such as pieces of fresh fruit or pickles, *cannot* be re-served to other customers.

Dairy Products

To protect dairy products from temperature abuse and airborne contamination, milk given to the customer should be supplied in commercially filled, unopened packages that are one pint in volume, or smaller. Milk may also be drawn from commercially filled, mechanically refrigerated, bulk dispensers. When bulk dispensers are not available, and portions of less than one-half pint are required, milk products may be poured from a commercially filled container of not more than one-half gallon in capacity. Cream or half-and-half should be provided in individual, unopened containers or in covered pitchers, or should be drawn from a refrigerated dispenser.

Packaged Foods and Condiments

Sealed packages of food and condiments such as wrapped crackers and breadsticks, as well as individual portions of condiments, such as ketchup and mustard, provide a level of protection from customer contamination as long as the packages are unbroken. Only those items that are wrapped and unopened may be re-served.

Bread, Rolls, Crackers, and Breadsticks

Unwrapped rolls, breadsticks, and crackers that are served but uneaten, may *not* be re-served to other customers.

Protecting Previously Prepared Food

Many foodservice operations find it necessary to prepare foods a day or more before they are served, while other operations must effectively use their leftover foods in order to control food costs and stay in business. Rapid cooling and protection from contamination are the keys to protecting food during the cooling process. Some highly perishable foods, such as custards, creamed casseroles, and gravies, which have already been held for service, cannot be held over for next-day service. Leftover products like these must be discarded.

Cooling Food Safely

In general, cooked, potentially hazardous foods should be chilled to an internal temperature of 45°F (7.2°C) or less in less than four hours. However, the allowable time for chilling must be reduced if food has already spent time in the temperature danger zone during other steps in the preparation process.

The chilling of food does *not* occur rapidly when a hot item is taken from the range or oven and placed in a refrigerator. Several variables influence the rate at which food cools. Managers who have a general understanding of these variables can experiment with various cooling methods to determine the best way to cool the foods for their individual operations. These methods should then be written into recipes, flowcharts, operation procedure manuals, and employee training programs.

In general, the thickness or distance to the center of the food mass is the greatest influence on the cooling rate. A large stockpot full of food may take four times as long or more to cool as a stockpot that is one-half its size. That is why putting a large stockpot full of hot food straight off the stove or out of the oven in the walk-in cooler is dangerous. The more the distance to the center of the food is reduced, the more rapidly it will cool.

Other factors to consider are:

- The nature of the food influences the cooling rate. A very thick, dense food like refried beans will take much longer to cool than a thin broth.
- Some food containers tend to transfer heat better and therefore, cool more rapidly than others. Stainless steel containers tend to transfer heat more quickly than plastic containers.
- Agitation or stirring of cooked foods being cooled will accelerate the cooling.
- The type and capacity of the refrigerator is also important. Most refrigerators are cold holding units designed to receive foods at 45°F (7.2°C) or below

and keep them cold. They do not have the capacity to receive large amounts of hot foods and cool them down. Special refrigerators with "pull down" capability may be necessary for operations that cool large amounts of hot foods on a regular basis. (For a complete discussion of different types of refrigerators refer to Chapter 9.)

- Uncovered foods will cool faster, but are more likely to be exposed to accidental contamination. The safest procedure may be to leave food uncovered on or near the top shelf until it is completely cooled, and then covering it.

Cooling Methods

Let's consider the beef stew that was discussed earlier (see Exhibit 8.8). Suppose that 12 gallons of stew were prepared for service at lunch the following day. Since it is not practical to hold the stew above 140°F (60°C) until the next day, the most acceptable practice is to chill the stew and hold it in the refrigerator.

Twelve gallons of stew will weigh about 93 pounds. If this stew is placed in a stockpot with a diameter of 16 inches, the stew will be about 13 inches deep. If this pot of stew were placed in a refrigerator that has an air temperature of 40°F

Exhibit 8.8 A HACCP flowchart for safe cooling procedures for cooked beef stew

Critical Control	Hazard	Standards	Corrective Action if Standard not Met
		Cooling	
Cooling for storage	Bacterial survival and growth	Cool rapidly in ice water bath and/or shallow pans (<4"deep)	Move to shallow pans
		Cool rapidly from 140°F (60°C) to 45°F (7.2°C) in four hours or less	Discard, or reheat to 165°F (73.9°C) and re-cool one time only
		Verify final temperature with a thermometer; record temperatures and times before product reaches 45°F (7.2°C) or less	If temperature is not reached in less than four hours, discard; or reheat product to 165°F (73.9°C) and re-cool one time only
	Cross-contamination	Place on top shelf	Move to top shelf
		Cover immediately after cooling	Cover
		Use clean and sanitized pans	Wash, rinse, and sanitize pans before filling them with product
		Do not stack pans	Separate pans by shelves
	Bacterial growth in time or after prolonged storage time	Label with date and time	Label with date and time or discard

Exhibit 8.9 Safe procedures for cooling hot foods

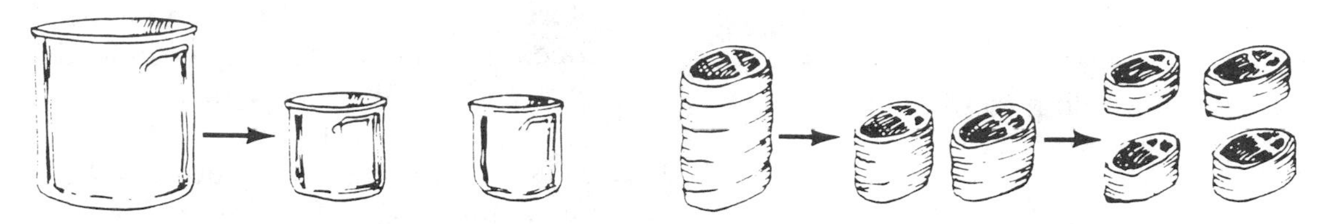

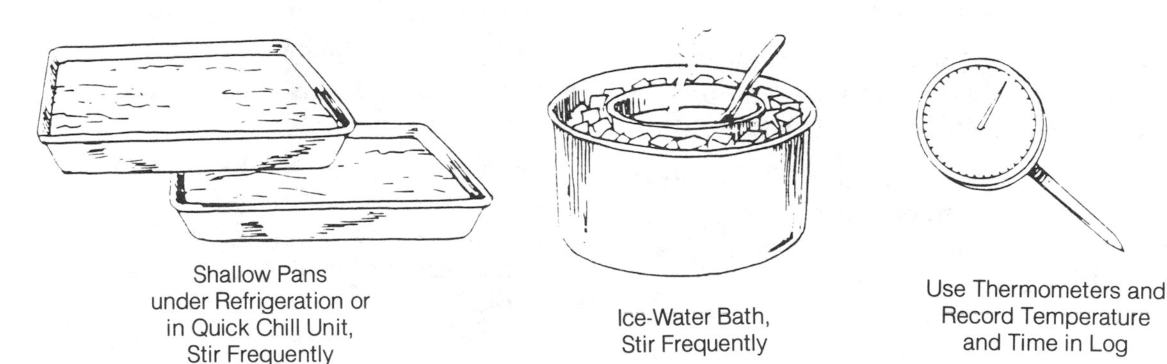

(4.4°C), it would require well over 36 hours to cool the center portion below 50°F (10°C). *That's too long.* There would be more than enough time for enormous bacterial growth, so a better cooling procedure must be used (see Exhibit 8.9).

Ice-Water Bath

Water is a much better heat conductor than air, so one way to achieve proper cooling would be to divide the stew into several one-gallon pots and put each pot in an ice-water bath to chill. To chill the stew uniformly in each pot, it must be stirred frequently. By using the ice-water bath and frequent stirring, the temperature of the stew can be brought down from 140°F (60°C) to around 75°F (23.9°C) in about two hours. The refrigerator can accommodate it much better now.

Shallow Pans

To speed up further cooling in the refrigerator, the stew should be put in *shallow* containers that are no more than four inches deep with a product depth of no more than three inches. It is necessary to frequently stir the product under refrigeration until the product temperature reaches 45°F (7.2°C) or below, being careful not to spill. Sufficient shelving should be available in refrigerators, so pans are not stacked. Hot food in shallow pans can also be refrigerated in a quick-chilling unit or tumbler-chiller to reduce its temperature.

Steam-Jacketed Kettles

Large quantities of soup or other highly liquid foods can be cooled quickly in a steam-jacketed kettle by introducing cold water, instead of steam, into the jacket. The cooling can be further accelerated with a mechanical agitator.

Guidelines for Cooling and Reheating Previously Prepared Food

The following guidelines are important in cooling, storing, and reheating previously prepared foods (see Exhibit 8.10).

- Always use thermometers to determine the internal temperatures of foods as they cook. Record the temperatures, as well as the amount of time it takes the food products to cool to 45°F (7.2°C) or lower. The internal temperature of the food cannot be identical to the air temperature of the refrigerator, and will be higher.
- Cut cooked turkey and roasts into smaller pieces and place them in shallow pans to facilitate cooling.
- Store uncovered cooked and cooled foods on the upper shelves of the cooler. Cover the food when it reaches 45°F (7.2°C) or lower. Cooked and cooled foods must never be stored beneath raw foods. Shallow pans must not be stacked on top of one another, but separated so that air can flow freely around the pans.
- Label cooled foods with preparation dates and times.
- Rapidly reheat previously cooked foods, such as beef stew that has been refrigerated, to 165°F (73.9°C) within two hours. (Refer to the discussion of hot holding earlier in this chapter.)
- Never mix leftover foods with fresh foods.
- Never reheat previously prepared food *more than once.*

Exhibit 8.10 A HACCP flowchart for safe reheating procedures for precooked and cooled beef stew

Critical Control	Hazard	Standards	Corrective Action if Standard not Met
		Reheating	
Reheat for service	Survival of bacterial contaminants	Heat rapidly on stove top or in oven to 165°F (73.9°C)	Reheat to 165°F (73.9°C) within two hours
		Maintain temperature at 140°F (60°C) or above; verify temperature with a thermometer	Transfer to preheated hot holding unit to maintain 140°F (60°C) or above
		Do not mix new product into old product	Discard product
		Do not reheat or serve leftovers more than once	Discard product if any remains after being reheated

PROTECTING FOOD IN CENTRAL KITCHENS AND IN MOBILE, TEMPORARY, AND VENDING UNITS

It is one of the paradoxes of modern life that though we are always on the go, we never want to go very far to meet our basic needs. One way the foodservice industry has kept pace with this phenomenon is through the development of fast-food outlets. But there are other, less obvious parts of the industry that help to bring food and beverages to consumers. Restaurants offer home delivery, and caterers and grocery store delis bring parties to private homes and supply food for company picnics. Temporary food stands supply the needs of audiences at sporting events and outdoor theaters. Pushcart operations sell frankfurters and soft drinks to crowds in city streets. Vending machines offer a wide variety of foods in numerous locations.

The primary rules of sanitation and HACCP apply to all mobile, temporary, and mechanical units. Selected menu items must support safe service. Facilities must be designed with sanitation and ease of cleaning in mind. Operators must follow rules of personal hygiene. Food must be protected from contamination and temperature abuse. Special requirements, however, are in order for each type of facility.

Central-Kitchen Food Preparation and Delivery

Central kitchens are usually used to provide meal service for large groups of people, such as elementary and secondary school students; for providing food for shut-ins, such as home-delivered meal programs for the elderly; or for off-premise catering. Even when safe conditions are maintained in storage and preparation, breakdowns that occur during distribution and service of the food can allow food to become contaminated or allow bacterial contamination to increase at the final stage. Also, if any contamination occurs in the earlier stages of preparation, the problem can be intensified by conditions in transporting the food.

Equipment that is used in transportation must be designed to hold food and keep hot foods hot and cold foods cold (see Exhibit 8.11). Safe food temperatures must be maintained regardless of delays caused by traffic, scattered stop sites, and weather conditions. Makeshift supplies like cardboard boxes are not designed to hold food for any period of time. Containers and vehicles specifically designed for the purpose are the only options.

Hot-food containers for transporting food should be: (1) rigid and sectioned so food items do not mix; (2) capable of being tightly closed to retain heat; (3) nonporous so there is no leakage; and (4) easy to clean or disposable. The carrier should be built to be stacked while the interior design should allow for air circulation around the containers during transporting.

Some organizations providing individual meal delivery services pack an extra meal to measure the temperature at the end of the delivery route to evaluate how well the system is working. Other operations include food safety guidelines for the customer. These guidelines state

Exhibit 8.11 Specially designed carriers for transporting hot foods

Photo courtesy of Cambro Manufacturing Company, Huntington Beach, California

which items should be eaten immediately and which items may be saved for later consumption. Since many of the customers of delivered meals are elderly or have weak immune systems, protecting them from foodborne illness is especially important.

Mobile Units

If a mobile unit serves only frozen dessert novelties, soft drinks, popcorn, candy, wrapped crackers or cookies, and similar confections, it will need to meet basic sanitation requirements. However, if a mobile unit prepares and packages food and serves this packaged food directly to customers, it will need to meet more complex sanitation regulations identical to those that apply to permanent foodservice operations.

These mobile units will need equipment for handling potentially hazardous foods, such as adequate cooking devices, hot-holding units, and mechanical refrigeration units (not ice). In addition, adequate facilities for cleaning, sanitizing, and especially hand washing are necessary. These facilities will need hot and cold potable water under pressure.

Arrangements must be made to provide refuse storage and disposal, insect control, satisfactory ventilation, and proper food storage. Mobile units should be returned daily to a central commissary for cleaning, maintenance, repairs, and waste disposal. These tasks should be performed in areas separate from the food preparation and loading areas of the commissary.

Temporary Units

A temporary foodservice unit is generally considered to be one that operates in a single location for a period of time, usually not more than 14 consecutive days, and in combination with a single event or celebration. This definition also refers to restricted operations, where the units may be set up for a longer period of time; for example, during the summer near a beach area. Definitions of, and requirements for, temporary units vary from location to location. For that reason, information on the requirements for temporary units should be obtained from the local public health authority.

In general, it is not advisable for restricted operations to serve any potentially hazardous foods other than those that require limited preparation, such as hamburgers and frankfurters, or foods that are preprepared and prepackaged. If local laws allow the service of potentially hazardous foods in a temporary installation, the following precautions should be observed:

- The foods must be *prepared* in a nontemporary facility, such as a central kitchen, in accordance with rules of sanitation.
- These foods must be prepackaged for individual service.
- These foods must be transported to and held in temporary installations at temperatures below 45°F (7.2°C) or above 140°F (60°C). No potentially hazardous foods, no matter how carefully wrapped, should be stored in direct contact with water or ice.
- Potable water must be available for cleaning and sanitizing, and hand washing. If cleaning and sanitizing of utensils is impossible, only single-service articles may be used.

Since temporary installations are often set up in warm weather at fairs or picnic grounds, control of insects can pose problems to food safety. (See Chapter 12 for a complete discussion of insect control.) It is essential to keep food covered at all times, and where practical, use prepackaged products. Ideally, all counter-service areas should be enclosed with tight-fitting solid or screened doors. Openings should be kept closed except when in actual use for service. Since temporary units are collapsible, the walls and ceilings should be made of materials that protect the interior from the weather. Single-service utensils should be made available for customer use.

Vending Machines

Coin-operated machines for vending of various types of food and beverages are a common sight today. Foods dispensed by these machines are usually prepared and packaged in commissaries that are inspected and controlled under the same sanitation regulations applied to foodservice operations. It is also important to prevent contamination of vended foods during delivery, storage in the machines, and during dispensing.

Potentially hazardous foods must be dispensed in their original containers. Fresh fruits, which have a peeling that will be eaten, should be washed and then wrapped before being placed in a machine.

The basic principles of sanitation apply to automatic cold and hot food machines. Temperature controls are required in machines to maintain potentially hazardous foods below 45°F (7.2°C) or above 140°F (60°C). In addition, machines vending potentially hazardous foods have temperature cut-off controls which inactivate the vending mechanism whenever the temperature rises above 45°F (7.2°C) or below 140°F (60°C). The cut-off controls prevent a customer from purchasing a product that has been subjected to adverse conditions.

Proper cleaning, sanitizing, pest prevention, easily cleanable location, storage and disposal of wastes, and safe water supplies are important considerations where vending machines are located. Good personal hygiene of service employees who handle food products during the servicing and cleaning of machines is a must. Hand-washing facilities at vending locations are used by service employees before servicing machines.

The food-contact surfaces of vending machines should be cleanable, corrosion-resistant, and nonabsorbent. Machines must be cleaned and serviced regularly to ensure that customers receive quality products.

Restaurant and foodservice operators may use vending machines as one form of service in their operation. Under such circumstances, responsibility for protection of the food in the vending machines rests with the restaurant operator. When installing machines, the foodservice operator must make certain that the machines are not placed directly under sewage pipes; adequate power and potable water are available; and cleanable floors, walls, and ceilings are present.

THE TEN RULES OF SAFE FOODHANDLING

We began our investigation of food contamination within the foodservice establishment with a presentation in Chapter 1 of "Critical Offenses." The following list takes a positive approach. It reviews the ten most important rules of safe foodhandling.

1. Require strict personal hygiene on the part of all employees, and relieve infected employees of foodhandling duties. Instruct employees not to touch cooked food or food that will not be cooked.
2. Identify all potentially hazardous foods on your menu and create a flowchart to follow these foods throughout the entire operation.
3. Obtain foods from approved sources.
4. Use extreme care in storing and handling food prepared in advance of service.
5. Keep raw products separate from ready-to-eat foods.
6. Avoid cross-contamination from raw to cooked and ready-to-serve foods via hands, equipment, and utensils. Clean and sanitize food-contact surfaces of equipment after every use to avoid cross-contamination.
7. Cook or heat-process food to recommended temperatures.
8. Store or hold foods at temperatures below 45°F (7.2°C) or above 140°F (60°C) during preparation and holding for service.
9. Heat leftovers quickly to an internal temperature of at least 165°F (73.9°C) within two hours.
10. Cooked food should be rapidly chilled in shallow pans in a refrigerator, in a quick-chilling unit, or in an ice water bath and stirred or agitated frequently during the chilling.

SUMMARY

Food must be protected at every step in the process of preparation. The most important guideline for protecting food in the kitchen and in the dining room is the time-and-temperature principle.

In the first stages of preparation, raw foods must be cleaned thoroughly; frozen foods must be thawed according to safe, accepted procedures. Special care must be exercised in the mixing of raw and cooked ingredients, especially if the product will not be cooked again.

Cooking can kill many, but not all, of the bacteria present in food if proper internal temperatures are reached. Hot foods must always be held at temperatures above 140°F (7.2°C). Leftovers and any other cooked foods that are to be refrigerated must be chilled immediately and according to proper procedures. Leftovers must be thoroughly and quickly reheated before they are served again. (The only exceptions are those leftovers that are used for items that are served cold, such as roast beef sandwiches and chicken salad. These foods should be served promptly or held at a temperature of 45°F (7.2°C) or less until serving.) Thermometers must be used throughout the processes of cooking, chilling, holding, and reheating.

Customers can be sources of contamination. At various self-service or buffet areas, particularly food bars, an operation must take precautionary measures to guard against this type of contamination. Employees and signs can be used to offer direction to customers.

Special precautions to be observed by operators of mobile units, temporary installations, satellite feeding systems, and vending machines include regulations covering maintenance, construction, waste disposal, and supplies, as well as restrictions on the types of food that can be kept safely in such facilities.

The Ten Rules of Safe Foodhandling present positive guidelines for achieving and maintaining high standards of sanitation in the foodservice industry.

A CASE IN POINT

Monday evening, Angie prepared the chicken breasts (see "A Case in Point," Chapter 7) for dinner. She also prepared tapioca pudding, a favorite dessert among the residents of the White Oaks Nursing Home. Because several residents were away from the home and Angie had not been informed, she prepared more chicken breasts and tapioca pudding than were necessary.

"No problem," Angie thought. "I'll just use the leftover chicken to make chicken salad. I can serve that for lunch later this week, and I'll re-serve the tapioca, too." She refrigerated the leftover chicken and tapioca pudding in five gallon containers at an air temperature of 40°F (4.4°C) until she was ready to prepare lunch for next week Monday.

Monday morning, Juanita, the manager, observed Angie readying chicken salad, fresh fruit, cole slaw, and tapioca pudding for lunch.

"Aren't those leftovers from last week?" Juanita asked. When Angie nodded, Juanita asked her to stop preparing lunch.

Why did Juanita stop Angie?

STUDY QUESTIONS

1. What is the time-and-temperature principle? Give one example of its application.
2. Name three permissible methods of thawing frozen foods.
3. What is the minimum required, final cooking temperature for poultry, stuffing, and pork?
4. Describe precautions that should take place in the preparation of battered and breaded food items.
5. What rules should apply to the use of eggs added to foods that will receive little or no additional heat treatment?
6. What precautions should be exercised in the preparation of salads and of sandwiches containing potentially hazardous ingredients?
7. What is the purpose of a sneeze guard or a food shield?

8. What are some ways in which customers might contaminate food in self-service areas?
9. What risks are present when stuffing is cooked in poultry?
10. Describe three methods of cooling cooked items before they are to be held over in a refrigerator.
11. How does the rapid cooling of food items provide safety measures?
12. When should buspersons be sure to wash their hands?
13. Under what conditions can potentially hazardous foods be served in a temporary foodservice unit?
14. What types of foods should not be prepared in a temporary unit?
15. If a foodservice operator provides a vending machine, what practices should be observed?

ANSWER TO A CASE IN POINT

Juanita stopped Angie because neither the chicken nor the tapioca pudding were safe to serve. The chicken served last Monday was already at risk, since it had been both improperly received and stored. Fortunately, the White Oaks Nursing Home experienced no foodborne illness when the chicken was served Monday. Even if the chicken had *not* been questionable, it should only have been stored for two days; then it should have been discarded. Puddings should be served only on the day they are prepared. Therefore, the leftover tapioca pudding stored in a five gallon container should have been discarded immediately. Had proper flowcharting been established by the manager of the White Oaks Nursing Home foodservice staff, and safeguarding procedures written into the recipes for each item, Angie would have known the proper policies for the preparation and storage of leftover food items.

MORE ON THE SUBJECT

LONGRÉE, KARLA and GERTRUDE ARMBRUSTER, *Quantity Food Sanitation.* 4th ed. New York: Wiley, 1987. 452 pages. Chapter 11 on "Multiplication and Survival of Bacterial Contaminants in Ingredients and Menu Items," and Chapter 13 on "Time-Temperature Control" provide many hints for implementing the rules for protecting food in preparation and service, as well as much discussion of the scientific evidence on which the rules are based.

NSF *INTERNATIONAL. Vending Machines for Food and Beverages. Foodservice Standard Number 25.* Ann Arbor, MI: NSF *International,* 1980. This publication covers the basic construction of vending machines.

U.S. DEPARTMENT OF HEALTH AND HUMAN SERVICES, PUBLIC HEALTH SERVICES, FOOD AND DRUG ADMINISTRATION, DIVISION OF RETAIL FOOD PROTECTION. *The Vending of Food and Beverages, Including a Model Sanitation Ordinance.* Washington, DC: FDA, 1978. DHEW Publication No. (FDA) 78-2091. This publication provides a recommended model sanitation ordinance regulating the sale of food and beverages through vending machines.

MARRIOTT, NORMAN G. *Principles of Food Sanitation.* 2nd ed. New York: Van Nostrand Reinhold, 1989. 387 pages. Chapter 15, "Foodservice Sanitation," discusses sanitary procedures for food preparation, among other related subjects.

PART III

Clean and Sanitary Facilities and Equipment

9

Sanitary Facilities and Equipment

Facilities and equipment are integral factors in a HACCP-based system because they make sanitation procedures more effective. Poorly designed equipment and facilities make the job of cleaning and sanitizing too difficult. Many breakdowns in sanitation, including pest infestation, are caused by facilities that are simply too hard to keep *clean,* which is free of visible soil and food waste.

The first requirement for sanitary design of a food service facility and its equipment is *cleanability,* which means exposing an item or surface for inspection and cleaning without difficulty. Both the facility and its equipment need to be constructed so that soil can be removed effectively by normal cleaning methods. The easier the establishment is to clean, the closer it will come to achieving an environment free of contamination.

Other factors enter into sanitary design; for example, the layout. *Layout* is the arrangement of equipment in each area of the foodservice facility, for example, the placement of the dishwashing machine and the food preparation counter in the kitchen. The layout of equipment must be such that it is not possible to contaminate food in preparation with refuse from dirty dishes or equipment.

As mentioned in Chapter 3, a *food-contact surface* is one that normally touches food, or one from which food could drip or drain onto surfaces that normally touch food. A *food-splash surface* is one on which food can splash or spill, such as a wall or the exterior rim of a mixing bowl. Toxic or potentially toxic materials must not be used for food-contact or food-splash surfaces.

The design of utilities is also important in creating an environment where cleaning and sanitizing become easier to accomplish.

This chapter focuses on the following four subjects of importance that make up the plan for a sanitary environment for food:

- Construction of walls, floors, and ceilings for easy maintenance and cleaning
- Arrangement and design of equipment and fixtures to comply with sanitation standards
- Design of utilities (for example, electric, gas, and water) to prevent contamination and to make cleaning and sanitizing easier
- Proper solid waste management to avoid contaminating food and attracting pests.

Designing for sanitation should begin when a facility is being planned. Operators or managers who inherit an established foodservice facility may have to work harder, but they can improve the food environment every time they remodel the facility, make repairs, or purchase new equipment. Built-in sanitation should be considered in every structural and mechanical feature.

SANITARY DESIGN AND THE LAW: PLAN REVIEW

In most communities, the sanitary design of a foodservice facility is governed largely by law. Public health, building, and zoning departments may all have power to regulate construction of a facility. Many jurisdictions provide for the review and approval of plans prior to new construction or extensive remodeling.

The U.S. Department of Justice and the Equal Employment Opportunity Commission, in accordance with the Americans with Disabilities Act (ADA), have specific guidelines for facility design that includes appropriate access to the facility by patrons and employees with disabilities. Local health authorities often provide checklists of features they consider desirable or necessary for good sanitation.

An operator should provide information on sanitation requirements to the architect, interior designer, and construction contractor, and check that these are met in the plan. Then the health department and building safety department should inspect the plans. *A plan review is essential.* Some health departments will provide guides for submitting plans and specifications, and the operator should ask for a copy.

Even if local laws do not require it, a manager should have plans reviewed by the local public health agency. In addition to ensuring compliance with sanitation requirements, such reviews can save time and money, not to mention reducing frustration during cleaning time. This holds true for remodeling or conversion of a facility as well as for new construction and installation of new pieces of fixed equipment.

The plan and specifications must include the proposed layout, the proposed arrangement, the mechanical plans, lists of the construction materials for work areas, and the types and models of proposed facilities and fixed equipment. The *mechanical plans* are drawings that include exact measurement specifications.

Factors that must be considered when preparing a design plan are the menu, the service methods and standards, the mood, the hours, and the rate of patron turnover. The building department will probably require the specifications for utilities, plumbing, ventilation, and lighting.

Before granting a permit to allow a facility to operate, the local health department will usually conduct a pre-opening inspection to make certain that all design and installation requirements have been met. (More information on plan inspections is provided in Chapter 15.)

INTERIOR CONSTRUCTION DESIGN

Materials for floors, walls, ceilings, and dry storage have to be selected for ease in cleaning, maintenance, and safety, as well as for appearance.

Flooring

The choices and price range of floor coverings are almost unlimited. Exhibit 9.1 lists the advantages and disadvantages of several common floor coverings. Floor coverings are usually classified on the basis of their *resiliency*; that is, on the basis of their ability to withstand shock. Resilient floors include asphalt, linoleum, vinyl, and sealed wood. Nonresilient coverings are concrete, marble, quarry tiles, terrazzo, and others.

Exhibit 9.1 Advantages and disadvantages of common flooring materials

Material	Description	Advantages	Disadvantages
Asphalt	Mixture of asbestos, lime rock, fillers, and pigments, with an asphalt or resin binder	Resilient, inexpensive, resistant to water and acids	Buckles under heavy weight; does not wear well when exposed to grease or soap
Carpeting	Fabric covering	Resilient, absorbs sound, shock; good appearance	Not to be used in food preparation areas. Sometimes problems with maintenance elsewhere
Ceramic tiles	Clay mixed with water and fire	Nonresilient; useful for walls; nonabsorbent	Too slippery for use on floors

(Continued)

Exhibit 9.1 *(Continued)*

Material	Description	Advantages	Disadvantages
Concrete	Mixture of portland cement, sand, and gravel	Nonresilient; inexpensive	Porous; not recommended for use in food preparation areas
Linoleum	Mixture of linseed oil, resins and cork pressed on burlap	Resilient; nonabsorbent	Cannot withstand concentrated weight
Marble	Natural, polished stone	Nonresilient; nonabsorbent; good appearance	Expensive; slippery
Plastic	Synthetics with epoxy resins, polyester, polyurethane, and silicone	Most resilient; nonabsorbent	Should not be exposed to solvents or alkalies
Rubber	Rubber, possibly with asbestos fibers. Comes in rolls, sheets, and tiles	Anti-slip; resilient	Affected by oil, solvents, strong soaps, and alkalies
Quarry tiles	Natural stone	Nonresilient; nonabsorbent	Slippery when wet, unless an abrasive is added
Terrazzo	Mixture of marble chips and portland cement	Nonresilient; nonabsorbent. If sealed properly, good appearance	Slippery when wet
Commercial-grade vinyl tiles	Compound of resins and filler and stabilized	Resilient; resistant to water, grease, oil	Water seepage between tiles can lift them, making floor dangerous and providing crevices for soil and pests
Wood	Maple, oak	Absorbent; sometimes inexpensive; good appearance	Provides pockets for dust, insects. Unacceptable for use in food preparation areas. Sealed wood may be used in serving areas

A more important consideration for the foodservice operator is the degree of absorbency or *porosity*; that is, the extent to which a floor covering can be permeated by liquids. When liquids are absorbed, the flooring can be damaged, and the elimination of micro-organisms becomes almost impossible. Nonabsorbent floor-covering materials should be used in all food preparation and storage areas.

In addition to these general requirements, each area of a food service has its own particular needs in floor coverings. In most operations, it is the kitchen that imposes the greatest demands on floor coverings. Materials, such as carpeting, which may be suitable for dining rooms, are prohibited from use in the kitchen because they would be difficult to clean and would not wear well under constant spilling and use of heavy-duty cleaners and degreasers. Nonslip textures must be used in traffic areas, as well as the entire kitchen because slips and falls are the most common type of foodservice injury. Prohibited floor coverings in some states include sawdust, wood shavings, or earth.

Kitchen floors may be of marble, terrazzo, natural quarry tile, asphalt tile, sealed wood, or other equally nonabsorbent, easily cleanable materials. Poured seamless concrete may be used if it has been adequately sealed to minimize porosity. Sealant must be renewed as needed to keep the floors nonabsorbent and durable, and to reduce possible health hazards. Regular paint does not work as a sealant because it chips and wears easily. Lifted or cracked tiles must be replaced immediately to avoid giving soil and pests a place to collect, as well as to prevent accidents.

Carpeting is limited to serving areas where it absorbs sound and shock. Carpeting should be closely woven to facilitate cleaning. Although it is available in fibers that are easy to clean and inexpensive to maintain, carpeting does require daily vacuuming and regular shampooing to remove soil.

The way the floor is constructed is also important for built-in sanitation. Coving is required in all new or extensively remodeled establishments that use terrazzo tiles, ceramic tiles, sealed concrete, or similar materials as a flooring material. *Coving* is a curved sealed edge between the floor and wall that eliminates sharp corners or gaps that would be uncleanable (see Exhibit 9.2). Make sure the

Exhibit 9.2 Coving is a curved, sealed edge between the floor and wall

Source: The Food and Drug Administration

coving tile or strip adheres tightly to the wall to eliminate hiding places for insects. Gaps between floors and walls must be less than $^1/_{32}$ of an inch.

Walls and Ceilings

Some of the same factors to be considered in judging floor coverings also apply in the selection of wall and ceiling materials. Cleanability and noise reduction are important. Color is also significant. Walls and ceilings in food preparation areas must be light in color in order to distribute light and make soil more visible for cleaning. Location is another consideration when selecting wall and ceiling materials. What may work well in one part of the facility may be disastrous in another part due to splashing and heat as well as other factors.

Ceramic tile is a popular wall covering in almost every area. The grouting should be smooth, waterproof, and continuously sealed to help cleaning, with no holes to collect soil or to harbor insects. Stainless steel, with its resistance to moisture and its durability, is often used in food-preparation areas where the humidity is high and wear and tear is considerable.

Painted plaster or cinder-block walls are appropriate for relatively dry areas if they are sealed with soil-resistant and easy-to-wash glossy paints (for example, epoxy or acrylic enamel), however, this kind of surface is not recommended in areas where food or grease splash on the wall. Toxic paints, such as those that have a lead base, must *never* be used in a foodservice facility. In addition, nontoxic paints that flake and chip can become physical hazards and result in food contamination, so they should not be used.

Like floors and walls, ceilings should be covered in smooth, nonabsorbent, easily cleanable materials. Coverings that also improve the distribution of light and absorb sound are the best selections. Smooth sealed plaster, plastic laminated panels, plastic coated tiles, or panels of other materials coated in plastic are all possible choices.

Studs, joists, rafters, and pipes should not be exposed in the dining room unless they are finished and sealed for easy cleaning. Drapes, sconces, and wallpaper in dining rooms, bars, and lobbies must also be chosen for cleanability, or they can become hiding places for pests.

Dry Storage

See Exhibit 9.3 for the proper layout of a dry storage area. Easy-to-clean materials should be used in the construction of dry storerooms. Floors should be of nonporous material such as sealed concrete or quarry tile. The ideal storeroom has:

- Walls that are covered with epoxy or enamel paint, stainless steel, or glazed tile
- Shelving and tabletops of corrosion-resistant metals
- Bins for dry ingredients that are easily cleanable, and made from corrosion-resistant metals or food-grade plastics. The bins must be covered to keep out moisture and vermin. These bins should be labeled with contents and a *use-by* date for proper stock rotation.

Using slatted shelves and avoiding over-crowding will help to improve air

Exhibit 9.3 An acceptable dry storage facility

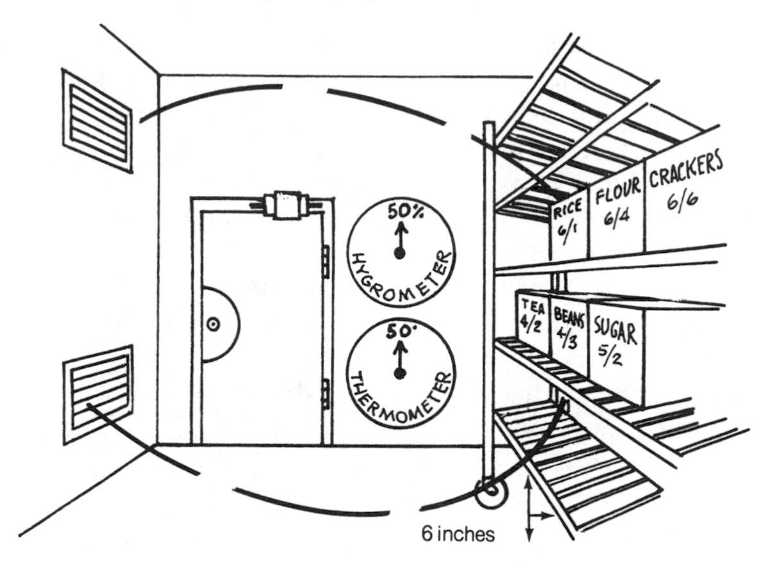

circulation. Food should be kept six inches away from walls and at least six inches above the floor. Keeping food above the floor eliminates hiding places for pests, makes cleaning easier, protects food from dampness, and helps keep food containers clean.

If there are windows in the storeroom, they should have frosted glass or shades, since direct sunlight can increase the temperature of the room and affect food quality. Exposure to bright light can change the colors of spices and chocolate and help turn cooking oils and fats rancid, if they are improperly packaged for storing.

Steam pipes, ventilation ducts, water lines, and other conduits have no place in a well-designed storeroom. Dripping condensation or leaks in overhead pipes can promote bacterial growth in such normally stable items as crackers, flour, and baking powder. Leaking overhead sewer lines are a highly dangerous source of contamination for any food. Hot water pipes and machinery could increase the temperature of the storeroom to harmful levels.

It should be emphasized that dry goods are especially susceptible to attack by insects and rodents. Cracks and crevices in the floor or walls should be filled. Outside doorways to the storage area should be closed off with solid or screened, self-closing doors to stop flying insects.

Toilet and Lavatory Accommodations

Local building and health codes usually specify how many sinks, stalls, toilets, and urinals are required in a foodservice facility to meet the needs of both diners and employees. These lavatories must also be accessible and functional for individuals with disabilities.

It is best if separate restrooms are provided for employees and diners. In any event, patrons must never pass through the food-preparation area to reach the restrooms, since they could contaminate food or food-contact surfaces both on the way to, and back from, the restrooms.

Restroom facilities must be convenient, sanitary, and adequately stocked with toilet paper, disposable paper towels, and trash receptacles. The inclusion of at least one hot-air hand dryer may save the establishment from a health violation should the disposable paper towels run out. Common towels are not permitted. Pull-down cloth roller towels should be used only if the cloths are checked and changed regularly, since they can transmit contamination from one patron's hands to others when not used or laundered properly. Individual cloth towels can be used if they are provided one per customer and then laundered.

(Requirements specifically for employee restrooms and hand-washing stations are covered in detail in Chapter 4.)

EQUIPMENT

Food-contact equipment can harbor bacteria and pests. When properly designed, equipment can prevent risks to food in a sanitary environment.

The task of choosing equipment designed for sanitation has been made simpler by a number of equipment standards provided by organizations and manufacturers, such as NSF *International,* formerly, the National Sanitation Foundation. Underwriters Laboratories, Inc. (UL), also provides sanitation classification listings that comply with those of NSF *International.* Before selecting equipment, it is helpful for the foodservice operator to be familiar with the NSF *International* standards or their equivalent.

Although it is up to the equipment manufacturer to know the standards thoroughly, the foodservice manager should know the general features on which NSF *International* bases its standards. These include:

1. Equipment must be easy to clean. *Easily cleanable* means it is possible to reach all food-contact parts by normal methods.
2. All materials that have food-contact surfaces must be nontoxic and impart no significant color, odor, or taste to food. They are to be nonabsorbent, corrosion-resistant, and stable, so they do not react in any way with food products or cleaning compounds.
3. Internal corners and edges exposed to food must be rounded off. Solder and caulking are not acceptable rounding materials. External corners and angles are to be sealed and finished smooth.
4. All food-contact and -splash surfaces should be smooth and free of pits, crevices, ledges, inside threads and shoulders, bolts, and rivet heads. Non-food-contact, food-contact, and food-splash surfaces must be easily cleanable and corrosion-resistant.
5. Coating materials—especially those on food-contact surfaces—must be

nontoxic and must resist cracking and chipping. Coating materials must also be easily cleanable.

6. Waste and waste liquid should be easily removed. Beverage dispensers, for example, should have trays that do not require tools in order to remove them for cleaning or to dispose of overflows and dripped liquids.

Foodservice managers should look for the NSF *International* mark or the UL sanitation classification mark on commercial foodservice equipment. Exhibit 9.4 shows facsimiles of the NSF *International* and UL sanitation marks. The regular UL mark indicates compliance of equipment to electrical safety standards (see Exhibit 9.5). In general, *only commercial foodservice equipment should be used in foodservice operations.* Household equipment is not built to withstand the heavy use found in food services, and it is not subject to NSF *International* standards or their equivalent.

Exhibit 9.4 Look for NSF and UL marks (facsimiles shown) for sanitary equipment

Source: Reprinted by permission of NSF *International*, Ann Arbor, Michigan; and Underwriters Laboratories Inc., Northbrook, Illinois

Exhibit 9.5 Look for the Underwriters Laboratories marking for equipment that meets safety standards.

Source: Reprinted by permission of Underwriters Laboratories Inc., Northbrook, Illinois.

Although *all* equipment used in a foodservice operation needs to meet the standards such as those set by NSF *International,* certain items require particular attention by the foodservice operator or manager.

Cutting Boards

Although it is a kitchen item as traditional as the chef's hat, the wooden cutting board can easily become a source of contamination. The wood can become crisscrossed with cuts that harbor microorganisms.

If wooden cutting boards and baker's tables are allowed and used, they must meet the NSF *International* listing standards. The wood in cutting boards and baker's tables must be nontoxic and impart no odor or taste to the food. The wood must be a nonabsorbent hardwood, such as maple. Food grade, hard rubber, or acrylic blocks are preferable to wood and are required in many areas because they can be cleaned and sanitized by immersion or in the dishwashing machine. All cutting boards must be free of seams and cracks.

Separate cutting boards need to be used for raw and cooked foods in order to prevent cross-contamination. They also need to be washed, rinsed, and sanitized between *every* use. Due to the high risk of cross-contamination, cleaning and sanitizing cutting boards should be included in the flowcharts and recipe directions for all potentially hazardous food items.

Dishwashing Machines

Dishwashing machines vary widely in size, style, and capabilities. Mechanical dishwashing machines also come in a variety of time and temperature combinations that achieve the same results. Selecting the size and type of machine depends upon the nature and volume of the utensils and other items to be cleaned, and upon dishwashing schedules. Since a machine is a big investment, the operator or manager needs to consult with manufacturer's representatives to learn which machine matches the needs of the establishment. Chemical-sanitizing machines usually require more drainboard space to air-dry utensils, but hot-water sanitizing machines have higher energy costs.

When planning the placement for a dishwashing machine, keep in mind that piping to and from the dishwashing machine should be as short as possible to prevent the loss of heat from the water entering the machine, and to avoid hard-to-clean surfaces. The main structure of the machine must be raised at least six inches off the floor to permit easy cleaning underneath.

Materials used in constructing dishwashing machines should be able to withstand wear, including the action of detergents, and not be entered easily by pests. A plumbing contractor or reputable supplier can help a foodservice manager evaluate these features as part of the plan review.

Proper water temperature, as well as water pressure and chemical concentration information, should be posted in a conspicuous place on or near the machine. The sensing element of the thermometer in the water tank is generally mounted on top of the dishwashing machine so that it can be read easily by the operator of the machine.

An adequate sanitizing cycle, either hot water or chemical, must be an integral part of the machine's operation. There are basically two types of dishwashing machines: high-temperature machines and chemical-sanitizing machines.

High-temperature machines include the following models:

1. *Single-tank, stationary-rack type with doors.* This machine will hold a rack of dishes that does not move. Dishes are washed by detergent and water from below—and sometimes from above—the rack. The wash cycle is followed by a hot-water final rinse.
2. *Conveyor machine.* The moving conveyor takes the dishes through the various cycles of washing, rinsing, and sanitizing. This type of machine may have a single tank or multiple tanks.
3. *Carousel or circular conveyor type.* This multiple tank machine moves a rack of dishes on a peg-type conveyor or in racks. In some models,

dishes must be removed after the final rinse, or they may continue to go through the machine. Other models are available that have an automatic stop after the final rinse.

4. *Flight-type.* This machine is a high-capacity, multiple tank machine with peg-type conveyor. It may also have a built-on dryer. It is commonly used in institutions and very large foodservice establishments.

There are three kinds of chemical-sanitizing dishwashing machines. These include two kinds of stationary racks:

1. *The batch-type, dump.* This machine combines the wash cycle and rinse cycle in a single tank. Each cycle is timed, and the machine automatically dispenses both the detergent and the sanitizing chemical.
2. *The recirculating, door-type, nondump.* This machine is not completely drained of water between cycles. The wash water is diluted with fresh water and reused from cycle to cycle.

The other kinds of chemical-sanitizing machines are the conveyor type *with* a power prerinse and the conveyor type *without* a power prerinse. Glassware washing machines are also chemical-sanitizing machines. (Chapter 10 has a complete description of the cleaning properties of both the high-temperature and the chemical-sanitizing dishwashing machines.)

Clean-in-Place Equipment

Some equipment, such as certain automatic ice-making machines, and soft-serve ice cream and frozen yogurt dispensers, are designed to be cleaned by having a detergent solution, hot-water rinse, and sanitizing solution passed through them. These machines must be constructed so that the cleaning and chemical sanitizing solution remains within a fixed system of tubes and pipes for a predetermined amount of time and that the cleaning water and solutions cannot leak into the rest of the machine. All food-contact surfaces must be reached by the solutions.

Clean-in-place equipment must be self-draining and capable of complete evacuation, to be sure that no cleaning solution is left in the machine that could come into contact with the food or beverages being dispensed. Some means of inspection, preferably exposure of a part of the machine by removing a panel, must be provided. Read and follow the manufacturer's instructions.

Refrigerators and Freezers

It is essential that refrigerators and freezers be easy to clean. Forced air circulating fans in service or upright units are essential to help provide a quick recovery time so refrigerator and freezer temperatures remain at the appropriate level.

There are several types of foodservice refrigeration and freezer units, two of which are: the walk-in and the reach-in, or upright, units (see Exhibits 9.6 and 9.7). Walk-in and reach-in refrigerators should be made of stainless steel or a combination of stainless steel and aluminum on the outside, and may have plastic or galvanized interior liners.

Exhibit 9.6 The commercial-type walk-in refrigerator

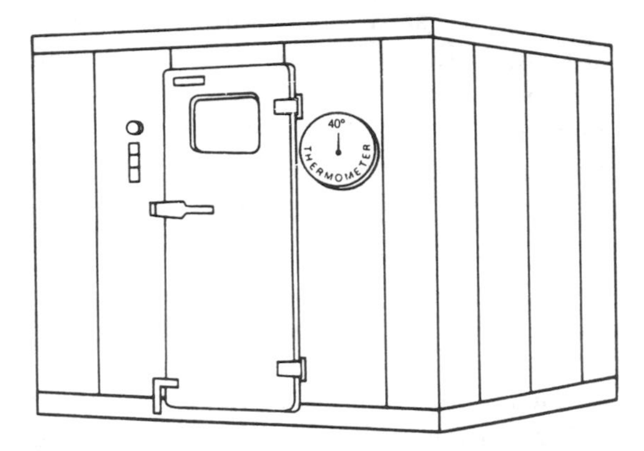

Source: U.S. Department of Agriculture

Exhibit 9.7 The commercial-type reach-in freezer

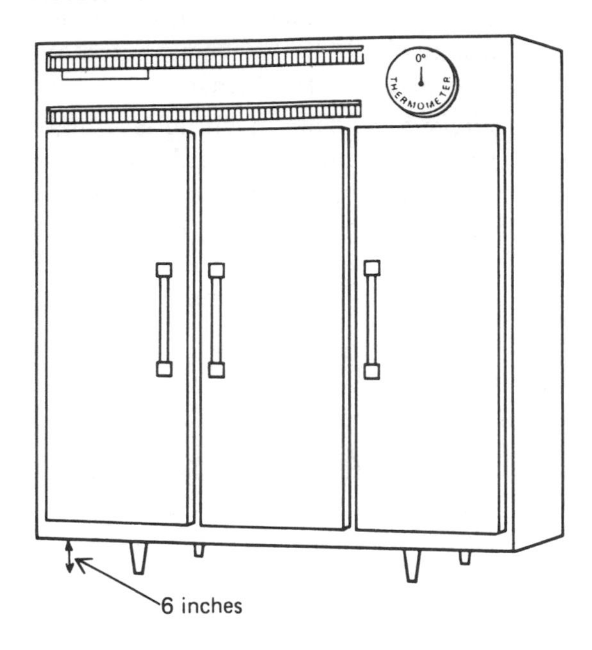

Source: U.S. Department of Agriculture

The doors should be constructed to withstand heavy use. The door gaskets can be fixed or removable for easy cleaning. Doors should close with a slight nudge.

Walk-in units must include an inside safety release to prevent employees from locking themselves in. Walk-in refrigerators and freezers that have windows in the door for observing the food items reduce unnecessary opening of the door to determine if the item is available.

Because condensation and defrost water are unsanitary, a drain outside of the unit must be provided and maintained. An optional device for both types of refrigeration units is a warning alarm that signals if the power is interrupted.

Other types of refrigeration and freezer units include under-the-counter units, open units, and display cases. When purchasing a refrigeration or freezer unit, consider the following:

- Sufficient storage space encourages frequent cleaning, allows proper cooling of foods, maintains required cold- and frozen-holding temperatures, and prevents moisture build-up. Breakdowns are more frequent when equipment is overloaded.
- Refrigerators and freezers must meet the temperature requirements for the foods they are expected to hold at safe temperatures. Ideally, separate refrigerators should be available for different food types.
- All surfaces of the unit should be made of easily cleanable, nonabsorbent materials.

- Lighting should be adequate to see items and read labels on items within the unit.
- All shelves inside the cabinet should be removable without tools so the unit may be cleaned easily on a regular basis.
- Interiors should be free of sharp edges and corners.
- Food-contact and -splash surfaces should resist corrosion, chipping, and cracking. Otherwise, cleaning will become difficult and loose particles may end up in stored food.
- Coils should be located so that condensation will not collect and drip onto food. They should also be located so food-contact surfaces are protected. Drip pans should be installed to catch and hold any condensates.
- Drains, except those for condensation, should not be located inside the refrigerator cabinet.
- Reach-in refrigerators and freezer units should be either elevated six inches off the floor on legs, or mounted on a masonry or curb base. Casters that make it easy to move the unit for cleaning are also acceptable.
- Walk-in units should be sealed to the floor and wall and offer no access to moisture or vermin. Floor materials

Exhibit 9.8 A blast-chiller unit

Photograph courtesy of Vulcan-Hart, Louisville, Kentucky.

must be able to withstand heavy impact.

- A walk-in unit's internal thermometers should be placed to make sure that the maximum safe cold- and frozen-holding temperatures are maintained, even in the warmer regions of the unit.

Cook-Chill Equipment

The process of cook-chill was mentioned in Chapter 8. A cook-chill kitchen requires special equipment to rapidly cool foods and then to reheat, or retherm, them. It works best for large-scale operations. Two options for quick chilling food are the blast chiller and the tumbler chiller. The interior and exterior construction of both should meet the same standards as regular refrigerators and freezers and be NSF *International* and UL listed, or the equivalent, for sanitation and safety.

The *blast chiller* resembles a refrigerator or freezer unit with a high-tech computerized control panel (see Exhibit 9.8). In addition, it must cool foods to 37°F (24°C) within 90 minutes. There should be some alert feature to let the operator know when the blast-chill

Exhibit 9.9 A tumbler chiller unit

Photograph courtesy of Vulcan-Hart, Louisville, Kentucky.

cycle is complete. Items that have been blast chilled can be cooked or rethermed in standard cooking equipment or a retherm unit.

Tumbler chillers are highly automated so there is little direct contact with food during the process. The *tumbler chiller* is designed as a total system consisting of steam-jacketed agitator kettles, pump/filler stations, conveyors, and tumbler chiller/cook tanks (see Exhibit 9.9).

Tumbler chillers work best with viscous or liquid food products, and should be chosen based on menu planning and space considerations for equipment layout. The tumbler chiller can double as a cook tank, eliminating the need for separate retherm equipment.

ARRANGEMENT AND INSTALLATION OF EQUIPMENT

Kitchens that work best in regard to economy, efficiency, and sanitation help establish a flow for quick, easy, and safe foodhandling. This efficiency minimizes the time the food is in the temperature danger zone and the number of times the food is handled. Layout must be determined so chances for food contamination or cross-contamination are eliminated, and equipment is easily accessible and cleanable (see Exhibit 9.10).

Portable equipment can be moved by being picked up and relocated or by being rolled on casters by one person to allow for easy cleaning of the equipment itself and of the walls and floors behind

Exhibit 9.10 Considerations in layout design

There are eight basic rules that should be remembered in establishing flow in work centers, sections, and the entire layout.

1. Functions should proceed in proper sequence directly, with a minimum of criss-crossing and backtracking.
2. Smooth, rapid production and service should be sought, with minimum expenditure of worker time and energy.
3. Delay and storage of materials in processing and serving should be eliminated as much as possible.
4. Workers and materials should travel minimum distances.
5. Materials and tools should receive minimum handling, and equipment should receive minimum worker attention.
6. Maximum utilization of space and equipment should be achieved.
7. Quality control must be sought at all critical points.
8. Minimum cost of production should be sought.

Source: *Foodservice Planning: Layout and Equipment,* Lendal H. Kotschevar and Margaret E. Terrell. Copyright 1985 by Macmillan Publishing Company. Reprinted by permission.

and under the unit. Any small spaces or crevices created by the improper placement of equipment in relation to the floor, walls, or other equipment will be difficult to keep clean and will prove attractive to pests. Food-preparation equipment should be kept far apart from nonfood-preparation equipment. For example, it is not wise to put the soiled-dish table next to the salad preparation sink.

Storage areas should be near the receiving area to avoid delays in storing perishable food items. An added bonus of a good equipment layout is increased productivity (see Exhibit 9.11).

Exhibit 9.11 Simplified foodservice floor plan. Arrows indicate normal work-flow patterns

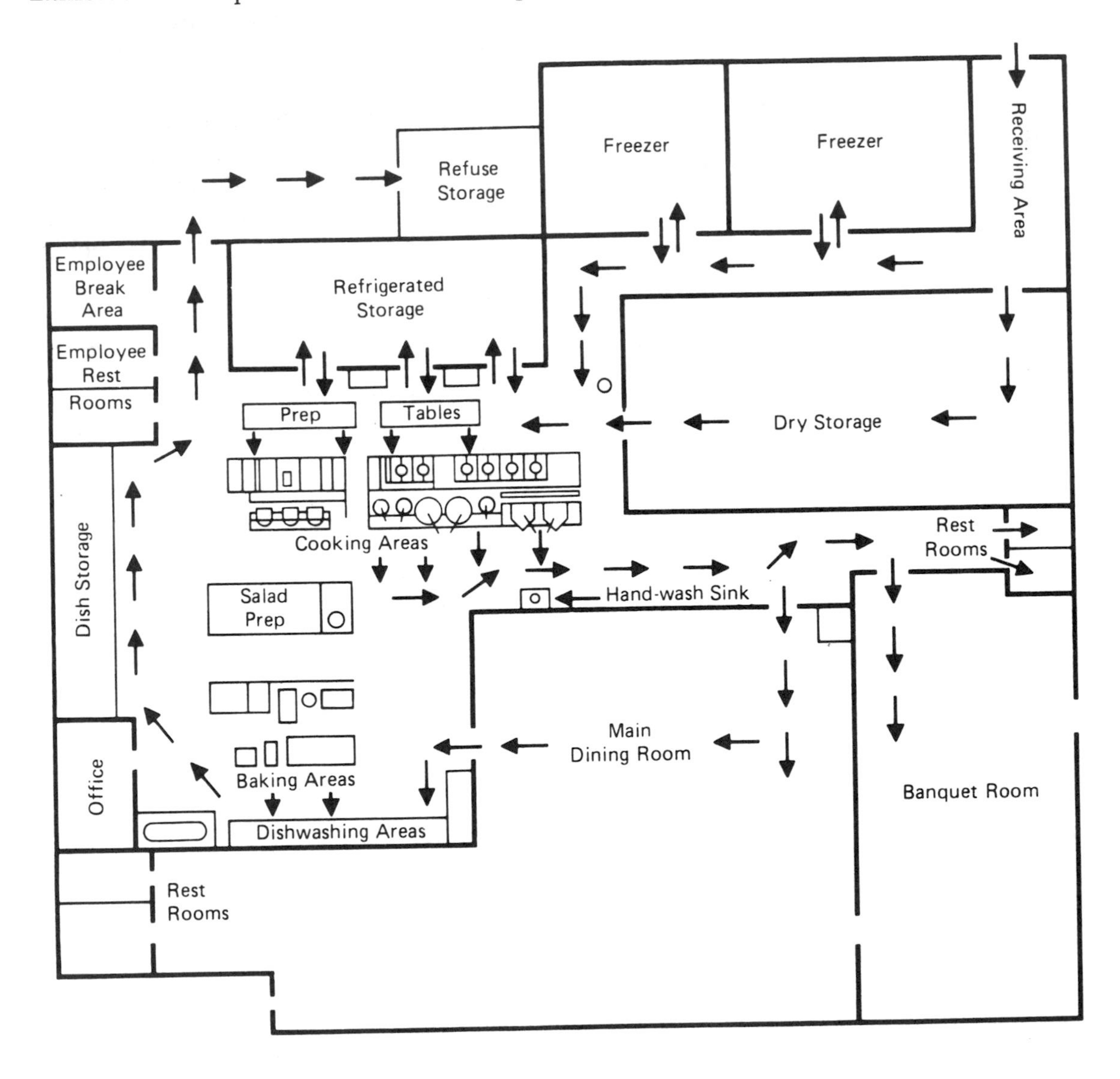

The primary sanitation goals in installing equipment are easy cleaning and elimination of hiding and breeding places for pests. *Immobile equipment,* which due to its large size cannot be moved for cleaning purposes, must be mounted on legs at least six inches off the floor, or sealed to a masonry base to avoid uncleanable spaces below and behind it. If sealed to the floor, a one- to four-inch toe space should be provided (see Exhibit 9.12). The recommended distance from the wall, or between pieces of equipment, depends on the size of the equipment and the amount of surface to be cleaned, so you should refer to the manufacturer's directions for exact specifications.

Countertop equipment can also pose cleaning problems if an item kept on a counter cannot be moved easily by one person. Immobile countertop equipment should be mounted on legs that will provide a minimum four-inch clearance between the base of the equipment and the countertop, or it should be sealed to the countertop with a nontoxic, food-grade sealant.

When equipment is sealed to the floor, wall, or counter, it is important that the sealant not be used to cover wide gaps caused by faulty construction or repairs. Renovation should correct any gaps, either by closing them entirely or providing an adequate space of six or more inches.

Some pieces of equipment may be *cantilever mounted,* in which a bracket attached to the wall supports the equipment (see Exhibit 9.13), to allow for cleaning behind and underneath. Tiltable equipment can be tilted to one side by a mechanical means to allow easy cleaning underneath it. If equipment is designed to be free-standing, however, problems may result from mounting it on the wall. Be sure to follow manufacturer's directions for installing all equipment.

Exhibit 9.12 Acceptable floor mounting of equipment for easy cleaning

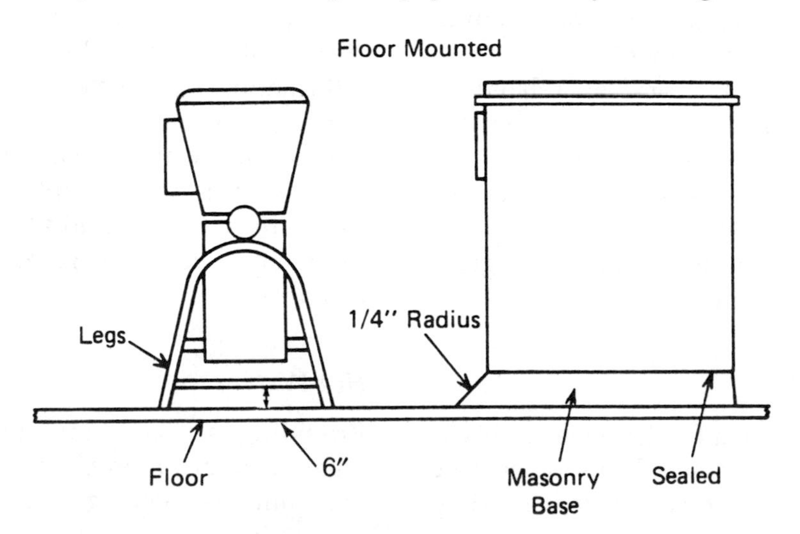

Exhibit 9.13 Wall mounting allows free access for cleaning under a large kettle

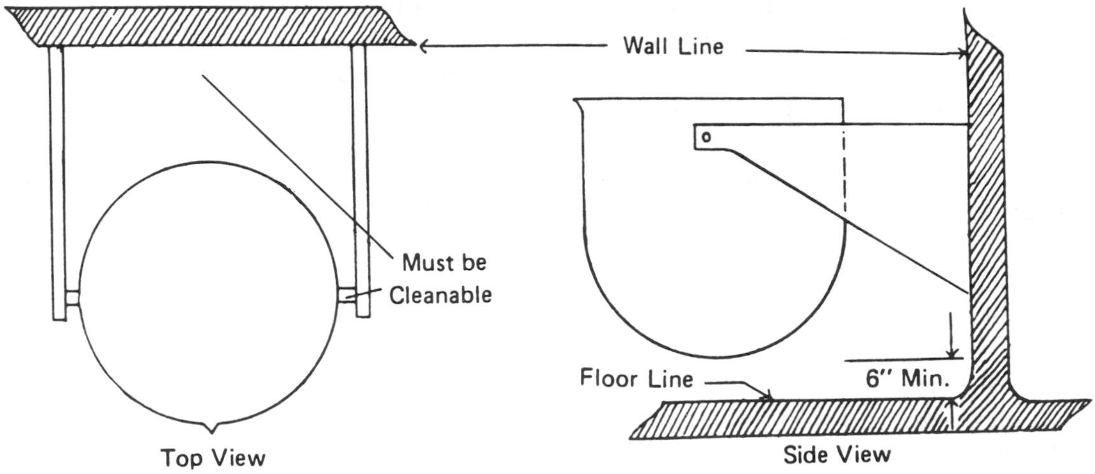

UTILITIES

Utilities are the water and electricity that a foodservice operator depends on to run the facility. Two goals must be met in the sanitary design of utilities: first, provision of utilities themselves must not contribute in any way to contamination within the food service; and second, the resources provided by utilities must be in sufficient quantity to meet the cleaning needs of the establishment.

Other necessary services related to the utilities include plumbing, sewage, garbage disposal, and trash recycling pick-ups. The provision of these services and other considerations, such as lighting and ventilation, requires a great amount of planning.

Water Supply

No matter where a foodservice facility is located, the water supply used must be potable. *Potable water* is safe for drinking. Variations in water pressure may occur at any time in a potable water system. A drop in water pressure can ruin the efficiency of an automatic spray-type dishwashing machine so that tableware, equipment, and pots and pans will not be cleaned, rinsed, and sanitized properly.

If your establishment uses a private water supply, such as a well, rather than an approved public source, you should check with your health department for information on inspections and other requirements.

If bottled water is used, it must come from sources that comply with federal food safety laws. In addition, the water must be dispensed from its original container.

Hot Water

Providing a continuous supply of hot water can be a problem for many establishments serving the public. New demands placed on the hot water supply through remodeling and expansion must not go beyond the water heater's capacity.

Water heaters should be evaluated routinely to ensure their effectiveness. Important factors to consider in evaluating a water heater include its *recovery rate,* which is the speed with which the heater produces hot water; the size of the holding tank; and the location of the heater in relation to the sinks or dishwashing machine. To maintain the right temperature for heat sanitizing by machine—180°F (82.2°C)—the installation of a *booster heater* near the dishwashing machine is usually necessary. If the booster is located more than five feet from the dishwashing machine, water will stand in the pipes and cool while the machine is off. In this instance, a recirculation device must be used to circulate the hot water through the heater to maintain the desired water temperature. Pipes may also need to be insulated.

Plumbing

In almost every community in the United States, plumbing design is regulated by law, and with good reason. In few areas of food service can improper design pose such serious potential dangers. Illnesses can result from contamination of the potable water supply. Improperly installed or maintained plumbing, which allows the mixing of potable and nonpotable system water, has been implicated in outbreaks of typhoid fever, dysentery, hepatitis A, Norwalk virus, and other gastrointestinal illnesses. Improperly installed water pipes can also lead to contamination from metals, such as copper poisoning from beverage dispensers, or contamination from chemicals used in the system, such as detergents, sanitizers, or drain cleaners.

The greatest challenge to water safety comes from cross-connections. A *cross-connection* is any physical link through which contaminants from drains, sewers, or waste pipes can enter a potable water supply. Water supply outlets submerged in an unapproved source, such as a faucet located below the floodline of a sink, are also cross-connections.

The danger from cross-connections rests in the possibility of backflow, or back siphonage. *Backflow* is simply the flow of contaminants from unapproved or undrinkable sources into potable water distributing systems. *Back-siphonage* is one form of backflow that can occur whenever the pressure in the water supply drops below that of the contaminating supply.

Backflow can also occur when a hose is attached to a faucet, as is sometimes done for cleaning purposes. For example, a hose attached to the faucet of a utility sink is used to fill a mop bucket to clean the kitchen floor. If the nozzle is left submerged in the bucket of dirty water, a cross-connection is established. If heavy usage elsewhere in the supply system causes the water pressure to be reduced below the atmospheric pressure—that is, the pressure of the air on the water in which the nozzle is submerged—contaminated water can then be drawn through the hose and into the potable water supply.

A hose should not be attached to a faucet *unless* a backflow prevention device, like a vacuum breaker, is attached. Threaded faucets and connections between two piping systems must have vacuum-breakers or other approved backflow prevention devices installed.

The only completely reliable backflow prevention device is the air gap. An *air gap* is an unobstructed, vertical distance through the air that separates an outlet of the potable water supply from any potentially contaminated source. For example, it usually separates the faucet of a sink from the contaminated water draining from the sink (see Exhibit 9.14). If a faucet has an opening with a diameter of one inch, the air gap between it and the floodline of the sink must be at least two inches. Other types of air gaps separate waste water sources from the potable water supply may be located below the sink (see Exhibit 9.15).

Exhibit 9.14 An air gap to prevent backflow

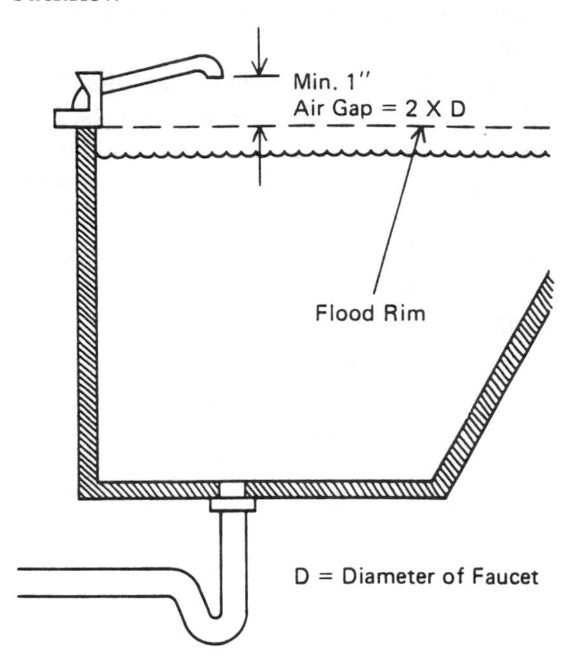

Exhibit 9.15 Sinks with air gap underneath

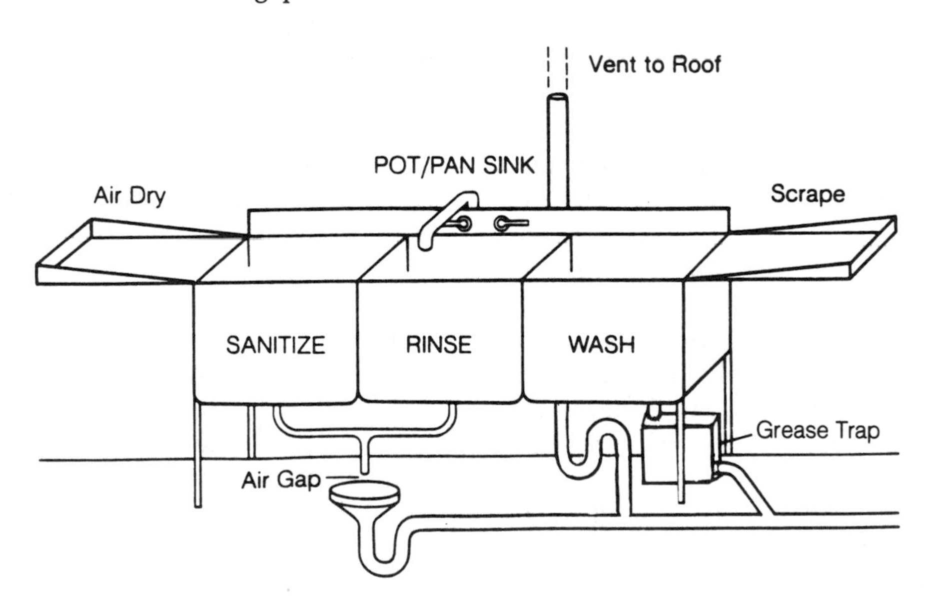

Source: Michigan Department of Public Health

While local foodservice operators should be able to recognize hazards, local plumbing codes vary widely regarding the types of connections permitted and what kinds of protection are required. If you are in any doubt about the plumbing in your establishment, check with your local health department or other agencies that have responsibility for the water supply. Have your licensed plumber upgrade your system to meet the required standards.

Overhead waste-water drain lines or sprinkler systems for fire safety can leak, and even overhead lines carrying potable water can be hazardous, since water can condense on the pipes and drip onto food. All piping should be serviced immediately when leaks occur.

Sewage

Sufficient drains must be provided to handle the waste water produced by the operation. Any area subject to heavy water exposure, such as a kitchen cleaned by hosing down, should have its own floor drain. Waste water coming from equipment and the potable supply should be channeled into an open, accessible waste sink or floor drain. The drainage system should be designed to keep floors from being flooded, which is both unsanitary and a safety hazard.

The piping of any nonpotable water system, such as that from toilets or sinks, should be clearly identified by a licensed plumber so that these pipes are readily distinguishable from those pipes carrying potable water.

Sewage and waste water are dangerous reservoirs of pathogens, soil, and chemicals in a food service. It is absolutely essential that there be no possible contamination of food or food-contact surfaces from this source.

Electricity

Electricity does not pose a threat to food safety unless it fails. However, electrical outlets and wiring should be adequate for the power needs of the facility, for operation and cleaning. If the dining room is carpeted, then enough outlets of the proper kind are needed to plug in vacuum cleaners and shampooers.

Electrical wiring on equipment must not be allowed to become frayed or to get wet. Plugs need to be checked to make sure they are not broken. Both plugs and cords should be checked at regular intervals by the staff that use the equipment, and repaired immediately by the manufacturer, supplier, or an electrician. Electric housings should be installed so that exposed conduits do not have to be run *over* the surface of interior walls at a later time. Exposed conduits would pose additional cleaning and safety problems.

LIGHTING

Good lighting in a foodservice facility generally results in improved employee work habits, easier cleaning, and a safer work environment.

Building and health codes usually set minimum acceptable levels of lighting, typically based on footcandles. A *footcandle* is a unit of illumination one foot from a uniform source of light. Other units of measurement for light include lumens, luxes, and luminaires.

Lighting should always be sufficient for the employee to be able to do his or her job efficiently and safely. Over food-preparation areas and tableware and equipment washing areas, light must be at least 20 footcandles. This level also applies to utensil and equipment storage areas and lavatories. For walk-in refrigerator and freezer units, dry storage, and the dining room during cleaning, the light level must be at least 10 footcandles.

Overhead or ceiling lights over work stations should be positioned so that the employee does not cast a shadow on the work surface. Good location of light fixtures prevents shadow formations that can hide dirt. The location of lighting fixtures is also important in minimizing physical contamination that may result from shattered glass bulbs or fluorescent tubes.

In food preparation, storage, service, refrigeration or display facilities, and in places where utensils and equipment are cleaned and stored, the location must be such that bulbs are not subject to impact breakage. If this is not possible, protective covers, sleeves, or shatter-resistant bulbs must be provided. Recommended devices are plastic panels, metal covers, and glass globes for incandescent lights; plastic sleeves with endcaps in good condition for fluorescent tubes; and shields for heat lamps.

VENTILATION

Ventilation is the removal of steam, smoke, grease, and heat from food-preparation areas and equipment. It helps to maintain indoor air quality for both safety and the comfort of patrons and employees by reducing heat and grease from the atmosphere. Ventilation has the following five functions:

1. To reduce the possibility of fires resulting from accumulated grease
2. To eliminate condensation and other airborne contaminants, which may drip from walls and ceilings onto food
3. To reduce the accumulation of dirt in the food-preparation area
4. To reduce odors, gases, and fumes
5. To reduce mold growth by reducing humidity.

Relying on open windows and doors for ventilation increases the risk of pests entering your establishment and is inadequate. Mechanical ventilation equipment must be used in areas for cooking, frying, and grilling. Exhaust hoods are frequently used over cooking equipment, steam tables, dishwashing machines, and pot washers. Hood filters or grease extractors, required when air is grease-laden, must be tight-fitting, easily removable, and cleanable.

Ventilation is measured by the rate at which air volume changes are made in a given time period. This rate depends on velocity, building air pressure, and the type of equipment being ventilated, among other factors (see Exhibit 9.16).

Since so much air is moved through exhaust hoods, provisions must be made to take in clean air to replace it. This replacement air is called *makeup air.* Air must be replaced, without creating drafts, either by the natural flow of air through a grating or by mechanical

Exhibit 9.16 Rate of ventilation for cooking range

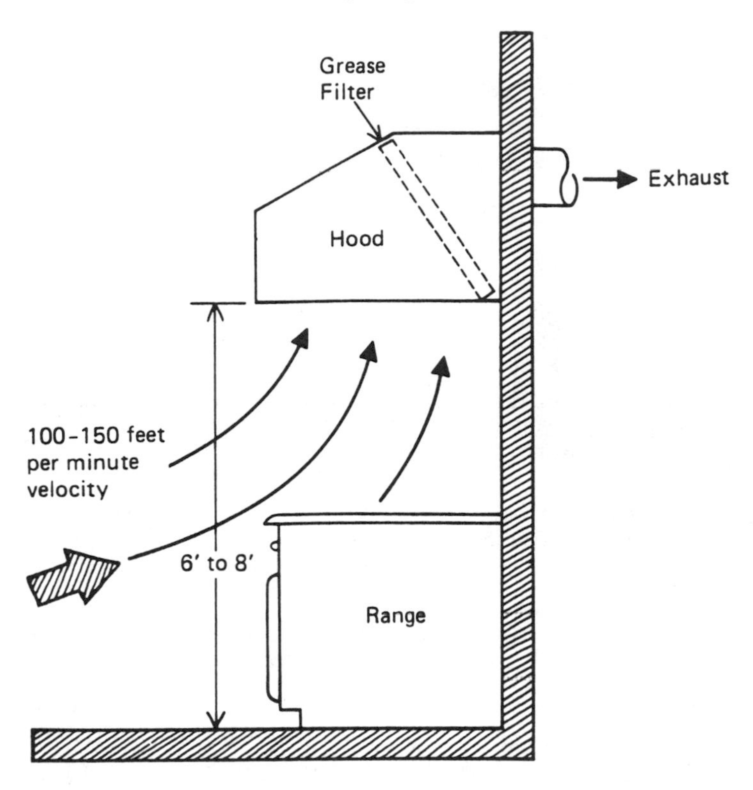

Source: Ohio Department of Public Health

air-changing devices (see Exhibit 9.17). All outside air intakes must be screened. Any ventilation equipment leading to the outside should be designed to keep insects and other pests out.

In many areas of the United States, clean-air ordinances restrict exhausts vented to the outside from retail and other establishments. Exhaust air from cooking areas, which is thick with food odors, smoke, and grease, may have to be purified by filters or other devices. It is the operator's responsibility to see that the installation is in compliance with applicable regulations, according to the plan review.

GARBAGE DISPOSAL

Garbage is wet waste, usually from food, that can result in a hazard to a foodservice operation. Food wastes attract pests and have the potential to contaminate food items, equipment, and utensils. A few general rules on handling garbage include:

1. Garbage containers must be leakproof, waterproof, easily cleanable, pest-proof, and durable. These containers can be made of galvanized metal or an approved plastic. Plastic bags and wet-strength paper bags

Exhibit 9.17 Fresh air supply installation

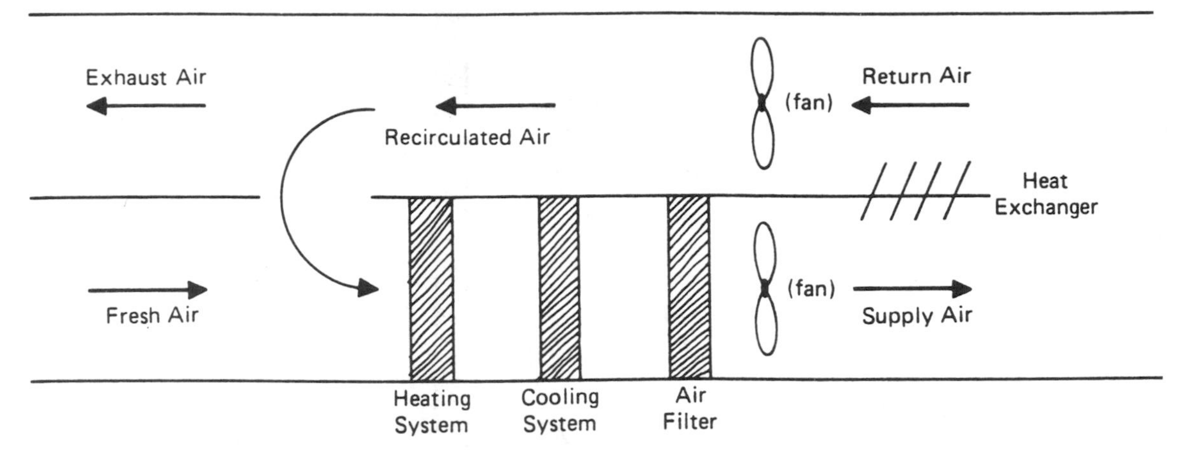

Source: *The Management of Maintenance and Engineering Systems in Hospitality Industries* by Frank D. Borsenik. Copyright © 1979 by John Wiley & Sons, Inc. Reprinted by permission.

may also be used to line these containers.

2. Garbage and refuse containers should be stored on or above a smooth surface of nonabsorbent material, such as sealed concrete.
3. Garbage must not be allowed to accumulate anywhere except in the designated garbage containers. Garbage storage areas, inside or outside, should be large enough for the amount of garbage that will accumulate, and should be provided with enough containers to hold this refuse. Inside garbage storage areas should be easily cleanable, pest-proof, and kept away from food-preparation areas.
4. Garbage should be removed from food-preparation areas as soon as possible and should be disposed of often enough to prevent the formation of odor and the attraction of pests. When removing the garbage, do not carry it above or across a food preparation area. Containers kept outdoors or temporarily near food-preparation areas must have tight-fitting lids, and must be kept covered when not in use.
5. Regardless of whether a garbage container is stored inside or outside the facility, each container must be cleaned frequently and thoroughly, both on the interior and exterior surfaces of the container.
6. An area equipped with hot and cold water and a floor drain must be provided for washing garbage containers. This area must be located so that food in preparation or storage will not be contaminated when the cans are washed and disinfected.

SOLID WASTE MANAGEMENT

As concerns over environmental issues have increased, the foodservice industry has come under close scrutiny because of the perceived volume of solid waste we are disposing. *Solid waste* is dry, bulky trash that includes glass bottles, plastic wrappers and containers, paper bags, and cardboard boxes. Although the industry's contribution to the problem is relatively minimal, the public perception is that the contribution is much greater.

As legislators and solid waste experts seek solutions to the solid waste problems in this country, foodservice operations can do their part by initiating activities within their facilities that will reduce the volume of waste generated for disposal.

The volume of garbage and trash found in a food service can be reduced through the use of pulpers or grinders. *Pulpers* grind refuse into small parts that are flushed with water. The water is then removed so that the processed solid wastes can be trucked away. *Mechanical compacting* is used for dry bulky trash, such as cans and cartons, and is particularly valuable in establishments that are cramped for space. This process reduces trash volume to as little as one-fifth of its original bulk. Use of a compactor requires access to a drain, water for cleaning, and a power source.

For dealing with trash, one activity is *source reduction*—decreasing the total amount received and the total amount disposed. For example, unnecessary packaging, like multiple layers of overwrap, can be eliminated.

Recycling is another activity, which involves separating and storing recyclable items for pick up. These items include paper, cardboard, and polystyrene packages. Also, glass bottles, aluminum and tin cans, and used cooking oil can be removed from your trash for recycling.

Waste-to-energy incineration, which is a municipal rather than individual approach, burns trash efficiently and cleanly, but requires air-pollution controls.

When these systems are used, the need for landfills is greatly reduced. Operators can take the following steps to help solve the solid waste problem.

- Train your employees to eliminate unnecessary waste.
- Evaluate the packaging of the products received by your operation.
- Initiate a source reduction recycling program in your operation.
- Participate in local recycling programs.
- Join with other operators to identify a hauler who will take recyclables to recycling areas.

When you recycle, you have to exercise caution in the storage of these materials. Check local regulations for recommendations or requirements for storage and handling of recyclables.

SUMMARY

The foodservice establishment that is difficult to clean will not be cleaned well. Sanitation efforts will be greatly facilitated if the establishment is designed and equipped with easy cleaning, or built-in sanitation, in mind.

The operator and manager must give sanitation a high priority whenever planning a new foodservice or remodeling or expanding an old one. In most communities, plans for new construction or extensive remodeling are subject to review and approval by local regulatory agencies.

The operator or manager can create built-in sanitation in three aspects of the facility: construction of floors, walls, and ceilings; choice and placement of equipment; and planning of utilities.

Do not select floor, wall, and ceiling coverings solely on the basis of good appearance. In the long run it will pay off to consider their cleanability and durability.

Much the same can be said of equipment used in a foodservice operation. Equipment must meet the sanitation standards set by NSF *International,* Underwriters Laboratories, Inc., or their equivalent. Electrical equipment should be listed by Underwriters Laboratories, Inc., for safety considerations as well.

Equipment should be placed so that all areas around and under it can be cleaned effectively. Crevices or surfaces that catch dirt are difficult to clean and can harbor micro-organisms and pests.

A foodservice establishment with built-in sanitation will also have utilities and related services that are designed to promote easy cleaning of facilities and equipment, and are not a source of contaminants. Included in this category are the water supply, plumbing and sewage installations, electricity, ventilation and lighting systems, toilet and lavatory facilities, and solid waste management.

A CASE IN POINT

In 1986, several people became ill shortly after drinking beverages at a local restaurant with a bar lounge. They all complained that their iced drinks tasted like chemicals. At this time in the lounge, the glassware washing machine had been out of service.

When interviewed, the manager explained that the machine's wash cycle was functional, but that the unit could not be used because the large volume of water discharged after each washload was worsening a recent drain blockage problem. A maintenance worker mentioned that there had been intermittent backups in the plumbing during the previous week and a large pool of water was found under the glassware washing machine. Drain cleaners had been used repeatedly with no change in the condition.

The beverage ice-making machine shared piping with both the glassware washing machine in the bar, and the grease trap on the sink in the restaurant. The manager revealed that he had installed the plumbing fixtures himself.

When ice cubes were removed from the beverage ice bin, congealed food grease and typical washing debris were found within it. Garbage debris also covered the bottom of the ice bin where the water drained out of the unit and into the pipes.

Why do you think the people became ill?

What should the manager do to correct this problem?

STUDY QUESTIONS

1. What is meant by *cleanability*? What is its importance in a foodservice establishment?
2. What is the most important sanitation requirement to consider in choosing a floor-covering material for food-preparation areas? Name several flooring materials that meet this requirement.
3. List at least three of the sanitation characteristics recommended by NSF *International* in its general standards for foodservice equipment.
4. How far off the floor should the main structure of an immobile item of equipment be? Why?
5. What are some of the sanitation factors that should be considered in selecting a mechanical dishwashing machine?
6. From a sanitation viewpoint, what features should be present in a well-designed refrigeration unit?
7. List at least three sanitation features that should be considered when purchasing equipment for frozen-food storage.
8. What are the desirable characteristics of a dry-storage area?
9. What factors should be remembered in the selection of clean-in-place equipment?
10. Define the terms *cross-connection* and *backflow.*
11. How can backflow be prevented?
12. List at least three functions of ventilation in a food service.
13. What danger can arise from overhead water pipelines or waste water pipelines?
14. What requirements must be met by a garbage or refuse container?
15. What are three activities to manage solid waste?

ANSWER TO A CASE IN POINT

The people who had beverages with ice became ill from the chemical drain cleaner that was used in the glassware washing machine. The pipe connection to the ice-making machine lacked the air gap underneath it that prevents backflow, like the one illustrated in Exhibit 9.15 in this chapter. With no protection device on the greasetrap in the restaurant sink *or* the glassware washing machine drain, which were both connected to the same pipe as the ice-making machine, back-siphonage of waste and drain cleaner was allowed into the ice cube bin, where the potable water supply became contaminated.

This case shows how important it is to know the potential hazards associated with plumbing connections, backflow, and the use of chemicals in potable water systems. Ice bins and water used for dispensing beverages must be protected by approved air gaps. The manager needs to consult a licensed plumber in order to correct his plumbing system in accordance with the regulations in his region.

MORE ON THE SUBJECT

NATIONAL RESTAURANT ASSOCIATION. *Facilities Operations Manual.* Washington, DC: The National Restaurant Association, 1986. #MG573. 252 pages. This manual outlines the details for facility construction, building structures and systems, and foodservice equipment.

NATIONAL RESTAURANT ASSOCIATION. *Managing Solid Waste: Answers for the Foodservice Operator.* Washington, DC: The National Restaurant Association, rev. 1990. #PB527. 48 pages. This booklet gives a national overview of the solid waste problem, and provides resource information for operators.

ELECTRICAL POWER RESEARCH INSTITUTE (EPRI). *The Revised National Electric Code for New Restaurants.* Palo Alto, CA: EPRI. 1990. This brochure gives detailed information on Section 220-36, which is used to determine the size of wire conductors, conduit, and main electric panels for new foodservice facilities. Under this new section, it is possible to use smaller sized panels and conductors than under other provisions of the National Electric Code, without reducing safety, but giving substantial dollar savings to operators. Data charts are used to illustrate potential savings. To obtain brochures, contact EPRI, 3412 Hillview Avenue, P.O. Box 10412, Palo Alto, California, 94303.

BIRCHFIELD, JOHN C. *Design & Layout of Foodservice Facilities.* New York: Van Nostrand, 1988. 272 pages. This book provides details on preliminary planning, principles of design, equipment layout, facilities engineering, and interior design for the foodservice industry. It includes a multitude of photos, specifications, and sample layouts, as well as a glossary. This book is part of The Education Foundation's "Management Development Diploma Program," and has companion student study and instructor guides.

NSF *INTERNATIONAL. Foodservice Equipment Standards.* Ann Arbor, MI: NSF *International,* 1987. 309 pages. All NSF standards concerning foodservice equipment are gathered together in this single volume. This book is the authoritative guide for managers and contains numerous helpful illustrations and photographs. Individual standards are updated every five years or less, and complimentary updated information on individual standards can be requested from NSF *International* by calling 1-313-769-8010. Catalogues of listings include all manufacturers of foodservice equipment that is NSF *International* listed and are published three times a year.

SCRIVEN, CARL and JAMES STEVENS. *Food Equipment Facts.* Troy, NY: Conceptual Design, 1980. 430 pages. A handbook and guide that gives general information and maintenance guidelines for equipment in all aspects of foodservice, including utilities. It includes reference charts, specifications, space requirements, and illustrations.

10

Cleaning and Sanitizing

At this point in our study of sanitation, we have learned how to set up a HACCP-based system by analyzing the menu, purchasing safe food, storing it out of the temperature danger zone, cooking it thoroughly, and serving it safely to appreciative customers in a sanitary, well-designed establishment. However, after the menu item has been served, the tableware and flatware, as well as the pots, pans, utensils, and equipment used have been soiled and must be cleaned and sanitized before re-use in order to avoid contamination problems.

Fortunately, a whole arsenal of cleaning and sanitizing products is available to remove this potential contamination problem. All we need to do is find out what the correct procedures are, what the products are, and how to apply them properly. The following topics are discussed:

- Cleaning with a detergent with sufficient agitation or under sufficient pressure, for an adequate period of time to remove food soil

- Thorough rinsing after cleaning for effective sanitizing
- Sanitizing by using heat or chemical compounds
- Mechanical and manual methods for washing, rinsing, and sanitizing
- Storage of clean and sanitary equipment and utensils
- Communicating precautions to employees using hazardous chemical products.

CLEAN AND SANITARY

Although the concepts of clean and sanitary have already been explained, the difference is worth repeating because of its importance. *Cleaning* is the physical removal of soil and food matter from a surface. *Sanitizing* is the reduction of the number of micro-organisms, such as bacteria, to safe levels on tableware, flatware, equipment, and any food-contact surface. The materials and procedures used in these two processes are different.

All surfaces coming into contact with food (whether plates, pots, tables, trays, or potato peelers) must be washed, rinsed, and sanitized after each use, after any interruption of service during which they may have become contaminated, or at regularly scheduled intervals if they are in constant use. This rule applies as well to equipment used in cleaning other food-contact surfaces.

CLEANING PRINCIPLES

Effective cleaning is more complicated than the simple combination of detergent, water, and muscle power. Cleaning takes place when a cleaning agent, such as a detergent, contacts a soiled surface under sufficient pressure, like that exerted by a brush, mop, cloth scrubbing pad, or water spray, and for a long enough time to penetrate the soil and loosen it so that it may be easily removed by rinsing. Cleaning involves two steps: (1) washing with a detergent solution; and (2) rinsing with water.

Factors in the Cleaning Process

Since many factors can affect a cleaning process, an assessment of these factors and possible techniques needs to be completed before starting.

1. *Type and condition of soil.* Soils and stains can be classified as protein-based (blood, egg); grease or oil (margarine, animal fat); water miscible, that is, dissolved in water (flour, starches, beverage deposits); and acid or alkaline (tea, dust, wine, fruit juice). In addition, the condition of the soil (fresh, ground-in, soft, dried, or baked on) affects the ease of its removal.
2. *Type of water.* Water hardness is the amount of dissolved minerals (calcium, magnesium, and iron) in the water, measured in parts per million (p.p.m.), or grains per gallon (see Exhibit 10.1). Cleaning in hard water can be more difficult, because the minerals present can reduce the effectiveness of some detergents. And where boiling or evaporation occurs, hard water may cause a *scale,* or lime deposit, to form on equipment, such as the dishwashing machine. Use of

Exhibit 10.1 Measures of water hardness

	Soft	Moderately Hard	Hard	Very Hard
Grains per gallon	0.0–3.5	3.6–7.0	7.1–10.5	10.6–
Parts per million or milligrams per liter	0.0–60	61–120	121–180	More than 180

Source: U.S. Geological Survey

a detergent containing "builders" that tie up the minerals can help prevent lime buildup, as well as keep the hardness from affecting detergents. If water hardness remains a problem, the operator may want to consult with a reputable water treatment firm who can provide answers.

3. *Temperature of water.* In general, the higher the temperature of the water in which the cleaning agent is dissolved, the faster the chemical reaction and the more effective the cleaning action. A practical limitation for manual cleaning is 120°F (49°C), which is the highest temperature in which most people can immerse their hands. Some chemical cleaning agents are designed to be used in cold water, but are too slow acting for cleaning greasy food soils.
4. *Surface being cleaned.* Different surfaces demand different cleaning agents. For example, aluminum surfaces can darken in highly alkaline or chlorinated cleaners; hard, crusted, baked-on soils may require use of an abrasive cleanser; coated, resilient flooring may require a less alkaline cleanser.
5. *Type of cleaning agent.* Soap may leave a greasy film under certain conditions; abrasives may scratch some surfaces; and some cleaning agents are only used for special purposes. For example, a mildly alkaline cleaning solution may be used for washing painted walls, while a highly alkaline cleaning solution may be used for hot-water dishwashing machines. Acid cleaners may be periodically needed to remove lime buildup from dishwashers or steam tables.
6. *Agitation or pressure to be applied.* In many cleaning operations, *agitation,* or scouring action, consists primarily of the friction from the proper brush, or another cleaning tool, such as a cloth or mop. The purpose is to remove outer layers of loosened soil from the surface of the object, so that the cleaning agent can penetrate deeper into the soil. This loosen-penetrate cycle continues until all soil is removed. The amount of agitation or pressure will affect the success of the outcome. Other forms of agitation include the pressure from water-spray action in mechanical dishwashing machines and the agitation created when water is forced through equipment with clean-in-place systems.

7. *Length of treatment.* All other factors being equal, cleaning effectiveness increases with the time the object is exposed to the cleaning agent.

Cleaning Agents

For our purposes, *cleaning agent* refers to a chemical compound specifically formulated to remove soil, or to remove mineral deposits. A particular cleaning agent should be selected for its specific cleaning properties. A compound that is powerful in one situation may prove totally ineffective in another. It must be effective, stable, noncorrosive, and safe for both employees and the surfaces when applied as directed. Since some cleaning agents are more effective than others, the quantities and labor required to do a job should be considered in making cost comparisons so that cleaning agents can be evaluated on a cost-effectiveness basis.

It's wise to check with suppliers to decide which chemical compounds are suitable for specific needs. The advice of a foodservice operator down the street may not be dependable. Cleaning agents that work well in one place may not be effective in another because of differences in soils, surfaces, and availability of tools.

Cleaning agents may be divided into four categories: detergents, which can be mildly or highly alkaline; solvents; abrasive cleaners; and acid cleaners. Some overlapping of categories is possible. For example, most abrasive cleaners contain detergents; many detergents contain solvents for full-strength use; and some acid cleaners use detergents.

Detergents

While soap is the oldest manufactured cleaning agent and is still one of the most effective in removing soil, its poor rinsability and sensitivity to water hardness limits its use to hand washing in the foodservice industry.

For most purposes, including the cleaning of food-contact surfaces, soap has been replaced by synthetic detergents. Detergents have been designed to meet almost any cleaning requirement. They are easily rinsable, and can be high- or low-sudsing (no sudsing is also available, but since loss of suds is the universal indicator that soil is attached to the detergent in the water and no longer effective, you probably would not want to use this type).

All detergents contain surfactants. *Surfactants* (surface active agents) reduce tension at the points where the detergent meets the soiled surface, allowing the detergent to quickly penetrate in order to loosen and disperse the soil.

Most synthetic detergents used in foodservice operations are alkaline because food soils are more easily dispersed in alkaline conditions. General-purpose detergents are mildly alkaline and are effective for removal of fresh soils from floors, walls, ceilings, and most equipment and utensils. Heavy-duty detergents are highly alkaline and are reserved for use in removing wax, aged or dried soils, and baked-on or burned-on greases. Detergents used in dishwashing machines are also highly alkaline.

Most detergents are effective if used as directed, but can be ineffective or even dangerous if misused. For example, handwashing soap will not delime a dishwashing machine no matter how much of it is applied. Always read and follow all manufacturer's instructions.

CAUTION: Under no circumstances should a foodservice manager or employee attempt to formulate his or her own detergents or mix chemicals, such as ammonia and chlorine bleach. The resulting chemical compound is potentially very dangerous to employees and customers. By the same token, one type of detergent should never be substituted for another, unless the intended use is clearly stated on the label. Using a dishwashing machine detergent for manual washing could result in serious damage to a person's skin.

Solvent Cleaners

Greasy soils that become extra tough in areas such as grill backsplashes, oven surfaces, and less frequently cleaned nonfood-surfaces, such as driveways, are quickly and easily loosened with solvent cleaners. *Solvent cleaners,* usually called de-greasers, are alkaline detergents containing an agent that dissolves grease, such as a diglycol butyl ether. These cleaning solutions are only effective at full or half strength, so their use must be limited to small areas or infrequent tasks to keep costs low.

Acid Cleaners

There are certain types of soil that are not affected by alkaline cleaners. The scaling in dishwashing machines, the rust stains that develop in washrooms, and the tarnish that darkens copper and brass are examples. Acid cleaners are used for these cleaning purposes.

The kind and strength of the acid varies with the purpose of the cleaner. Whatever their strength, acid cleaners must be selected and applied carefully to avoid damaging the surface being cleaned or the health of the user.

Abrasive Cleaners

Sometimes soil is attached so firmly to a surface that alkaline or acid cleaners will not work. In these cases, a cleaner containing a scouring agent is used to attack the soil. Badly soiled floors benefit from the use of abrasive cleaners.

Abrasives should be used with caution in a foodservice facility since they may mar smooth surfaces, especially plexiglass or plastic, and make cleaning more difficult.

SANITIZING PRINCIPLES

After a surface has been thoroughly cleaned and rinsed, it is ready to be sanitized. In the case of a food-contact surface, it *must* be sanitized. Sanitizing is not a substitute for cleaning, however, and is only effective if items are already clean. Caked-on soils not removed by cleaning, for example, may shield bacteria from a sanitizing solution. Sanitizing is a step beyond *clean,* which is merely the absence of soil, and a step below *sterile,* which is the absence of all living micro-organisms.

Why isn't everything sterilized to be on the safe side? The answer is that total sterilization is expensive and would be extremely difficult to do in a foodservice establishment. It is also unnecessary.

Sanitizing can be accomplished in two ways: either the object may be immersed in water with a temperature of at least 170°F (76.7°C), which is high enough to kill micro-organisms; or it can be treated with a chemical sanitizing compound. *You cannot sanitize without washing and rinsing first.*

Heat Sanitizing

Exposing a clean object to adequately high heat for a sufficient length of time will sanitize it. Generally, the higher the heat, the shorter the time required to kill harmful organisms. The purpose of heat sanitizing is to bring the food-contact surface to a temperature that will kill most harmful micro-organisms, usually 162° to 165°F (72.2° to 73.9°C) or higher. Heat sanitization can be used in either mechanical or manual warewashing operations (see Exhibit 10.2).

One method for heat sanitizing in a manual system is by immersing an object in water maintained at a temperature of not less than 170°F (76.7°C) nor higher than 195°F (90.6°C) for at least 30 seconds. *NOTE:* Some states currently require 180°F (82.2°C) *minimum*

Exhibit 10.2 The minimum and maximum required washing, rinsing, and sanitizing temperatures and times for manual and machine methods

Temperature	Procedure
75°–120°F (24°–49°C)	Range for chemical sanitizers to be effective
80°–110°F (26.7°–43.3°C)	Pre-wash cycle for high-temperature machine
120°F (49°C)	Wash and rinse water temperatures for manual immersion of tableware and equipment
120°–140°F (48.9°–60°C)	Range for sanitizing (final) cycle in chemical sanitizing machine
150°F (65.6°C)	Wash cycle for single-tank, stationary rack and multiple-tank, conveyor models of high-temperature machines
160°F (71.1°C)	Wash cycle for single-tank, conveyor model of high-temperature machine
162°–165°F (72.2°–73.9°C)	Range that kills most micro-organisms
165°F (73.9°C)	Wash cycle for a single-temperature model, high-temperature machine
170°F (76.7°C)	Heat sanitizing for manual immersion
180°F (82.2°C)	Heat sanitizing for manual immersion in some local regulations
180°F (82.2°C)	Final rinse cycle for high-temperature machine at the manifold
195°F (90.6°C)	Upper limit for heat sanitization by machine or manual process
200°F (93.3°C)	Heat sanitization using live, additive-free steam

on the machine manifold for heat sanitization, so be sure to check local regulations.

Manual immersion may be accomplished by use of racks to lower the items into the water so that all items are completely covered by the water. A thermometer to check the water temperature in the sink compartment must be accessible and used often.

The most common method of hot-water sanitizing is in a high-temperature dishwashing machine with a manifold temperature of 180°F (82.2°C). Still another means of heat sanitization is through the use of live, additive-free steam. This method may be used on equipment that is too large to be immersed, provided that the equipment is capable of confining the steam. Steam used on assembled equipment should be at a temperature of at least 200°F (93.3°C).

It is important to note, however, that the temperature at the *surface* of the object being sanitized is what counts, not the temperature of the steam in the line. Water or steam is hot when it leaves its source, but it will cool very quickly on contact with a cool object.

The effectiveness of heat sanitization can be checked in one of several ways. One method to indicate surface temperatures is water-proof, pressure-sensitive labels that are applied directly to tableware and equipment. The labels change color irreversibly when sanitization temperatures are reached.

Another method is using heat-sensitive tape that contains or consists of paraffin wax with a specific melting point that indicates whether the required surface temperature was reached. Do not use your bi-metallic stemmed food thermometer in a dishwashing machine because it can melt and will be ruined.

Chemical Sanitizing

Sanitizing can also be achieved through the use of chemical compounds capable of destroying disease-causing micro-organisms. Due to energy costs and ease of use, chemical sanitizers have found wide acceptance in the foodservice industry.

Chemical sanitizers are regulated by the U.S. Environmental Protection Agency (EPA), as well as state EPA agencies, which register them as pesticides. Consequently, any product that uses the word *sanitize* on its label and bears an EPA registration number will, if used as directed, reduce micro-organisms to levels acceptable to most local regulatory agencies. EPA labeling requirements are strict, so the information on the label will include what concentrations to use, data on minimum effectiveness, and warnings of possible health hazards.

Chemical sanitizing is done in two ways: either by immersing an object in a specific concentration of sanitizing solution for one minute; or by rinsing, swabbing, or spraying with another specific concentration of sanitizing solution.

Since the sanitizer is depleted in the process of killing bacteria, the strength of the sanitizing solution must be tested frequently. The solution also gets bound up by leftover food particles and detergents not adequately rinsed off. Make sure that you use a test kit designed for the sanitizing solution you are using. Test kits are usually available from the manufacturer of the particular sanitizing agent you use, or can be obtained from your restaurant supplier. The solution must be changed when it is dirty or when the concentration is less than the required parts per million (p.p.m.).

Chemical sanitizers are most effective in temperatures between 75°F (23.9°C) and 120°F (48.9°C). Temperatures that are in the lower part of this range can extend the life of the sanitizing solution.

Three of the most common chemicals used for sanitizing are *chlorine* as hypochlorites, *iodine* as iodophors, and *quaternary ammonium compounds* (quats). The properties of these agents differ somewhat, as shown in Exhibit 10.3. Chlorine and iodine compounds have some properties in common.

- They will kill most bacteria if used correctly.
- They are less effective in water that is alkaline from detergent carryover, making it necessary to *thoroughly rinse* items that have been cleaned with general-purpose or heavy-duty detergents before applying either of these sanitizers.
- They are relatively non-irritating to the skin at specified sanitizing strengths.

There are also some differences between chlorine and iodine compounds.

- Chlorine compounds are more likely than iodine compounds to tarnish and damage metal, such as pewter, stainless steel, aluminum, or silverplate utensils, as well as rubber products.
- Some chlorine compounds may leave a mild odor on dishes if the concentration is too strong.
- Iodine-based compounds should be used only in solutions that have a pH of 5.0 or less, unless the manufacturer's instructions specify a higher pH limit of effectiveness. To maintain the pH of the water, be sure you rinse thoroughly, which will minimize the amount of detergent carry-over.

Quats can be effective in both acid and alkaline solutions. They are generally non-irritating to the skin when used at proper dilutions as recommended by the supplier. Certain quats may not be compatible with some soaps or detergents, which is why rinsing is so important. For some quaternary ammonium sanitizers, the mineral deposits in hard water may interact with certain bacteria and make the bacteria less susceptible to the action of the sanitizing solution.

Quats can only be used in water that has a hardness of 500 p.p.m. or less (which is the level of virtually all municipal water supplies). For those quats that are accepted by the FDA, a concentration of 200 p.p.m. is specified. Since sanitizers are manufactured according to many different formulas, it is necessary to read the label to make sure that the particular formula in question suits the sanitizing needs of the operation and are accepted by the appropriate agencies.

A number of factors affect the action of chemical sanitizers and should be considered by management before use.

1. *Contact.* In order for the sanitizing solution to kill a sufficient number

Exhibit 10.3 Chemical sanitizing agents

	Chlorine	**Iodine**	**Quaternary Ammonium**
Minimum Concentration			
—*For Immersion*	50–100 parts per million (ppm)	12.5–25.0 ppm	200 ppm
—*For Power Spray or Cleaning in Place*	100–200 ppm	25–50 ppm	
Temperature of Solution	75°F/23.9°C+	75°–120°F/ 23.9°–48.9°C Iodine will leave solution at 120°F/48.9°C	75°F/23.9°C+
Time for Sanitizing			
—*For Immersion*	1 minute	1 minute	1 minute; however, some products require longer contact time; read label
—*For Power Spray or Cleaning in Place*	Follow manufacturer's instructions	Follow manufacturer's instructions	
pH (Detergent Residue Raises pH of Solution So Rinse Thoroughly First)	Must be below pH 10	Must be below pH 5.5	Most effective around pH 7 but varies with compound
Corrosiveness	Corrosive to some substances	Noncorrosive	Noncorrosive
Response to Organic Contaminants in Water	Quickly inactivated	Made less effective	Not easily affected
Response to Hard Water	Not affected	Not affected	Some compounds inactivated but varies with formulation; read label. Hardness over 500 ppm is undesirable for some quats
Indication of Strength of Solution	Test kit required	Amber color indicates effective solution, but test kits must also be used	Test kit required. Follow label instructions closely

of micro-organisms, the sanitizing solution must make contact with the item for an amount of time specified by local regulations, usually one minute immersion. Short cuts *cannot be taken* with label instructions no matter what the method of application.

2. *Selectivity.* Chemical sanitizers, particularly quats, may be selective in their capacity to kill certain micro-organisms.

3. *Concentration.* Proper concentration of sanitizers is a critical factor. Concentrations below the legally required minimum could result in a failure to sanitize equipment and utensils. Concentrations higher than necessary may create a safety hazard. High concentrations can cause taste and odor problems, corrode metals and other materials, leave residues, and waste money, besides being a health department violation (refer to Exhibit 10.3). Test kits are available for all chemical sanitizers. *It is essential to follow the manufacturer's instructions for each type of sanitizer and to check the strength often in order to meet your local regulations.* To test your sanitizing solution using test paper, tear off a small strip or use a pre-cut portion, dip it in the solution or touch it to a treated plate, and compare the color to the chart on the test kit. The minimum concentration of chlorine should be at least 50 p.p.m., iodine should be 12.5 p.p.m., and quats should be the concentration listed on the manufacturer's label, or one that is required by local regulations. In order to maintain the minimum required concentrations, you should mix solutions of chlorine to 100 p.p.m. and iodine to 25 p.p.m.

4. *Temperature.* The temperature of the solution is crucial. A minimum temperature of 75°F (23.9°C) is required. Chlorine and iodine compounds are less effective at temperatures lower than 75°F (23.9°C). At temperatures higher than 120°F (48.9°C), chlorine compounds may corrode some metal items and both chlorine and iodine compounds may evaporate from the solution. Generally, all sanitizers work best at temperatures between 75° and 120°F (23.9° and 48.9°C).

Some sanitizers are blended by manufacturers with detergents to make *detergent-sanitizers.* These products will sanitize effectively, but sanitizing must still be a separate step from cleaning. In other words, the same product must be used twice—once to clean, and once to sanitize. In general, detergent-sanitizers are more expensive than regular detergents and sanitizers and are more limited in their applications than detergents. They were created for two-compartment sink operations, which are no longer common. Some are not meant to be used on food-contact surfaces, but can be used on floors or the exteriors of equipment; check your local regulations and labeling for EPA approval for use on food-contact surfaces.

Scented or oxygen bleaches are not acceptable as sanitizers. These bleaches are inappropriate for use on food-contact surfaces because they leave a residue on equipment, utensils, and in the water. All bleaches are not sanitizers; check labeling for EPA approval.

MACHINE CLEANING AND SANITIZING

Dishwashing machines, properly operated and maintained, can be reliable in removing soil and bacteria from tableware and kitchen implements. Because of the needs of high-volume operations, the foodservice industry has increasingly moved to the use of dishwashing machines, both hot-water and chemical-sanitizing types. (The purchase of and planning for dishwashing equipment was covered in Chapter 9.)

A successful machine cleaning and sanitizing program requires:

- Sufficient water supplies for the entire washing operation, including hot-water sanitizing (unless chemical sanitizing is chosen)
- An efficient layout in the dishwashing area to utilize employees and machines to the best advantage and to prevent recontamination of clean dishes by soiled ones (see Exhibit 10.4)
- Employees who know how to operate and maintain the equipment, and who are knowledgeable about the correct use of required detergents and other chemicals
- A protected dish-handling and storage area to contain sanitary dishware
- Regular inspections of the washed and sanitized items to make sure that the machine is performing up to par
- Regular inspections of the whole operation to make certain that correct procedures are being followed.

Be sure to read the manufacturer's data plate, but whatever the type of dishwashing machine selected, the following requirements should ordinarily be met.

- Dish tables of adequate size should be provided to handle both soiled and clean equipment, tableware, and utensils.
- Unless the machine has a prewash cycle, items to be cleaned must be scraped or soaked to remove food debris before being washed in the machine.
- Dishwashing machines should come equipped with a device that allows the operator to check on the pressure of the wash and final rinse water. A pressure between 15 and 25 pounds per square inch (psi) is required for good spray action.
- All dishwashing machines, especially the scrap trays, must be cleaned as needed, and at least once a day.

High Temperature Machines

Temperatures during the wash cycle vary based on which model of machine is being used, but typically they should be one of the following: 150°F (65.6°C) for the single-tank, stationary rack model and the multiple-tank conveyor model; 160°F (71.1°C) for the single-tank conveyor model; and 165°F (73.9°C) for the single-temperature unit. These are the temperatures necessary to dissolve grease and loosen soil.

For high-temperature machines with a prewash cycle, the temperature during that cycle should be between 80° and 110°F (26.7° and 43.3°C). These are the

Exhibit 10.4 Efficient working space for mechanical dishwashing function

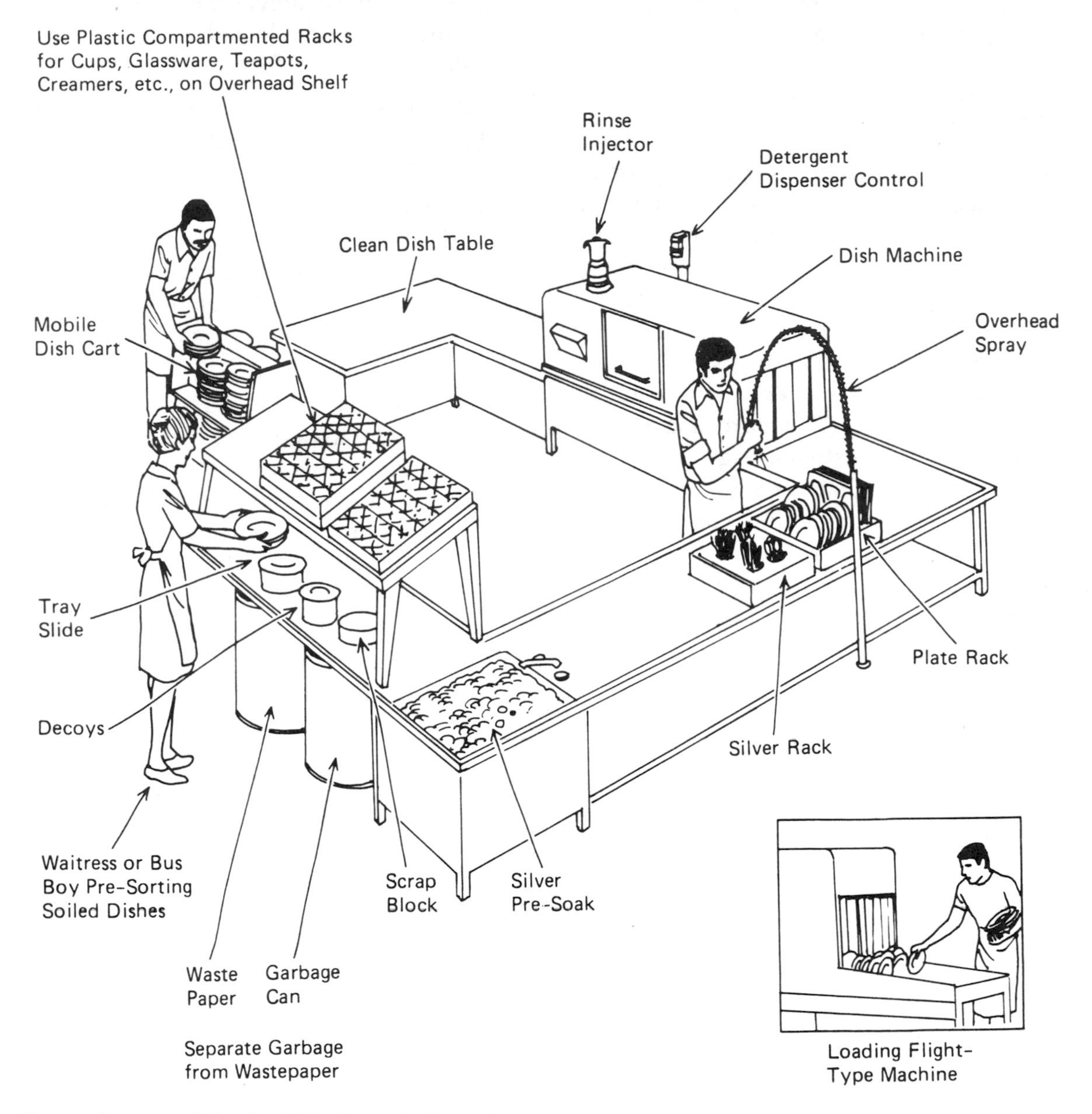

Source: *Programmed Cleaning Guide.* Copyright © 1984 by The Soap and Detergent Association. Reprinted by permission.

minimum temperatures needed to remove soil and particles of food, and which provide accumulative heat sanitization.

Effective hot-water sanitizing by a machine depends on several factors. The temperature of the water for the final rinse must be at least 180°F (82.2°C), but no more than 195°F (90.6°C). At higher temperatures, the water will vaporize before sanitizing the dishes. High temperatures will also bake food particles onto dishes, making their removal difficult. Since most general-purpose water heaters do not bring water to such high temperatures, a *booster heater* is usually required.

An in-line thermometer, which measures the hot water temperature at the manifold where it sprays out into the machine, is the method used to verify the effectiveness of heat sanitizing. In order for sanitization to work, all steps in the clean, rinse, and sanitize process must be followed correctly.

Chemical-Sanitizing Machines

The water temperatures for chemical sanitizing are typically between 120° and 140°F (48.9° and 60°C). It is important that the recommended sanitizing chemical be used in the proper concentration. The chemical solution must be dispensed automatically into the final rinse water.

Problems and Cures

Although dishwashing machines can be the most reliable way to clean and sanitize tableware and utensils, they can also be a source of many problems if installed or operated incorrectly. Exhibit 10.5 summarizes some of these problems, with possible causes and corrective actions.

MANUAL CLEANING AND SANITIZING

Washing, rinsing, and sanitizing of equipment, utensils, and food-contact surfaces can also be done manually.

Work Area

The first requirement for washing, rinsing, and sanitizing most portable food-contact items is a washing area, preferably away from the food preparation area. This work station must be equipped with a minimum three-compartment sink, separate drainboards for clean and for soiled items, and an area for scraping and rinsing food soil into a garbage container or grinder for disposal (see Exhibit 10.6). Requirements for cleaning and sanitizing equipment vary from locality to locality, so you should check regulations as they apply in your area. In some areas, two-compartment sinks are permitted for manual cleaning and sanitizing of pots and pans, and in other areas four-compartment sinks are required.

If hot water will be used to sanitize, the third sink or compartment of the sink must be equipped with a heating unit to maintain water at the desired minimum temperature of 170°F (76.7°C). For either hot water or chemical sanitizing, a thermometer must also be mounted in each sink compartment to accurately indicate

Exhibit 10.5 Some chemical and physical factors affecting the dishwashing process

Symptom	Possible Cause	Suggested Cure
Soiled Dishes	Insufficient detergent	Use enough detergent in wash water to ensure complete soil removal and suspension.
	Wash water temperature too low	Keep water temperature within recommended ranges to dissolve food residues and to facilitate heat accumulation (for sanitation).
	Inadequate wash and rinse times	Allow sufficient time for wash and rinse operations to be effective. (Time should be automatically controlled by timer or by conveyor speed.)
	Improperly cleaned equipment	Unclog rinse and wash nozzles to maintain proper pressure-spray pattern and flow conditions. Overflow must be open. Keep wash water as clean as possible by pre-scraping dishes, etc. *Change water in tanks at proper intervals.*
	Improper racking	Check to make sure racking or placement is done according to size and type. Silverware should always be pre-soaked and placed in silver holders without sorting. Avoid masking or shielding.
Films	Water hardness	Use an external softening process. Use proper detergent to provide internal conditioning. Check temperature of wash and rinse water. Water maintained above recommended temperature ranges may precipitate film
	Detergent carryover	Maintain adequate pressure and volume of rinse water, or worn wash jets or improper angle of wash spray might cause wash solution to splash over into final rinse spray.
	Improperly cleaned or rinsed equipment	Prevent scale buildup in equipment by adopting frequent and adequate cleaning practices. Maintain adequate pressure and volume of water.
Greasy Films	Low pH; insufficient detergent; low water temperature; improperly cleaned equipment	Maintain adequate alkalinity to saponify greases; check detergent, water temperature. Unclog all wash and rinse nozzles to provide proper spray action. Clogged rinse nozzles may also interfere with wash tank overflow. Change water in tanks at proper intervals.

Exhibit 10.5 *(Continued)*

Symptom	Possible Cause	Suggested Cure
Streaking	Alkalinity in the water; highly dissolved solids in water	Use an external treatment method to reduce alkalinity. Within reason (up to 300–400 ppm), selection of proper rinse additive will eliminate streaking; above this range external treatment is required to reduce solids.
	Improperly cleaned or rinsed equipment	Maintain adequate pressure and volume of rinse water. Alkaline cleaners used for washing must be thoroughly rinsed from dishes.
Spotting	Rinse water hardness	Provide external or internal softening. Use additional rinse additive.
	Rinse water temperature too high or too low	Check rinse water temperature. Dishes may be flash drying, or water may be drying on dishes rather than draining off.
	Inadequate time between rinsing and storage	Allow sufficient time for air drying.
Foaming	Detergent; dissolved or suspended solids in water	Change to a low sudsing product. Use an appropriate treatment method to reduce the solid content of the water.
	Food soil	Adequately remove gross soil before washing. The decomposition of carbohydrates, proteins or fats may cause foaming during the wash cycle. Change water in tanks at proper intervals.
Coffee, tea, metal staining	Improper detergent	Food dye or metal stains, particularly where plastic dishware is used, normally requires a chlorinated machine washing detergent for proper destaining.
	Improperly cleaned equipment	Keep all wash sprays and rinse nozzles open. Keep equipment free from deposits of films or materials which could cause foam buildup in future wash cycles.

Source: Reprinted by permission of NSF *International* from "Recommended Field Evaluation Procedures for Spray-Type Dishwashing Machines." Copyright © 1982 by NSF *International.*

Exhibit 10.6 A three-compartment sink for manual washing, rinsing, and chemical sanitizing

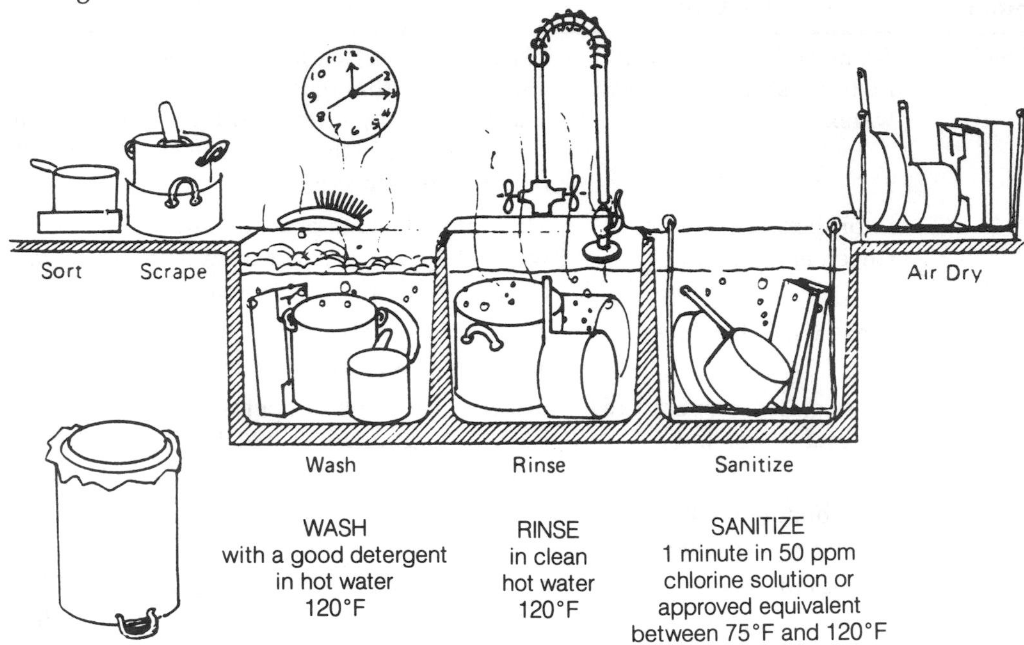

temperatures. A clock with a second hand should be easily visible so that employees do not have to estimate the time of immersion. Long-handled tongs, hooks, baskets, or racks will also be needed to dip clean items into the sanitizing bath of hot water or into the chemical sanitizing solution. These tools keep the foodservice employee from handling the items, which not only prevents injuries to the employee, but helps the items remain sanitary.

Tableware and Portable Equipment

Whatever the item, immersion washing, rinsing, and sanitizing involves six steps.

1. Clean and sanitize the sinks and the work surfaces before each use.
2. Scrape and pre-soak items to be cleaned to remove food soil that may deactivate the detergent. Garbage should be scraped and disposed without contaminating clean equipment and utensils. Items to be cleaned should be sorted. Silverware should be pre-soaked in a solution designed for that purpose.
3. Wash in the first sink using a clean detergent solution at 120°F (48.9°C). Use the proper brush or a cloth to loosen and remove any remaining soil.
4. Rinse in the second sink using clear, potable water at 120°F (48.9°C) to remove all traces of food soil and detergent that may interfere with the sanitizing agent.

5. Sanitize in the third sink by immersing items in hot water at 170°F (76.7°C) for 30 seconds, or in a chemical sanitizing solution at a minimum of 75°F (23.9°C) for one minute, or according to the manufacturer's instructions. Be sure that all surfaces contact the sanitizing chemical or hot water for the prescribed length of time. In particular, beware of air bubbles inside inverted containers that might shield the interior from the sanitizer.
6. Air dry. *Do not wipe dry.* Wiping can recontaminate all your newly sanitized utensils and equipment, and can remove the sanitizing solution from the surfaces before it has worked completely.

When the soap suds disappear in the first compartment, when soap suds remain in the second compartment, and when water temperature cools or water becomes dirty and cloudy in any of the compartments, it is necessary to empty all three compartments and refill them for proper cleaning and sanitizing to take place. Use the test kit periodically to check the third compartment for chemical sanitizing, in order to make sure that it remains at the appropriate level designated by the manufacturer, or according to local regulations.

Wooden surfaces, such as cutting boards, wood handles, and baker's tables, are an exception to the above steps. They must be scrubbed with a detergent solution and a stiff-bristled nylon brush, rinsed in clear, potable water, and wiped down and swabbed with a sanitizing solution after every use. Wooden cutting boards should *never* be submerged in detergent or sanitizing solutions.

Clean-in-Place Equipment

Clean-in-place items are designed to have detergent and sanitizing solutions pumped through them. For items that are equipped for cleaning in place, the manufacturer's instructions should be followed.

Immobile Equipment

Immobile food preparation equipment comes with manufacturer's instructions for disassembly and cleaning, which should be followed. If these instructions are not available, the following general instructions may be applied.

1. For general equipment cleaning, unplug the unit if it is electrically powered. For example, you need to unplug a meat slicer before disassembling it for cleaning, but you do not need to unplug a refrigeration or freezer unit.
2. Remove fallen food particles and scraps that have collected around and under equipment.
3. Remove whatever parts you can. Wash, rinse, and sanitize the removable parts following the steps for manual immersion.
4. Wash remaining food-contact surfaces, such as the base of the equipment where the removable parts attach, and rinse with clean water. Swab with a solution of chemical

sanitizing solution mixed to manufacturer's directions.

5. Wipe down the nonfood-contact surfaces with a sanitized cloth, and allow all parts to air dry before reassembling.
6. Resanitize the external food-contact surfaces of the parts that were handled when the equipment was reassembled.

Equipment can also be cleaned and sanitized using spray equipment. To be sanitized in this way, the object must be sprayed for two to three minutes. Use caution, though, especially if the equipment is electrical, and follow manufacturer's directions, because not all equipment can be spray cleaned. Check labeling on the sanitizer to determine whether spray application is allowed and what the correct chemical sanitizer concentration needs to be.

Refrigerated Units

Although refrigerators slow the growth of most micro-organisms, they may become homes for certain kinds of bacteria and fungi unless cleaned properly. All interior parts should be cleaned and sanitized on a regular basis—as often as once a day or once a week—depending on use. Spills should be cleaned up immediately. Shelves need to be removed as part of this cleaning process, so food already in the refrigerator will need to be moved to another unit during the cleaning and sanitizing process. Cleaning should be done before new food shipments are received, so less food will require moving. The presence of debris, mold, or objectionable odors indicates that immediate cleaning is required. Gaskets on unit doors need routine scheduled cleaning for removal of mold and mildew, as well as any collected food particles. Clean and maintained gaskets last longer and provide a good seal to maintain cold or frozen holding temperatures. Defrosting on a scheduled basis may also be necessary to maintain temperatures and for cleaning purposes.

The Facility

Nonfood-contact surfaces in an establishment include the floors, walls, ceilings, and drains. The frequency with which they are cleaned will depend upon the type of food served, the kinds of surfaces to be cleaned, and factors such as the rate of ventilation in the kitchen, but all should be cleaned on an as-needed basis. Floors present a unique problem because they can become a safety hazard if soil is allowed to accumulate or if they are not rinsed adequately. Sanitizing nonfood-contact surfaces is not required, but is a good idea.

Always use caution when preparing the detergent and sanitizing solutions and follow manufacturer's directions. Never use food utensils or baking measuring cups to measure chemicals for cleaning and sanitizing. Any tool used for measuring chemicals should be purchased for that purpose only, and stored with the chemicals away from food in storage and preparation.

Dry Storage

The dry storage areas must be swept and scrubbed regularly. Walls, ceilings, floors, shelves, light fixtures, and racks also require cleaning on a routine basis.

Walls can be cleaned by either the spray or swab methods, depending on which methods you use for other cleaning jobs. If you use the spray method, be sure that the spray will not damage the walls or nearby equipment or supplies. You will need to be careful of wall seams as well, since soaking them can cause damage behind walls.

Floors

Floors can be cleaned by spraying with a pressure nozzle on a hose or by mopping. Whichever method you choose, work toward the floor drain, or from the outer edges toward the middle of the floor. Sweep floors before cleaning, then apply detergent, scrub, and rinse. To remove excess dirt and grease, spray or pour the cleaning solution on the floor and use a floor or deck brush to scrub along the grouting or floor surface before regular wet and damp mopping.

Prepare the mop by swishing it up and down in the mop bucket to get it thoroughly wet and to further dissolve the cleaning agent. Twist the mop in the wringer on the bucket and then press it. Use a "figure 8" motion to cover an area of about ten feet by ten feet. Flip the mop at least twice in this area, then return the mop to the bucket to start again on a new section of floor. Use a wet mop for the first run, and then a damp mop over the same area.

Remove excess water with a clean squeegee or a damp mop, again working toward the drain. Rinse floors thoroughly so residue from the detergent will not make the floors slippery when they dry.

Ceilings

Ceilings may not need to be cleaned as frequently as floors and walls, but they should be checked daily to be sure cobwebs, dust, dirt, or condensation cannot fall from the ceiling and contaminate food-contact surfaces or food below. To clean ceilings use the swab method, rather than spraying, to keep water away from lights and ceiling fans. The swab method involves using a sponge or cloth to apply the cleaning solution, which will eliminate excess water and cleaning agents from dripping.

Floor Drains

Cleaning floor drains should be the last step in the daily cleaning schedule. For cleaning floor drains, workers should wear heavy-duty rubber gloves. Remove the drain cover or grate and clean out debris with a drain brush. Replace the cover or grate and flush it with a pressure sprayer or hose through the drain. Be sure it is flushed thoroughly and that the water does not splash back up through the drain. Pour cleaning detergent down the drain, following the manufacturer's directions for preparing the solution. Wash with the pressure sprayer or the drain brush, and rinse thoroughly. As an optional step, pour in a sanitizing or disinfecting solution. Drains under refrigerator cases and freezer units also need to be cleaned on a regular basis.

Public Restrooms

An unsanitary restroom can harbor dangerous micro-organisms and often has unpleasant odors. Public and employee restrooms must be checked on an hourly basis and cleaned thoroughly at least once a day, or more frequently if necessary to keep them clean, sanitary, and restocked.

All walls, counters, sinks, mirrors, dispensers, toilets, urinals, partitions, and waste receptacles must be kept thoroughly clean. Toilet bowls and urinals must also be cleaned and disinfected at least once daily, both inside and outside the bowl. Regular checks throughout the day should include refilling toilet paper, soap, and sanitary supplies, as well as removing any trash out of urinals.

Do not underestimate your cleaning needs in this area; a restroom hasn't been invented that can be too clean. Keep in mind that many patrons tend to associate cleanliness of your public restroom with their perceived impression of the cleanliness of your kitchen.

Cleaning Tools

Tools used for cleaning should be kept separate from those used to sanitize, and nonfood-contact tools should be kept separate from those that are for food-contact surfaces. Items such as cloths, sponges, and scrubbing pads must be clean, rinsed frequently, and should be stored in separate containers with sanitizing solution between uses, or allowed to air dry. Cloths should be laundered daily, or more often if necessary.

Brushes and mops need to be washed, rinsed, and sanitized as well. They should be hung up by their handles in order to air dry. Do not rest brushes on the bristles, since this will cause unnecessary wear and damage. Do not leave mops in buckets or resting on the floor, as both of these practices can harbor pests or lead to the growth of micro-organisms. All buckets and mop pails need to be emptied, washed, rinsed, and disinfected or sanitized as needed, and at least once a day.

STORAGE

When all tableware, utensils, and equipment are clean and sanitary, they must be kept that way. Cleaning supplies also need to be stored so they remain sanitary and do not pose safety or physical hazards to food or people.

Clean Tableware and Equipment

Utensil drawers and clean-dish shelves must be cleaned and sanitized *before* the newly sanitized utensils and dishes are brought in from the dishwashing room. Wheeled dish tables or carts, if they are easy to move, generally provide the best means of transportation. These dish carts must be clean and sanitized as well.

The clean tableware storage area must be more than six inches off the floor and protected from splash, dust, and contact with food. All items must be accessible without the necessity of touching surfaces that will contact the food or the customer's mouth. For example, glasses and cups should be stored upside-down, and utensils should be stored handles up.

The food-contact surfaces of clean-in-place equipment must also be covered or otherwise protected when not in use.

Cleaning Supplies

Cleaning and sanitizing tools and chemicals should be kept in a dry and locked cabinet or area, which is away from food supply areas and other chemicals. Where portion-control systems are used, product supplies must be connected to the portion control devices so that proper dilutions are readily available.

When you have selected your cleaning tools, see that they are kept in a well-lighted storage area where they are easily identifiable and accessible. This area should be in a part of the facility away from food storage and preparation. This area should include hooks for hanging mops, brooms, and brushes.

There should be a separate sink to fill and empty mop buckets, to rinse and clean mops, and to clean brushes and sponges. Hand-washing, food-preparation, and equipment-cleaning sinks must *never* be used for cleaning mops and brushes. A floor drain in this storage area is necessary to rid the area of waste water or spillage (see Exhibit 10.7).

USE OF HAZARDOUS MATERIALS

Management must ensure that employees are protected within an operation. Employees must be warned whenever they are at some safety risk in performing a task; therefore, they should be well-informed about the hazardous materials that are used in an operation. The OSHA Hazard Communication Standard (HCS), also known as "Right-To-Know," sets forth OSHA regulations that require employers to develop and implement a hazard communication program within their operations.

A current inventory of the hazardous chemicals used in your establishment must be maintained. Included in the inventory should be all cleaning agents, chemical sanitizers, and chemical pesticides if they are used.

Manufacturers carry the responsibility of seeing that all hazardous chemicals are properly labeled with the chemicals' identities, with appropriate hazard warnings, and with the name and address of the manufacturer, importer, or other responsible party. A Material Safety Data Sheet (MSDS) must be supplied by the manufacturer for each hazardous chemical and must be kept on file in the establishment. Exhibit 10.8 illustrates a sample MSDS.

The information that needs to be supplied to the employees by the manager includes the common name of the product, where the product is used, and the foodservice employees authorized to use the product. The manager must also communicate to employees the information contained on the MSDS. This information covers the chemical name of the

Exhibit 10.7 Cleaning supplies storage area

product; physical hazards of the chemical, such as its potential for explosion or fire, toxicity, and whether it is caustic or irritating; the health hazards, such as its potential to aggravate certain medical conditions; and emergency procedures to take in case of exposure.

In addition, information must be presented dealing with the protective steps necessary during regular use, as well as procedures to deal with spills and leaks. All MSDSs must be readily available to both employees and to regulatory personnel who might request them.

Management's responsibility is to make sure that all labels are legible and clearly displayed on containers. If the original label is not on each container, the identity of the hazardous chemicals and the hazard warnings must be communicated by other means. For example, signs, placards, or operating procedures can be used. Secondary containers must be labeled as well.

Management must provide training that applies to the use of the specific hazardous chemicals being used in their establishment, and must offer information

Exhibit 10.8 A sample OSHA standardized MSDS

Material Safety Data Sheet
May be used to comply with
OSHA's Hazard Communication Standard,
29 CFR 1910.1200. Standard must be
consulted for specific requirements.

U.S. Department of Labor
Occupational Safety and Health Administration
(Non-Mandatory Form)
Form Approved
OMB No. 1218-0072

IDENTITY *(As Used on Label and List)* High Alkline Oven Cleaner/Degreaser
Extra Strong Degreaser

Note: Blank spaces are not permitted. If any item is not applicable, or no information is available, the space must be marked to indicate that.

Section I

Manufacturer's Name Chemicals Unlimited	Emergency Telephone Number 1-800-000-1234
Address *(Number, Street, City, State, and ZIP Code)* Tanktown, USA, 00001	Telephone Number for Information 1-123-000-4567
	Date Prepared February 31, 1989
	Signature of Preparer *(optional)*

Section II — Hazardous Ingredients/Identity Information

Hazardous Components (Specific Chemical Identity; Common Name(s))	OSHA PEL	ACGIH TLV	Other Limits Recommended	% *(optional)*
Sodium Hydroxide (Caustic Soda) 1310-73-2	2		2 C	5
Butoxyethanol (Butyl Cellosolve) 111-76-2	240		120	10
(Skin) — This product contains no other component considerd				
hazardous according to the criteria of 29 CFR				
1910.1200.				

Section III — Physical/Chemical Characteristics

Boiling Point 212°F	Specific Gravity (H_2O = 1) 1.05-1.08
Vapor Pressure (mm Hg.) N/A	Melting Point N/A
Vapor Density (AIR = 1) N/A	Evaporation Rate (Butyl Acetate = 1) N/A

Solubility in Water
99%

Appearance and Odor
Opaque red/purple liquid: slight glycol ether odor.

Section IV — Fire and Explosion Hazard Data

Flash Point (Method Used)	Flammable Limits	LEL	UEL
N/A	N/A	N/A	N/A

Extinguishing Media
N/A

Special Fire Fighting Procedures
Product does not support combustion.

Unusual Fire and Explosion Hazards
None

(Reproduce locally) OSHA 174, Sept. 1985

Exhibit 10.8 *(Continued)*

Section V — Reactivity Data

Stability	Unstable		Conditions to Avoid: Stable under normal conditions of handling.
	Stable	X	Do not mix with anything but water.

Incompatibility *(Materials to Avoid)*
Reacts violently with acids, reacts with soft metals such as aluminum and zinc.

Hazardous Decomposition or Byproducts
Do not use on surfaces above 130°F, irritating vapors may occur, keep from freezing.

Hazardous Polymerization	May Occur		Conditions to Avoid
	Will Not Occur	X	

Section VI — Health Hazard Data

Route(s) of Entry:	Inhalation?	Skin?	Ingestion?
	X	X	X

Health Hazards *(Acute and Chronic)*
SKIN — Causes severe chemical burns. EYES — May cause blindness. INGESTION — Harmful or fatal, causes chemical burns of mouth and throat and stomach. INHALATION — Damages airways and lungs, depending on amount and bronchitis and pneumonia. Those with asthma or other lung problems may be more susceptible.

Carcinogenicity:	NTP?	IARC Monographs?	OSHA Regulated?
N/A			

EMERGENCY AND FIRST AID PROCEDURES: EYES — Flush immediately with plenty of cool running water, continue flushing for at least fifteen minutes, holding eyelids apart. Call doctor immediately. SKIN — Flush skin immediately with plenty of cool running water for at least fifteen minutes while removing contaminated clothing and shoes. Wash clothing before reuse. IF SWALLOWED — Rinse mouth at once; then drink one or two large glasses of water or milk. DO NOT induce vomiting. Never give anything by mouth to an unconscious person. IF INHALED — Move immediately to fresh air. Call a poison control center or physician immediately.

Section VII — Precautions for Safe Handling and Use

Steps to Be Taken in Case Material Is Released or Spilled
CLEANUP: Rinse small amounts to drain where possible. Dike or dam large spills; pump to containers or soak up on inert absorbent. Flush residue to sewer with plenty of water; rinse area thoroughly.

Waste Disposal Method
RCRA corrosive waste (D002). Consult state and local authorities for restrictions and disposal of chemical waste.

Precautions to Be Taken in Handling and Storing
RESPIRATORY: Ventilate to maintain exposure below TLV(S). Use a NIOSH/MSHA approved organic vapor respirator in high concentrations of butoxyethanol. Avoid inhalation of vapor or mist. Wear rubber gloves or gauntlets; splashproof glasses, goggles

Other Precautions
or faceshield (do not wear contact lenses); rubber apron or other protective equipment to prevent contact with skin, eyes or clothing. Avoid contact with used solutions of this product as these may also be hazardous. KEEP OUT OF REACH OF CHILDREN.

Section VIII — Control Measures

Respiratory Protection *(Specify Type)*
As above.

Ventilation	Local Exhaust: X	Special
	Mechanical *(General)*: X	Other

Protective Gloves	Eye Protection
X	X

Other Protective Clothing or Equipment
Rubber apron.

Work/Hygienic Practices
As above.

The above information is believed to be correct with respect to the formula used to manufacture the product. As data, standards and regulations change, and conditions of use and handling are beyond our control, no warranty express or implied, is made as to the completeness or continuing accuracy of this information.

★ U.S.G.P.O.: 1986-491-529/45775

about steps that employees can take to see that they remain protected from the existing dangers. Training should include discussing the information given in the MSDS and demonstrating, for example, appropriate work practices, the use of personal protective equipment, and emergency procedures.

SUMMARY

All the work that goes into preparing and presenting wholesome and sanitary food can be undone by the failure to keep utensils and equipment clean and sanitary. All food-contact surfaces must be cleaned and sanitized after every use, or interruption of use, or at regular intervals of every four hours if in constant use.

Cleaning is the application of a cleaning agent at the proper temperature and with proper agitation for a sufficient time to remove soil from the surface of an object. Detergents are formulated to meet every cleaning need. *Sanitizing* is the application of hot water or a chemical solution that reduces micro-organisms to an acceptable level. Cleaning and rinsing must be done before sanitizing.

Dishwashing machines can be the most reliable tools for cleaning, rinsing, and sanitizing utensils and equipment if the machines are used properly.

Manual cleaning and sanitizing can be as effective as machine cleaning *if it is done carefully.* Items may be manually washed, rinsed, and sanitized in one of two ways: (1) either by heating them to 170°F (76.7°C) for 30 seconds, which is enough time to kill bacteria; or (2) by immersing them in the proper concentration of a chemical sanitizing solution. The six steps in manual cleaning are: (1) pre-cleaning the wash compartment or work surface; (2) pre-soaking or pre-flushing the items to be cleaned; (3) washing; (4) rinsing; (5) sanitizing; and (6) air drying.

Nonfood-contact areas, such as the public restrooms, floors, ceilings, walls, and floor drains, must also be cleaned on a regular basis, and disinfected where appropriate. Cleaning tools, such as brushes, mops, and buckets, need to be cleaned after every use and stored so they will air dry and remain clean and in good working order.

A protected storage area is necessary to keep clean and sanitary utensils and equipment from becoming contaminated before use.

The OSHA Hazard Communication Standard (HCS) requires that employers develop and implement a training program for all employees to communicate to them the hazards of any chemical used in the facility. The program should address the issues of labels and other forms of warning, material safety data sheets (MSDS), and employee precautions.

A CASE IN POINT

The student dishwashing employees in a university cafeteria had scraped every dish and utensil and placed all of them in the racks that automatically fed into the high-temperature dishwashing machine. When the wash and rinse cycles were complete, one of the employees noticed that the dishes were spotted. The thermometer that registered the final rinse temperature for the sanitizing cycle had a reading of 140°F (60°C), rather than the required 180°F (82.2°C).

The employee went to inform the manager. The manager called the manufacturer's representative to come and examine the dishwashing machine.

What alternative methods can be used to clean and sanitize the tableware since the machine is not functioning properly?

STUDY QUESTIONS

1. What is the fundamental foodservice rule about cleaning and sanitizing food-contact surfaces?
2. In selecting a cleaning agent, what requirements should a foodservice manager consider?
3. Why should foodservice managers not formulate their own detergents?
4. For what cleaning purposes should acid cleaners be used?
5. When should abrasive cleaners be applied?
6. Do the terms *sterile* and *sanitary* have the same meaning? If not, what is the difference?
7. What are the heat requirements for sanitizing with hot water using manual methods?
8. Describe the differences and similarities among chlorine, iodine, and quaternary ammonium (quat) sanitizers.
9. What is a detergent-sanitizer? How may it be used?
10. Describe the six steps in manual cleaning and sanitizing.
11. How often should dishwashing machines be cleaned?
12. How should the facilities be cleaned? How often?
13. How should clean and sanitary tableware, utensils, and equipment be stored?
14. What is the Hazards Communication Standard (HCS)?

ANSWER TO A CASE IN POINT

The tableware can be washed, rinsed, and sanitized in a three-compartment sink using a chemical sanitizer. After cleaning and sanitizing all the compartments in the sink, the tableware can then be pre-soaked and washed in the first compartment and thoroughly rinsed in the second compartment of the sink. The temperature in each compartment must be 120°F (48.9°C).

Chemical sanitizing can then be accomplished in the third compartment using the correct concentration at a temperature between 75° and 120°F (23.9° and 48.9°C). Tableware and equipment should always be air-dried and should not be handled until ready for use. Towel drying tableware can remove the chemical sanitizer before it has fully worked to kill micro-organisms, and unsanitary towels and hands can directly contaminate clean and sanitary dishes and utensils.

MORE ON THE SUBJECT

NATIONAL RESTAURANT ASSOCIATION. *Sanitation Operations Manual.* Washington, DC: National Restaurant Association, rev. 1984. 390 pages. This handy reference, three-ring binder has sections on cleaning and sanitizing equipment and facilities, as well as many other sanitation topics.

NATIONAL RESTAURANT ASSOCIATION. *"Right to Know"—A Foodservice Operator's Guide to the OSHA Hazard Communication Standard Program.* Washington, DC: National Restaurant Association. 1988. #PB100. 21 pages. A booklet that explains HCS, both what the standard is and what the foodservice operator must do to comply with it.

U.S. DEPARTMENT OF HEALTH AND HUMAN SERVICES. *Food Service Sanitation Manual.* Washington, DC: Food and Drug Administration, 1978. Chapter 5 of the "Model Food Service Sanitation Ordinance" deals with cleaning, sanitizing, and storage of equipment and utensils.

THE SOAP AND DETERGENT ASSOCIATION. *Programmed Cleaning Guide for the Environmental Sanitarian.* New York: The Soap and Detergent Association, 1984. This guide provides a look at overall cleaning and sanitizing procedures including floor care, warewashing, food sanitation, and laundering. Also, *Understanding Automatic Dishwashing,* 10 pages, offers a brief survey of detergent ingredients and water action requirements for automatic dishwashers.

SPARTA BRUSH COMPANY. *Wide World of Food Service Brushes.* Sparta, WI: Sparta Brush Company, 1990. 15 pages. This manual and study guide accompany an 18-minute videotape that covers basic foodservice cleaning and sanitizing, and how to select and use brushes for specific cleaning jobs. The manual includes charts, checklists, and a glossary.

11

Organizing a Cleaning Program

Effective and efficient cleaning begins with the manager's deliberate efforts to understand the precise cleaning needs of the foodservice establishment. The manager must provide the necessary equipment and cleaning agents to fill these needs, and use available employees and materials to the best advantage. In addition, the manager is responsible for continuously training and supervising employees.

It is important that the cleaning program is properly supervised and follows a schedule. Only a manager who is knowledgeable and alert can prevent breakdowns in good sanitation practices. He or she must know how a cleaning and sanitizing job should be done, instruct employees accordingly, and then see to it that instructions are carried out.

After foodservice operators have selected the necessary tools, proper brushes, cleaning agents, and chemical sanitizers they need for their equipment, tableware, and facilities, and have learned the principles of cleaning and sanitizing, emphasis needs to be placed on daily preventive maintenance.

In this book, we give the name *cleaning program* to the system that the operator or manager devises in order to organize all the cleaning and sanitizing tasks in the establishment.

A well-designed cleaning program is an integrated part of a HACCP-based food safety system. Among other things, having an organized cleaning program:

- Helps the manager to plan ahead and make the best use of supplies and employee resources
- Helps to distribute the work load fairly among all employees and shifts
- Pinpoints responsibility for specific tasks, thereby making it more likely that work will be done conscientiously and that efforts will not be duplicated
- Identifies the need for providing substitutes for ill or vacationing employees, and identifies contracting needs for cleaning by outside specialists
- Establishes a logical basis for supervisory tasks, such as inspections
- Saves employee time that might be spent organizing a job, or in deciding what to do next
- Provides a useful tool for familiarizing new employees with cleaning routines and with the management's approach to sanitation.

Chapter 10 presented the principles involved in cleaning and sanitizing. This chapter discusses ways of applying those principles using an organized and well-planned cleaning program. The basic steps in the design and implementation of a cleaning program are to:

- Survey cleaning needs.
- Devise a master cleaning schedule.
- Choose and collect cleaning materials.
- Introduce the program to employees.
- Supervise program implementation.
- Monitor the program.

SURVEYING CLEANING NEEDS

If you are a manager, one way to begin designing your cleaning program is to walk through every room of your establishment, clipboard or tape recorder in hand, and write down or record each area, surface, and piece of equipment that needs to be cleaned.

Try to examine your whole establishment, from the dining room to the back-of-the-house, with a fresh viewpoint, looking for spots that might be neglected simply because you are so familiar with your layout that you do not notice them.

While you are surveying your cleaning needs, you should closely examine the ways in which the cleaning is currently being done. Are there any procedures that could be improved? People tend to continue to perform their regular tasks according to habit, even if possible improvements are obvious.

Your next step is to estimate the amount of time needed for the specific cleaning and sanitizing procedures. Decide whether each job can be done most efficiently by one, two, or more people. Identify and describe all the equipment and materials that are required to do each job.

Keep in mind that many items require several different kinds of cleaning. Steam tables, for example, need periodic deliming in addition to cleaning and sanitizing. Kitchen floors require removal of dust, debris, and grease after each meal period, and deep scrubbing every few days. Include the responsibility for seasonal cleaning tasks whether you use your own employees or outside cleaning specialists. Also include the cleaning requirements outside the building, such as garbage and recycling areas, grease storage areas, drive-through lanes, and parking lots.

DEVISING THE MASTER CLEANING SCHEDULE

Now tabulate all the information from the survey in the form of a master cleaning schedule. The *master cleaning schedule* is a summary of all the cleaning operations in the establishment. For each item there should be an entry indicating:

- *What* is to be cleaned
- *Who* is to clean the item (an individual by job title, not a shift)
- *When* it is to be cleaned
- *How* the job is to be done.

Exhibit 11.1 presents a sample cleaning schedule for food preparation areas. Your master cleaning schedule can follow the same general outline.

What Is to Be Cleaned

The master cleaning schedule should be arranged in a logical form so that nothing will be overlooked. For example, all the cleaning jobs in one room can be placed together on the list. Alternatively, the schedule can list jobs in the order in which they are to be performed, according to the person responsible for them, or in another form that works well for the manager. The master schedule should be flexible enough so substitutions and changes can be made easily; for example, magnetized name labels or dry-erase chart boards can be used for posting daily cleaning schedules.

Who Is to Clean It

A specific individual should be responsible for each item on the list of jobs. In developing the schedule, include the employees in decisions. Since the employees regularly clean the items, they may have suggestions to increase efficiency. In addition, it will be their program, not just yours, so including their ideas will increase its success. One of the advantages of making up a schedule is that conflicts can be detected and worked out before they result in serious problems.

In general, employees should be held responsible for continuous cleaning of their own work areas, following a *clean-as-you-go* policy. Rotating cleaning assignments helps to prevent people from becoming bored. It will also distribute the unpleasant tasks, for example, cleaning the steam table, more evenly among employees.

Distribution of larger jobs will depend in part on the number of employees available. If your staff is limited, then it becomes necessary to schedule large jobs into the employees' slack periods. Where high turnover is a problem, have clearly defined roles or job titles, and schedule cleaning jobs by a particular job title.

When It Is to Be Cleaned

Major cleanup tasks should be scheduled at a time when contamination of foods is least likely to occur and interference with service is minimized. Obviously, the kitchen should not be mopped during the preparation of food, unless a major spill has occurred, because the splash raised by mopping can contaminate food, and wet floors pose a safety hazard.

Exhibit 11.1 Sample cleaning schedule (partial) for food preparation area

Item	What	When	Use	Who
Floors	Wipe up spills	As soon as possible	Cloth, mop and bucket, broom and dustpan	______
	Damp mop	Once per shift, between rushes	Mop, bucket	______
	Scrub	Daily, Closing	Brushes, squeegee bucket, detergent (brand)	______
	Strip, reseal	January, June	See procedure	______
Walls and ceilings	Wipe up splashes	As soon as possible	Clean cloth, detergent (brand)	______
	Wash walls	February, August		______
Work tables	Clean and sanitize tops	Between uses and at end of day	See cleaning procedure for each table	______
	Empty, clean and sanitize drawers, clean frame, shelf	Weekly, Sat. closing	See cleaning procedure for each table	______
Hoods and filters	Empty grease traps	When necessary	Container for grease	______
	Clean inside and out	Daily, closing	See cleaning procedure	______
	Clean filters	Weekly, Wed. closing	Dishwashing machine	______
Broiler	Empty drip pan; wipe down	When necessary	Container for grease; clean cloth	______
	Clean grid tray, inside, outside, top	After each use	See cleaning procedure for each broiler	______

Schedule the workshift to include enough time for cleaning tasks. A rushed employee may not do a thorough job, especially in areas that do not show, for example, proper sanitizing. If late night banquets and irregular closing times are the rule or the exception to the rule, the cleaning schedule should be adjusted accordingly. Avoid scheduling too much cleaning at closing time where corner-cutting is likely. You may consider scheduling a crew or crew member to come in near closing time to perform cleaning or oversee other employees.

Other goals to consider when scheduling cleaning operations include balance of periodic cleaning and arranging jobs in order of priority, or order to be done. For example, it is not wise to schedule all weekly jobs on the same day of the week, or to do semi-annual floor resealing in January and March.

How It Is to Be Cleaned

It is very important to provide clearly written and detailed procedures for cleaning all areas and items of equipment. These written procedures should lead the employee, whether a novice or a veteran, step-by-step through the cleaning process. Manufacturer's instructions for cleaning pieces of equipment must always be followed.

Specify tools by name and by color code, if colors are used to distinguish food-contact from nonfood-contact items, and cleaning from sanitizing items. List cleaning agents by brand name. Exhibit 11.2 presents a sample procedure form for cleaning a food slicer.

For practical reasons, the manager should post concise cleaning instructions near the item described so that employees do not have to run back and forth to the master cleaning schedule.

Any unusual hazards or special precautions involved must be clearly stated for each job. Have employees refer to the Material Safety Data Sheet (MSDS) for each chemical product. Keep manufacturer's operation manuals, cleaning directions, and MSDS for all equipment in one location for future reference for the manager and all employees.

CHOOSING CLEANING MATERIALS

Once the master cleaning schedule is developed, the manager has a reasonable idea of the kinds of cleaning products and tools needed for the program. Referring to Chapter 10, the manager can estimate the kinds of soils and the cleaning challenges each creates to better judge what the needs are.

Cleaning Products

To select an array of cleaning products that will efficiently provide the cleaning power needed for the kinds of soils and equipment in your establishment, work with a knowledgeable cleaning products supplier. Most cleaning products, with the exception of sanitizers, can be used for several different jobs, so it is reasonable to handle most cleaning tasks with fewer than ten different products. A typical cleaning and sanitizing product lineup is:

- Handwashing soap
- Floor and all-purpose detergent
- Abrasive cleaner (may also be a disinfectant)
- Solvent cleaner (degreaser)
- Glass cleaner
- Manual warewashing detergent
- Automatic dishmachine detergent
- Automatic dishmachine rinse aid
- Sanitizer (with test kit)

Exhibit 11.2 Sample cleaning procedure: How to clean a food slicer

When	How	Use
After each use	1. Turn off machine.	
	2. Remove electric cord from socket.	
	3. Set blade control to zero.	
	4. Remove meat carriage. (a) Turn knob at bottom of carriage.	
	5. Remove the back blade guard. (a) Loosen knob on the guard.	
	6. Remove the top blade guard. (a) Loosen knob at center of blade.	
	7. Take parts to pot-and-pan sink, scrub.	Manual detergent solution, gong brush.
	8. Rinse.	Clean hot water, 170°F (76.7°C) for 1 minute. Use double S hook to remove parts from hot water.
	9. Allow parts to air dry on clean surface.	
	10. Wash blade and machine shell by swabbing. CAUTION: PROCEED WITH CARE WHILE BLADE IS EXPOSED.	Use brush dipped in detergent solution or use a bunched cloth, folded to several thicknesses. Wear steel-reinforced gloves.
	11. Rinse by swabbing.	Clean hot water, clean bunched cloth.
	12. Sanitize blade, allow to air dry.	Clean water, chemical sanitizer, clean bunched cloth.
	13. Replace front blade guard immediately after cleaning shell. (a) Tighten knob.	
	14. Replace back blade guard. (a) Tighten knob.	
	15. Replace meat carriage. (a) Tighten knob.	
	16. Leave blade control at zero.	
	17. Replace electric cord into socket.	

Source: Adapted by permission of Macmillan Publishing Company from *Sanitary Techniques in Food Service* by Karla Longrée and Gertrude G. Baker. Copyright © 1982 by Macmillan Publishing Company.

Additional specialty products may be needed, and could include delimer, utensil/flatware pre-soak, drain cleaner, grill and deep fryer cleaner, oven cleaner, and stainless steel or silver polish. However, the functions of additional products can usually be handled by the basic products along with instructions on procedures from your supplier.

Keeping a simple group of basic products that are effective and versatile, combined with good procedures and tools that are appropriate and well-kept can provide the safest and most cost-effective route to a clean and sanitary establishment. This concept is shown in Exhibit 11.3. The inner circle represents the tangible elements that provide the means to

Exhibit 11.3 Wheel representing the controls needed to reach cleaning objectives and to improve the cleaning program

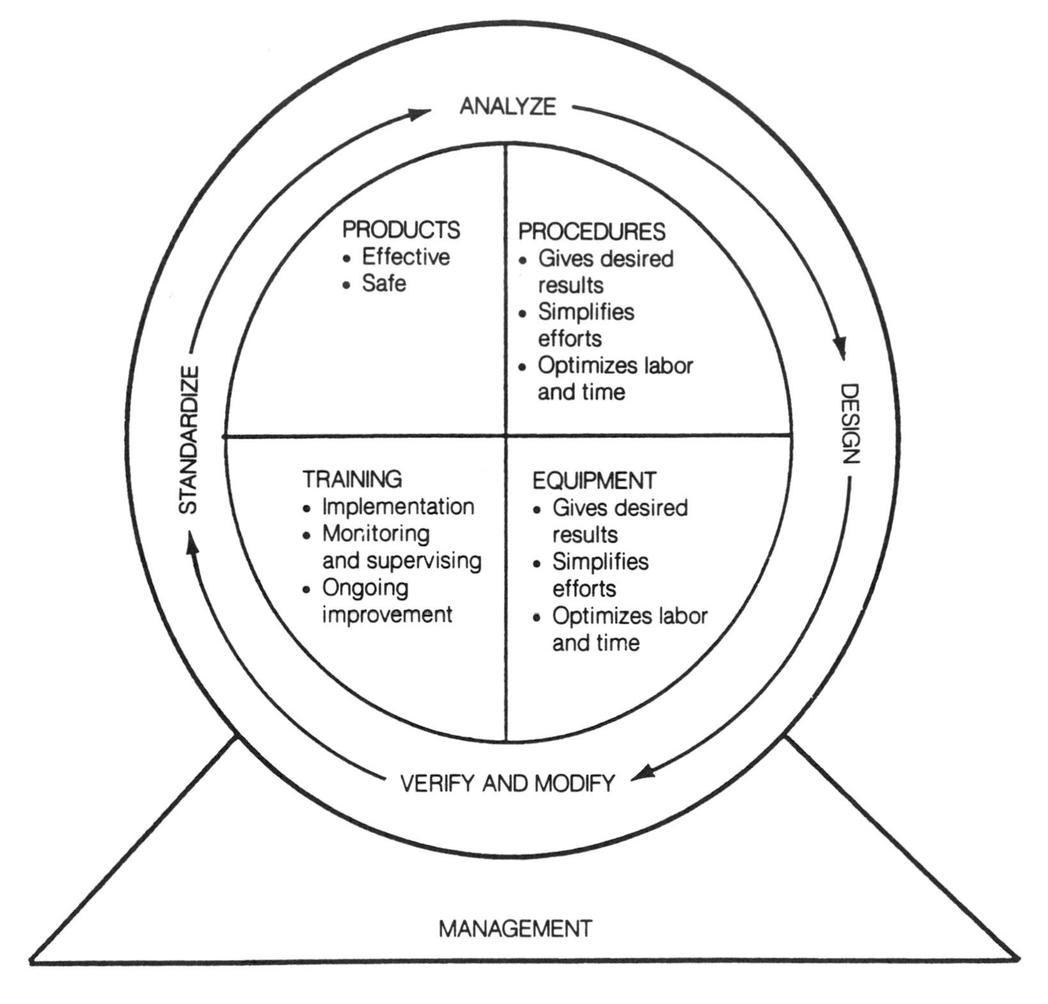

Source: Adapted with permission of Procter & Gamble, Cincinnati, Ohio.

meet cleaning and sanitation objectives. These are things the manager can control. The outer ring is the on-going process to analyze and improve cleaning effectiveness for total quality. This wheel rests on a foundation of good management.

Cleaning Tools

Providing adequate and appropriate sanitation tools is the next step in guaranteeing efficient cleaning. Metal scrapers, heavy rubber gloves, clean cloths, brushes, brooms (vertical and push-type), wet mops, dust mops, two-compartment hand buckets, mop buckets with presses, squeegees, and dust pans are examples of cleaning tools that will need to be available for use in the food-service establishment (see Exhibit 11.4).

Brushes of nylon, plastic, or natural fibers are common cleaning tools. Some wood-backed natural fiber brushes, though, are hard to keep clean because moisture and cleaning compounds are drawn into the wood backing by the bristles. The moisture in the wood causes small cracks to form where bacteria can enter and grow, and the absorbed chemicals can be a potential safety or chemical hazard. Nylon and plastic brushes with composition backing, in contrast, are easily cleaned and

Exhibit 11.4 Cleaning tools

General Supplies

Paddle scrapers
Window squeegees
Floor squeegees
Putty knives
Hoses
Dustpans
Mops (light and heavy duty)
Mop bucket and wringer
Buckets (various sizes)
Brooms (light and heavy duty)

Brushes

Glass washer
Tableware
Bottle
Tube
Scrub
Plumbing
Toilet
General cleaning
Drain
Grill (steel)

Cloths

Dishcloths
Counter cloths
Dustcloths
Special cheesecloths

Sponges

Varying grades and sizes
For non-stick surfaces
Nylon scrubbing pad

Heavy Equipment

Vacuum cleaner
Carpet sweeper
Wax applicator
Floor scrubber/polisher
Carpet shampooer

Worker Supplies

Plastic disposable gloves
Steel reinforced gloves
Goggles
Aprons

ideal for many hand- or mechanical-cleaning operations. They have strong and flexible bristles that wear well, are nonabrasive, and do not absorb water or chemicals. Different brushes need to be used for food-contact and for nonfood-contact surfaces.

Do not use steel wool, coarse abrasives, or anything else that will mar the surface being cleaned. Steel wool and other metal pads can break apart, leaving physical contaminants on the surfaces on which they are used, which can then get into food. An alternative to these products would be nylon scrub pads.

It is also a good idea to periodically test new materials and equipment that might be suitable for the establishment. A wet/dry vacuum with attachments could be versatile and economical. You may also wish to investigate the wide variety of power cleaning tools and look into new chemical cleaning agents as potential ways of improving cleaning efficiency. If you select novel materials and tools, though, you need to make sure the items are approved for use in a foodservice establishment. You will also need to train employees in their use and maintenance and provide competent supervision.

The equipment you keep on hand should match the needs you have identified in your survey of cleaning jobs and the master schedule. If you have scheduled two people to mop the floor, you must provide two mops. If you want the dining room vacuumed, make sure that you have the correct replacement bags for the vacuum.

When equipment becomes worn or too soiled to properly clean and sanitize, management must replace it. Not only are dirty and worn out cleaning tools ineffective, they can actually spread bacteria.

THE MANAGER'S RESPONSIBILITY

It is the manager's responsibility to establish a cleaning program *and* to make it work. Placing emphasis on effective two-way communication, the manager must see that all employees understand the importance of the program and stress preventive maintenance to avoid breakdowns in sanitation. The manager must provide employees with adequate training and reference materials to assist them as they perform their cleaning and sanitizing duties. The manager must also provide all employees with safety information on the chemicals and equipment used, as discussed in Chapter 10, in the section on the OSHA Hazard Communication Standard (HCS).

By evaluating the cleaning program and its participants regularly, the manager can identify any ineffective procedures and make the appropriate adjustments to increase efficiency. Diligent self-inspection provides the key to making the program operate at its best.

INTRODUCING THE PROGRAM TO EMPLOYEES

Preparing a cleaning program and writing schedules and procedures takes substantial time and effort. Unfortunately, the effort is wasted if only the manager works to carry the program out. The manager must devote time and energy to acquaint the employees with the schedule and to enlist their support for the program. Make sure employees are rewarded for a job well done, and that they know they are an integral part of the cleaning program and its success.

The manager should divide the master schedule into several smaller schedules for each room or area, for each individual employee, or for immediate tasks. These smaller schedules should refer to employees by name, therefore heightening their sense of responsibility and participation. The master schedule should use employee titles to establish its permanence. The master schedule should be used to introduce new employees to the cleaning program. Exhibit 11.5 illustrates a schedule of one employee's duties. Note that a server is assigned to several different stations during one week.

When posting the cleaning schedules for employees, consider using pictures of the products and tools needed for the job, as well as illustrations of the procedures. Posters that include both are usually available from suppliers.

A kickoff meeting offers an opportunity for the manager to explain the reasons behind the cleaning and sanitizing program, and to discuss the importance of following the procedures exactly as they are written. If people understand *why* they are being asked to do something a certain way, they will be more likely to follow all specified procedures. They will also feel that the manager has more confidence in their intelligence than if they are simply told, "Do it this way." They can offer some new suggestions that are helpful. By including sanitation in *all* employee meetings, the manager can demonstrate any new procedures or cleaning tools being introduced and the safety procedures that are required during cleaning. The sales representative or distributor may also be available to demonstrate cleaning products.

Exhibit 11.5 A sample individual cleaning schedule

Schedule for ______(Name)______.
Week of ____________________.

Monday	Tuesday	Wednesday	Thursday	Friday
Server station #1	*Server station #2*	*Server station #3*	*Server station #1*	*Server station #2*
Juice machine	Ice cream freezer	Coffee machine	Juice machine	Ice cream freezer
Refill paper supply	Cocoa machine	Catsup, mustard containers	Refill paper supply	Cocoa machine
Dust and wipe shelves under counter	Dust dining room window sill, booth partitions	Sugar, salt, pepper containers	Dust and wipe shelves under counter	Dust dining room window sill, booth partitions

SUPERVISING IMPLEMENTATION OF THE PROGRAM

One of the most important factors in keeping people aware of the cleaning program is the manager's continued interest and follow-up. The manager should let people know that the schedule is expected to function on a permanent basis. Changes in the menu, the employees, and the hundreds of other variables that affect a foodservice will necessitate changes in the schedule. The manager must make the schedule flexible, include the employees in decisions, and keep the posted schedules up to date.

The master cleaning schedule should be used for making self-inspection sheets for individual employees, and a master inspection sheet for the manager. By making use of these inspection sheets, the manager can successfully move into monitoring the program.

MONITORING THE PROGRAM

A manager should regularly review the cleaning program and consider the following questions:

1. Is the program working for the facility or should it be revised?
2. Do I have all the tools needed for the staff to do each task according to the specified procedures?
3. Does the schedule give enough time for employees to clean during their shifts and after closing?
4. Do I have enough employees for cleaning or do I need to hire outside contractors for certain jobs?
5. Are employees properly trained in cleaning, sanitizing, and chemical safety procedures?

SUMMARY

Clean and sanitary conditions in a foodservice operation are the product of an organized and well-planned *cleaning program.*

The first step in designing the program is for the manager to survey the establishment, noting all its cleaning requirements. Cleaning procedures, materials, and equipment must all be examined in order to reveal possible areas for improvement.

Second, to help ensure that every job is completed on time and as it should be, a *master cleaning schedule,* as well as cleaning procedures for each item on the schedule, should be written and prominently posted. The master schedule and individual procedures should state clearly and in detail *what* is to be cleaned, *who* is to clean it, and *when* and *how* the cleaning is to be done.

Third, the necessary cleaning materials should be available in appropriate quantities. These cleaning agents and tools must be kept in a safe, well-lighted, and convenient storage area.

A special effort must be made to introduce the schedule to the employees who will have to carry it out. Taking the time to enlist their support for the cleaning program will pay off in greater success.

After the cleaning program has been implemented, the manager must constantly monitor its functioning.

Supervising will make sure that the correct cleaning procedures are followed, and demonstrates the manager's continued interest in the program. Monitoring allows the manager to adapt the program as needed.

A CASE IN POINT

It was only 9:05 A.M., but already Tim, the day shift manager, could tell it was going to be a bad day. While the executives from American Widget munched unenthusiastically on complimentary doughnuts, they threw annoyed looks at Tim and the busboy, who were cleaning the private room that American Widget had hired for 9:00 A.M. Tim had opened the banquet room at 8:50 A.M. to find, to his dismay, that the remains of the Pine Valley Martial Arts Club Annual Banquet were still strewn about the room.

Later, Tim sat down with his cleaning schedule and tried to determine what had gone wrong. The banquet was supposed to have ended at 11:30 P.M. the previous night. The assistant dishwasher was scheduled to clean the room at midnight when he got out of the dishroom and before he went home at 12:30 A.M. But a note from Norman, the night shift manager, told Tim that the banquet had been a wild one and that the last guest had not staggered out until long after the 1:00 A.M. closing time. The dishwasher had punched out at 12:30 A.M. as he always did. Tim sighed.

What do you think went wrong?

How can Tim prevent this from happening again?

STUDY QUESTIONS

1. List five advantages of developing an organized cleaning program.
2. What are the basic steps to be followed in designing and implementing a cleaning program?
3. What are the four categories of information that should be present on the master cleaning schedule?
4. Give three ways of listing cleaning jobs on an individual cleaning schedule.
5. What is the difference between a *cleaning schedule* and a *cleaning procedure?*
6. What is the chief advantage to developing written cleaning procedures?
7. What are the manager's responsibilities in establishing the cleaning program?
8. What steps must be taken after the cleaning program is introduced to employees?

ANSWER TO A CASE IN POINT

Schedules don't clean dining rooms, people do. Tim had made his schedule too rigid and had failed to monitor it after its introduction. If he had, he would have been able to make the necessary adjustments to take late banquets into account. Shifts should be scheduled to include all cleaning duties, so if the banquet room closes at 1 A.M., then the shift should extend beyond closing time to take cleaning into account.

In the future, Tim should encourage his employees to follow a clean-as-you-go approach. If they had, the soiled tableware no longer being used by the banquet attendees could have been brought to the dishwasher before midnight.

Enlisting the cooperation of Norman, the night shift manager, to make sure the cleaning program is followed and to bring to Tim's attention any problems that need adjustment is essential. Two-way communication must be used between shift managers, as well as between employees and managers.

Also, if Tim had properly introduced the cleaning schedule and procedures to the employees, the assistant dishwasher would have realized one of his responsibilities was cleaning the room no matter how late he had to stay.

MORE ON THE SUBJECT

NATIONAL RESTAURANT ASSOCIATION. *Sanitation Operations Manual.* Washington, DC: National Restaurant Association, rev. 1984. 390 pages. This handy reference, three-ring binder has sections on cleaning and sanitizing equipment and facilities, as well as many other sanitation topics.

LONGRÉE, KARLA, and GERTRUDE G. BAKER. *Sanitary Techniques in Food Service.* 2d ed. New York: Macmillan, 1982. 271 pages. Sections A and B of Part III present outlines of procedures for cleaning facilities, equipment, and utensils.

FELDMAN, EDWIN B. *Housekeeping Handbook for Institutions, Business, and Industry.* New York: Frederick Fell, 1974. 423 pages. Though primarily concerned with the cleaning of industrial and high-rise structures, this book still offers much useful and practical information on cleaning methods and on training employees.

12

Integrated Pest Management

What can a foodservice manager do to overcome pest infestations that seem like an insurmountable problem? The answer is to develop an integrated pest management (IPM) program. *Integrated pest management* is a system that combines preventive tactics and control methods to reduce pest infestations.

The IPM approach involves both the foodservice manager and a licensed or registered pest control operator (PCO) working together. It goes hand-in-hand with good sanitation and food safety practices. This chapter will:

- Define an IPM program
- Identify methods to prevent the entry of pests in the facility and around the grounds
- Identify the most common pests
- Identify methods for detecting the presence of pests
- Identify methods to control pest infestations
- Describe how to work with a PCO.

DEFINING THE IPM PROGRAM

Rather than relying solely on the application of pesticides, the IPM approach requires the foodservice manager and PCO to follow a detailed monitoring program and to make decisions on what preventive and control measures to take based on a variety of nonchemical and chemical options.

The goals of IPM are long-term elimination of pests and reduced pesticide exposure for customers. Three commonsense rules used in developing an IPM program are:

1. Deprive pests of food, water, and shelter by following good sanitation and housekeeping practices.
2. Keep pests out of the foodservice facility by vermin-proofing the building.
3. Work with a licensed PCO to rid the operation of pests that do enter.

PREVENTING PEST INFESTATIONS

A HACCP-based program in place will help make sure that potential contamination from pests does not threaten food safety. By practicing good housekeeping and sanitation, an operation takes a major step toward the prevention of an infestation. A daily cleaning and sanitizing program, discussed in earlier chapters, is the foodservice manager's first line of defense in pest prevention. It denies food, water, entry, and *harborage,* which is shelter, to pests and it saves money over the long run if an infestation should happen to occur. If a foodservice is kept clean and sanitary, there will be fewer pests, and a PCO's materials will work more efficiently.

General Preventive Practices

The following is a list of general preventive practices that applies to keeping all pests out of a foodservice operation.

1. Use a reputable and reliable supplier. Check all incoming supplies outside on your receiving dock. Refuse any shipment of food, linen, or paper goods in which cockroaches or mice are found, even if there is only one. Roaches and mice can be carried into the facility on or in crates or cases. Since roaches like potatoes, inspect bags of potatoes carefully before storing them. Roaches and ants are often present in paper goods, such as napkins, cups, bags, and cardboard boxes. They can also be found in clean linen. Reject incoming goods if they contain any roach egg cases. The egg case is a dark brown or black capsule, about one-half inch (13 mm) long with a leathery or smooth and shiny appearance.
2. Dispose of garbage properly and promptly.
3. If you recycle trash, be careful how you store these materials because recyclables, such as bottles, cans, and paper and cardboard packaging cartons, can harbor pests if they are left dirty and remain in one place for extended lengths of time. Store all recyclables as far away from your establishment as local ordinances allow while awaiting pick up. (For a complete discussion of solid waste management refer to Chapter 9.)
4. Store all food and supplies properly. Keeping all food and supplies at least six inches off the floors and six inches away from walls is one of the more effective controls. Other controls are by maintaining the humidity of dry storage at 50 percent or

less, by providing good ventilation in storerooms and in food preparation areas, and by storing foods on movable carts. If you have inadvertently brought in roach egg cases, they are less likely to hatch with low humidity. Since most insects, including cockroaches, become inactive at temperatures below 40°F (4.4°C), refrigerating items that are frequently contaminated, such as cocoa, powdered milk, and nuts, can help. The FIFO method of stock rotation also helps to eliminate harborage in dry food and nonfood supplies in storage, since rotating stock disrupts insect breeding habits.

5. Dispose of mop and cleaning bucket water properly, and wipe up spilled water immediately. Keep cleaning supplies clean, dry, and properly stored. Wet mops are a favorite harborage for American roaches, who are extremely fond of dampness. Since rodents also need water, standing water in buckets or on floors encourages their presence.
6. Clean and sanitize your operation thoroughly. (For information on proper cleaning techniques refer to Chapter 10.) Food and beverage spills must be cleaned up immediately. Pick up crumbs and other scraps of food as quickly as possible. If baits are used as a control measure they will not be effective if pests have access to food other than the bait. Roaches can and will eat practically anything. A single crust of bread can support an entire roach population. Careful cleaning reduces the food supply for vermin, destroys many insect eggs, and may reveal new infestations before they become serious. It may also make IPM treatments more effective. If you have a break room or locker room for employees, make sure that the area is cleaned, maintained, and monitored. Inform your employees that they should not store food or soiled clothing in their lockers or on the floor under their lockers. Unsanitary conditions in lavatories and toilet areas will also attract pests and must be corrected.

Building and Grounds Maintenance

Structural defects, particularly in older buildings, allow pests entrance and provide pests with plenty of hiding places. Repairing, maintaining, and remodeling old facilities will lessen the cost of pest control.

Doors, Windows, and Vents

Use screens on all outside doors, windows, and other openings, such as transoms, if applicable, and keep all screens in good condition. According to the FDA's *Foodservice Sanitation Manual,* screens of not less than 16 mesh per inch must be used. Arches or framing around windows and doors should be checked regularly for cracks, and the cracks need to be filled or caulked. Outside maintenance and repair checks on windows, doors, and other entries to the facility should be included on the master cleaning schedule.

The use of *air curtains,* also called *air doors* or *fly fans,* can be helpful in preventing access to flying insects. The air

Exhibit 12.1 An air door must be of adequate width and provide sufficient air velocity to cover the entire door area

Photograph courtesy of Berner International Corporation, New Castle, Pennsylvania.

curtains form a shield of air that flying insects avoid. These air curtain units must be of adequate width and provide sufficient air velocity to cover the entire door area (see Exhibit 12.1). Check with local health departments on code requirements for the air velocity of air curtains. Using positive air ventilation that forces air out of storage areas or out of the building is also helpful.

In order to reduce the opportunity for pest entry under doors, door thresholds should not be removed. If the thresholds interfere with the movement of wheeled carts through doorways, they need to be replaced, rather than removed. All doors should be self-closing and should fit their frames flush, without leaving gaps beneath them or on either side. Install door sweeps on all outside doors as well. When receiving supplies, only leave doors and screens open for the shortest time possible.

Pipes

Weak masonry around external pipes and holes near pipes can be covered with sheet metal or filled with concrete (see Exhibit 12.2). Rats need a hole only the size of a quarter, and mice need a hole only the size of a nickel to gain entrance to buildings. Spaces around internal pipes also provide shelter for roaches. To protect ventilation pipes on the roof from rats, cover them with "hardware cloth."

Floors and Walls

Decaying masonry in building foundations allows rats to burrow into buildings. Seal all cracks in floors and walls with a suitable material that will

Exhibit 12.2 Stoppage of openings around pipes

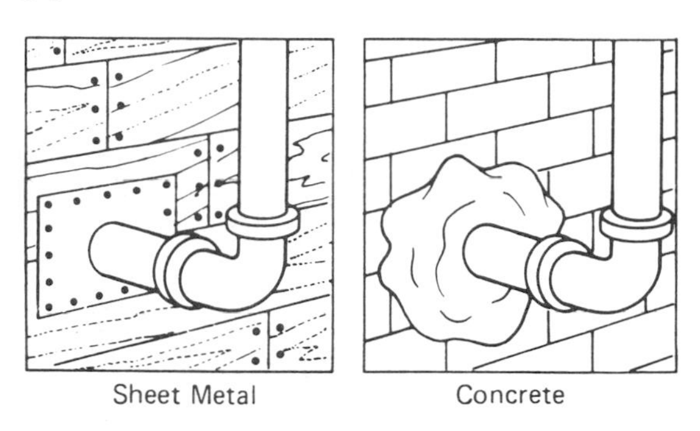

Source: U.S. Department of Health and Human Services, Centers for Disease Control, Atlanta, Georgia.

be permanent and keep out pests as recommended by your PCO, your representative from the health department, or a building contractor.

Be sure to close off spaces where large immobile equipment is fitted to its base or to the floor. Use a suitable material, such as silicone sealant or concrete, depending on the size of the spaces.

Painting a white strip around the edge of the floor of a storeroom will not only encourage workers to stack material away from the walls but will also make identification of rodent signs (tracks, droppings, hairs) much easier.

Basement and other floor drains should be covered by a perforated metal cap with a removable hinge (see Exhibit 12.3) since it is possible for rats to gain access to buildings through drains.

Grounds and Outdoor Serving Areas

Keep the grounds mowed, control weeds, and eliminate any stagnant or standing water on your property. In addition, make sure area residents are not leaving

Exhibit 12.3 Floor drain covered by a perforated metal cap with a removable hinge

Source: U.S. Department of Health and Human Services, Centers for Disease Control, Atlanta, Georgia.

litter or wastes from their pets on your property.

Since dining al fresco (in the fresh air) has gained popularity, it has opened a whole new area of concern for the food-service manager. Sanitation and good housekeeping outdoors takes on a new meaning. Pest control becomes even more critical. Outside dining requires you to expand your control boundaries further, so that your preventive measures include indoor and outdoor vermin-proofing.

Outdoor *zappers* are electrocutor traps used to control flying insects. Zappers should carry the Underwriters Laboratories (UL) seal for electrical safety, and should be maintained properly. Safety is a big issue. If a customer or employee somehow touches the electrical grid, the individual could be severely injured. In addition, these devices must be properly grounded, and the arcing equipment must be isolated from combustible materials.

Since the light they produce actually attracts insects, zappers should be placed away from entrances, exits, and eating areas.

Bees and wasps can also pose problems. The yellow jacket is a particularly aggressive wasp that feeds on the same garbage as flies. In June, the yellow jacket feeds on high-protein foods, which can be purchased at any fast-food or full-service restaurant. By mid-July, it prefers the high-carbohydrate foods that can be found in bakeries. Another outdoor pest that has received a great deal of press coverage is the Africanized "killer bee," which has migrated as far north as Texas. Early indications are that this bee is no more aggressive than the yellow jacket.

Keep all outdoor garbage containers and dumpsters tightly covered to prevent flies, bees, and wasps from feeding. Spills should be cleaned up immediately. Also have servers remove soiled tableware and uneaten food from tables as soon as diners are finished with them. Another measure for the safety of the customers is to avoid serving cold drinks in cans. Bees and wasps can often enter a can and hang underneath the drink opening, where they sting the unsuspecting drinker.

When hives are near or on the building, rely on your PCO for their removal. If hives are not visible, but bees or wasps are, it may be that the establishment has become part of a flight pattern for these insects, a problem that requires expertise in correcting and which should also be left to your PCO.

In addition, birds can be a problem. Make sure that customers are not feeding the birds, and therefore encouraging them to remain on your premises.

Indoor Garbage Storage Areas

Another area that requires special consideration is garbage storage areas *inside* the facility. Garbage should not be allowed to accumulate where it can give off odors or spill out of its container. Sanitation is the key to good insect control. All garbage should be sealed in plastic bags before it is placed in tightly covered containers. When the garbage containers are emptied, the containers should be rinsed with water and sprayed with a solution of 75 percent water and 25 percent chlorine bleach.

If inside zappers are used, they must be kept at least five feet away from food preparation and serving areas. The proper use of these devices requires that only wall-type devices be used; ceiling-type devices are not acceptable. When zappers are installed, the centers should be no more than three feet (one meter) above the floor. Zappers located close to the floor are more effective since the typical flight pattern of flies and other insects is at that level. Correct location near the garbage storage area is critical in order to avoid contamination of food or food-contact surfaces.

Zappers must also have a tray to contain the dead insect parts, so that they cannot fall out. The trays also need to be emptied and cleaned regularly, daily or as often as needed, depending on how bad the problem is. When the trays are not cleaned, the dead insects can serve as a food source for live insects.

DETECTING THE PRESENCE OF PESTS

Even if a foodservice manager is totally diligent concerning good sanitation and housekeeping practices, there is always the possibility that some pests will get into the establishment. By knowing how to detect pests and working with a PCO, the foodservice manager will be able to monitor the establishment and implement the appropriate control measures.

Cockroaches

Cockroaches head the list of the most frequent and bothersome insect pests. Efforts to control them are complicated by population dynamics and their ability to adapt to chemicals. In the lifetime of a female cockroach, which is between 60 days and two years depending on the species, she may produce offspring that number in the *millions.*

Cockroaches are known to be carriers of disease-causing organisms, such as *Salmonella,* fungi, parasite eggs, and viruses. Recent research indicates that many people are allergic to the residue left on surfaces and food by cockroaches. The disgusting appearance and vile odor of cockroaches also puts their elimination high on the list of priorities for a foodservice manager.

Cockroaches may be found virtually anywhere in a foodservice establishment. They tend to hide and to lay eggs in places that are dark, warm, moist, and hard-to-clean. Some of these areas are behind stoves and refrigeration units, under electrical burners in stovetops, inside water drains in sinks and on the floor, spaces around hot water pipes, inside the motors of electrical equipment, in cracks between ceiling tiles, under shelf liners, and under rubber mats on beverage dispensers.

If you see a cockroach in the daylight, it is an indication that you probably have a major infestation. Contact your PCO as soon as possible. Cockroaches usually prefer to search for food and water in the dark and only the weakest members of the colony will search for food in daylight. If you suspect you have a problem with roaches, check for these signs:

- A strong, oily odor
- Feces, which look like large grains of pepper
- Brown, dark brown, dark red, or black egg cases, which may appear leathery, smooth, or shiny, and are capsule-shaped.

The simplest way to identify a cockroach infestation is to use glue traps. These are cardboard containers that are open at both ends, allowing roaches to enter and become trapped by the sticky glue on the bottom of the container. These traps should be placed at an intersection of three surfaces, such as a corner, in order to be most effective where infestation is suspected.

The location of each trap should be recorded and monitored for 24 hours. The traps will reveal which species of cockroach are present, whether they are nymphs or adults, and how large the infestation is. This information will be particularly helpful for the PCO to determine what steps need to be taken.

While any species of cockroaches can be found anywhere in a foodservice establishment, knowing characteristics of certain cockroach species can help you identify and control infestations.

German cockroaches are pale brown or tan in color and are about one-half inch (13 mm) in length. This species is especially common in restaurants and can be found at any level in a room, from floor to ceiling. They are particularly fond of foods like potatoes, onions, and sweet beverages. They prefer warm crevices near stoves and other sources of heat, including the motors of refrigerators and soft-drink machines. They drink more water than any other species and reproduce more rapidly.

A new species, the *Asian cockroach,* has recently been identified. While it resembles the German cockroach in size and color, the Asian cockroach can fly, is attracted to light, and tends to hide in tropical plants.

Oriental cockroaches are shiny black and about 1 to 1½ inches (25 to 38 mm) long. Their favorite hiding places are in basements, water pipes, and indoor incinerators.

The *brown-banded cockroach,* which is about the same size as the German cockroach, is found in the kitchen, dry storage areas, and in dining areas under chairs and tables.

Finally, the *American cockroach,* which is reddish-brown in color and about 1$^1/_2$ inches (38 mm) long, feeds on starch found in food or wallpaper and loves water. This species is found in drainage and sewer areas, restrooms, heating pipes, and damp, open places.

Flies

The common housefly is an even greater menace to human health than is the cockroach. Flies can transmit a variety of foodborne illnesses because they feed on human and animal wastes and garbage. The dangerous bacteria present in these wastes stick to the mouth, footpads, and hairs of flies and may then be deposited in food intended for human consumption (see Exhibit 12.4). Contamination of food products and surfaces by fly feces is also likely.

Moreover, flies have no teeth and must take their nourishment in liquid form, so they *regurgitate,* or vomit, on solid food and let the food dissolve before consuming it. Fly vomitus is swarming with bacteria and contaminates food, surfaces, and utensils.

Exhibit 12.4 Transmittal of disease-causing organisms by flies in an al fresco setting

The following is a list of habits and characteristics of houseflies.

1. Flies can enter a building that has openings about the size of the head of a pin.
2. Flies are lured to the sources of attractive odors.
3. One female fly can produce thousands of offspring in a single breeding season.
4. Flies are especially fond of places protected from the wind, and of edges, such as garbage can rims.
5. Moist, warm, decaying material protected from sunlight is required for fly eggs to hatch and larvae, or *maggots,* to grow. Warm summer weather may allow flies to mature from larvae to adults in six days. Flies also favor animal and human waste areas for depositing their eggs.

Other Insect Pests

In addition to flies and cockroaches, the insects that plague foodservice operations include beetles, moths, and ants, which all thrive on very small amounts of food. Flour moth larvae, beetles, and other similar insects are found in dry storage areas. Evidence of their presence includes finding insect bodies, webbing, clumped-together food particles, holes in food, and holes in packaging. They can be controlled by storing food items in tightly covered containers, by using FIFO stock rotation, and by keeping all food storage, preparation, and serving areas clean and sanitary.

Ants sometimes make their nests in the walls and floors of establishments, especially in the vicinity of stoves or hot water pipes. They often enter from the outside. Ants are partial to sticky sweets and to grease. To control ants, don't allow food residue to collect anywhere, and clean up all food and beverage spills immediately.

Rodents

The economic loss from rats and mice is estimated in the millions due to consumption and contamination of food and structural damage to property, including damage from fires caused by rats gnawing electrical wiring. More important to the foodservice manager than economic problems is the serious health hazard posed by rodent contamination of food and utensils.

Rodents can be directly or indirectly involved in the transmission of such diseases as salmonellosis. One rat pill, or fecal dropping, can contain several million bacteria. Rats and mice have simple digestive systems, with very little bladder control. Therefore, they will urinate and defecate *as they move around the facility.* The waste particles may then be blown or carried into food.

Rats and mice have highly developed senses of touch, smell, and hearing, and can identify new or unfamiliar objects in their environment. Rats have a reach of up to 18 inches. They can climb vertical brick walls and jump up to three feet vertically and four feet horizontally. Rodents are also prolific breeders. A single pair of rats can produce 50 offspring in their life span of one year.

Because rodents tend to hide during the day, they may be present in your establishment even if you do not see them. Rats and mice may inhabit the same facility, although each will establish its own territory. Do not assume that if you have one type of rodent, you do not have the other. The following signs can help you determine whether you have an infestation and, if so, how extensive and recent it is.

1. *Droppings.* The presence of droppings, or feces, is one of the telltale signs of rodent infestation. If the droppings are black and shiny, a rodent has recently been in the vicinity. Older feces are brownish or grayish in color.
2. *Gnawing.* Rats gnaw to reach sources of food and drink and to keep their teeth short. Their incisor teeth are so strong that rats have been known to gnaw through lead pipes and unhardened concrete, as well as sacks, wood, and paperboard. If gnawings are recent, teeth marks should be visible.
3. *Tracks.* Rat and mouse tracks can be seen most clearly on dusty surfaces by a light source shining from an acute angle.
4. *Nesting materials.* Scraps of paper, hair, or other soft materials in a clump are often mouse nests.
5. *Holes.* Rats only nest in burrows, usually located in dirt, rock piles, or along foundations (see Exhibit 12.5).

Control of rats and mice is best left to the licensed professional PCO. Rats in particular may exhibit several behaviors that require the knowledge and experience of professionals to achieve successful control. For example, rats are highly territorial, and may exhibit "bait shyness," avoiding traps, baits, or other new items in their environment.

Exhibit 12.5 Rat burrows

Source: The Food and Drug Administration

PCOs have the experience necessary to effectively deal with these and other habits of rodent pests. PCOs also have access to a wider variety of equipment and poisons (called rodenticides), that are not available to foodservice managers. In addition, rats can be dangerous to humans. They will bite if they feel cornered or threatened, and even when dead they carry the potential to spread fleas and disease. These are formidable pests. *Their control must be entrusted to professionals only.*

Birds

The foodservice manager can gain insight on what bird control measures should be used by recognizing some characteristics of birds that commonly bother foodservice facilities.

Common pigeons, especially familiar to the citydweller, usually nest in poorly maintained buildings and in spaces in gutters, on roofs, on ledges, and under roof overhangs. *Starlings* and *sparrows* nest in holes or places of concealment.

These three kinds of birds feed on animal and vegetable matter, such as seeds and bread. Gathering in large flocks and roosting during the fall and winter months at your establishment poses health hazards because their droppings carry fungi that can cause illness in humans.

The most effective procedure in dealing with these birds is to employ a PCO who specializes in bird control. The PCO has the knowledge and equipment required for the safe use of chemicals and other techniques in combatting bird problems, and knows which birds are protected by law.

Some of the measures the PCO might use include pastes that repel birds and wires that administer a mild electric shock. Electric wires that produce mild shocks are generally safe, although measures should be taken to protect customers from any risks. This method, however, is fairly expensive and requires frequent inspection and maintenance. Netting may also be used around outside displays, such as signs or statues, to keep birds from roosting. Removing nests, spraying the birds with water, or simple trapping may be successful if carried out repeatedly.

PESTICIDES

For some operators, an immediate reaction upon seeing a few cockroaches is to run to the hardware store for a can of insecticide and to bombard the facility with it. This is a faulty approach for a number of reasons.

1. *Chemical pesticides are no substitute for good sanitation.* No matter how effective a pesticide program may be, rodents and insects will return when unsanitary conditions prevail.
2. *Any time pesticides are used or stored in the establishment, a potential hazard exists to the food supply, customers, and employees.* Pesticides may be toxic to humans, and some may also cause fire or explode. Since the manager is ultimately responsible for the safety of the entire operation, he or she must always see to it that hazards are minimized.
3. *Pesticide use is not a simple procedure.* Determining the most appropriate pesticide for a particular pest, where it is to be applied, and the most effective method of application often requires information and training not readily available to the foodservice manager. Because immunity and resistance are complicated problems varying from pest to pest and with each section of the country, it is best to work with a PCO who knows the proper methods to rid an operation of pests.

4. *Pesticide use is regulated by federal, state, and local governments.* Numerous regulations govern the use of pesticides. These regulations prohibit the use of any pesticides in a manner inconsistent with label restrictions. Read and follow all label directions.

Keep a written record on file of all pesticides used and treatment locations. It is illegal in some areas to use any pesticide in a foodservice establishment unless it is applied by a licensed PCO. In addition, some areas now require that you post a sign indicating that a pesticide is being used on the premises.

Pesticide Use

Although pesticide use can be effective against a variety of pests, it should be used only after preventive and nonchemical methods have been tried. For safety and effectiveness, rely on your PCO's knowledge of pesticides as part of an IPM solution to a consistent pest problem. Pest control methods, such as misting, dusting, and spraying should always be applied by a licensed PCO.

If the infestation is minor, for example, as indicated by glue traps for cockroaches, the foodservice operator may choose to use over-the-counter pesticides. In that event, the operator must use the following precautions in their application:

1. Cleaning should be performed *prior* to pesticide application.
2. If necessary, be sure the person applying the pesticides is protected by heavy work gloves, eye goggles, or a nose/mouth mask.
3. Read and follow package directions carefully. Use the pesticide *only* for the purposes for which it is designed.
4. When it comes to pesticides, it's dangerous to reason that if one dose is good, a double dose will be better. Effective results occur because of *thorough treatment,* not a double dose of pesticides.
5. Avoid any possibility of contaminating food-contact surfaces, equipment, and utensils with pesticides.
6. Thoroughly wash hands and clothing after using pesticides. Reclean, rinse, and sanitize all food-contact surfaces and equipment after any pesticide application.

Precautions in Storage and Disposal

Your PCO should store and dispose of pesticides used on your premises. However, if over-the-counter pesticides are stored in a food service, the manager should consult with the PCO about where to store them and how to dispose of them. The following list itemizes precautionary guidelines to be taken in storing pesticides.

1. Never allow a pesticide to be transferred from its original labeled package to any other kind of storage container. One potential cause of chemical poisoning is storing pesticides in empty food containers.
2. Store pesticides in a *locked* cabinet well removed from foodhandling and storage areas. Pesticides should

be stored separately from cleaning detergents or sanitizing chemicals.

3. Aerosol "bombs" or other pressurized spray cans should never be stored in an area where they may become overheated or leak. If they reach too high a temperature, they could explode violently. Pressurized pesticide containers should be stored in a cool place and should not be exposed to temperatures higher than 120°F (48.9°C).

Pesticides fall under new hazardous waste disposal regulations. Here are some precautionary guidelines for disposing of pesticides.

1. Containers for pesticides are a potential hazard even when they appear empty. Since particles of chemicals are still present, the containers should be disposed of according to manufacturer's directions as soon as they are empty.
2. Rinse bottles and nonaerosol or unpressurized cans three times in a disposal drain or a utility sink—never in a sink used for food, dishware, or utensils.
3. Break bottles and crush unpressurized cans, wrap the remains in paper, and place them in trash containers. Keep the glass separate from other trash as a safety precaution. If your local laws consider pesticides as hazardous waste, follow their guidelines for disposal.
4. Do not crush or burn empty aerosol cans, because of the likelihood of explosion. Pressurized spray cans must not be punctured or crushed in a compactor.

WORKING WITH A PEST CONTROL OPERATOR

While management can do its part to make sure that proper sanitation practices exist within an operation and reduce the occasion for infestation, the actual process of pest control should be entrusted to professionals. Management must obtain a licensed, certified, and reputable PCO to provide regular service. The PCO can provide an independent inspection and assessment of the conditions that exist in an operation. As a team, the manager and the PCO can rid the operation of pests and keep them from returning.

The following factors should be considered:

1. A skilled PCO can recommend an integrated pest management approach that includes all or a variety of control methods, such as correcting sanitary procedures, nonchemical solutions, sealing cracks and crevices to keep pests away, and chemical treatment. Pest problems are rarely solved by simplistic measures, such as random spraying of pesticides. A PCO who evaluates an operation's sanitation procedures and points out operational gaps is aware of this.
2. Pest control technology is changing and PCOs keep up with new developments.
3. A PCO works when no food is being prepared, for example, at night or

before the operation opens. Although it is necessary for the manager to know where the PCO is and what is being done, the manager is still free to concentrate on other tasks.

4. PCOs provide emergency service.

How to Select a PCO

The selection of a PCO merits serious consideration. The following are guidelines to help you make your decision.

1. Talk to others in the foodservice business. Find out who they are using or have used and what kind of job they felt a particular PCO did. Ask about return visits. Did the PCO offer sanitation advice or just spray corners and baseboards? (Remember you're looking for an *integrated* approach.) Any reputable PCO should be able to provide names of current clients as references.
2. Make sure that the PCO is licensed or certified by your state. Certification is required by federal law, especially for a PCO to supervise other employees. Certification usually indicates that the PCO has demonstrated at least minimal competency by passing an exam on pesticide use and other control procedures.
3. Ask about membership in professional organizations. Membership in a state or local pest control association or the National Pest Control Association (NPCA) is a positive sign. This professional membership is an indication that the PCO is receiving the most up-to-date information on control methods and procedures.
4. Be sure that any PCO you employ has insurance on his or her work in order to protect your establishment, employees, and customers.

A word of caution: Don't use price as the only deciding factor in choosing a PCO. Expertise is what you're looking for. The lowest bid may prove more costly in the long run.

The Teamwork Approach

Management must learn how to work effectively *with* a PCO; cooperation through two-way communication is key to a successful IPM program. Both management's expertise in the foodservice facility and in regard to sanitation, and the PCO's expertise in IPM techniques and pest behavior are valuable in reducing infestations.

The Contract

A contract is an essential element in the business relationship between the foodservice manager and PCO. A PCO may have a standard contract. Read it carefully. If possible, have a lawyer look at it. It should include a warranty of what is to be done, the period of service, legal liability, emergency service, and the obligations of the foodservice management.

The contract should also provide for detailed recordkeeping by the PCO. These records should include a log book or file kept at your facility with number of pests sighted or trapped, location of pests, pest species sighted, sanitation or structural defects noted, and any action taken or chemicals applied per visit.

In addition, these records should include maps of the facility that show the location of glue traps and other baits or traps, and problem spots for sanitation and pests; a calendar that notes things such as when to clean and replace glue traps or bait stations, when to replace bulbs in zappers or other electrocutor traps, when to re-apply chemicals, and when to check for seasonal pests; photos of sanitation and structural problems, along with sanitation notice reports for each sanitation or structural problem; the technician's report for each visit telling what was checked, found, and done; and, if applicable, a summary report from the PCO's supervisor each month, quarter, or year, that will complement past inspections and records, as well as indicate progress and success of the IPM approaches used.

The Inspection

You should have building plans and layouts of the facilities and equipment available for the PCO's initial inspection. Employees should be told about the inspection, and should be ready to answer any questions the PCO may have regarding sanitary procedures. The PCO must have complete access to the facility. If you have an idea where a problem is, point it out.

After the inspection, the PCO should outline *in writing* a treatment program for your establishment. The program must include the materials that will be used, the dates and times of the treatment, and any steps the foodservice manager should take to help eliminate pests.

At this time, the PCO should point out problem areas in sanitation and structural defects that invite pests. While you can expect and demand thoroughness and communication from your PCO, especially after every visit, you also need to do your part by listening and correcting sanitation or operational breakdowns. IPM is a shared responsibility between you and the PCO.

Treatment Procedures

Ask about proposed chemical methods of pest control, especially in areas where chemical substances may be hazardous, for example, boric acid and silica are relatively nontoxic but can be used only in areas that will not contaminate food areas. It is essential that foodservice managers know which chemical pesticides, if any, will be used throughout the operation.

As part of the inspection process, local health departments often require written records of when and where specific hazardous chemicals are used. The foodservice operator must know about the risks involved with pest control products and devices. These risks must be communicated to all employees. Training employees to recognize and minimize risks will help ensure that their safety, and the safety of customers, is not compromised (see Chapter 10 on the OSHA Hazard Communication Standard).

Mutual cooperation will speed the pest control process. The manager should accompany the PCO during his or her visits; or if that is not possible, talk to the PCO before, during, or afterward. Some PCOs will reserve the right to terminate the contract if clients ignore advice on sanitation or cancel scheduled treatments. On the other hand, the foodservice manager must make it clear that the unique nature of the foodservice business requires special care be taken in the IPM treatment plan.

Follow-Up

Pest control services must include follow-up. The PCO should return after all treatments and periodically check for pests. A manager should regularly check the facility as well, and contact the PCO if pests reappear.

PEST CONTROL OPERATOR METHODS

PCOs use a variety of methods to control pests that are not only environmentally sound, but safe for foodservice operations. In addition, they are trained to know the different techniques for controlling and monitoring different species of pests. Some of the available techniques that a PCO may use against specific pests are discussed here.

Controlling Insects

Many nonchemical methods for eliminating cockroaches and flies are commonly being used. It is best to consider *both* safety and effectiveness when checking out insect control methods from your PCO.

Repellents

Repellents are powders, liquids, or mists that work to keep bugs away from a specific area, but not necessarily to kill them. Repellents can sometimes be applied by the PCO in spaces inside and around the outsides of walls or other areas that are difficult to reach. Repellents can be effective when used in conjunction with glue traps, contact sprays, or residual sprays.

Sprays

Insecticide sprays are used frequently in the control of cockroaches and flies. A *residual spray* is applied to cracks and crevices and leaves a deposit that kills insects as they crawl across it. Residual sprays are applied near the hiding places of roaches and on the favorite resting and breeding places of flies. To be effective, the spray must form a thin, uniform layer on the treated surface.

A *contact spray* must directly touch the insect to kill it. This type of spray is usually applied to a group of insects, such as a cluster of cockroaches in a corner or crack, or a nest of ants. Contact sprays do not have residual action.

All foods and food-contact utensils must be removed from an area being sprayed to avoid contamination. Any immovable equipment, such as counters and ovens, must be covered *and* cleaned, rinsed, and sanitized after spraying.

Controlling Rodents

What can be done if preventive measures didn't solve the problem of rats or mice?

The following paragraphs list techniques that PCOs may employ for killing rodent pests.

Trapping

The use of traps is a slow, but generally safe, method of killing rats and mice. Traps should be placed at right angles to rodent runways, with the baited or trigger end toward the wall. Traps can be baited with any one of a number of different kinds of food, such as fresh ground meat or peanut butter. Traps should be checked and baits should be kept fresh. Since rodents are accustomed to human smells, odors from hands will not keep rats away from traps.

Glue Boards

Mouse-sized glue boards, strategically placed, can be an effective method in controlling mice as part of an overall IPM program. Glue boards do not contain poison and are safe for use in foodservice areas. Like traps, glue boards are placed where mice run through rooms or areas.

It may take several hours for a trapped mouse to die from suffocation, dehydration, or exhaustion. The dead mice and traps should be disposed of as soon as it is safe to do so. Glue boards can be an option for controlling rats, but are less effective than other methods and require special caution in handling.

Poisoning

Since rodent poisons are dangerous to humans, poison baits should be used with extreme caution, only as a last resort, and only under the direct control of a licensed PCO.

All poisons, baits, and traps require careful application and maintenance. Baits must be prepared by the PCO and placed along rodent runways or near breeding sites in tamperproof, locked bait stations. Fresh baits need to be put out every day for at least two weeks to be effective in reducing rodent infestation. If a bait remains undisturbed after several inspections, it should be moved to another location by the PCO. Baits must never be placed where they may be confused with food items or may be dropped onto food.

If several types of baits are distributed at the same time, a different-colored paper wrapper may be used for each type to help determine which type the rodents prefer.

Baits should be generous. Rodents travel limited distances from their burrows or hiding places—usually a maximum of 150 feet for rats, and 30 feet for mice. If baits are too few in number, or poorly placed, rodents may not find them. The PCO can bait liberally where signs of rodent activity are numerous and recent. The baits should be inspected by the PCO regularly and fresh ones provided when needed.

SUMMARY

From sanitary and economic standpoints, insect and rodent pests must be controlled and kept from entering foodservice facilities. Pests continue to be a menace in the foodservice operation because they are carriers of foodborne illness. Having a HACCP-based program in place will help ensure that potential contamination from pests does not threaten food safety.

To guard the establishment against pests, a foodservice manager should use integrated pest management (IPM)—a system that combines preventive tactics and control methods used to reduce pests. A foodservice manager must concentrate on the following two areas:

1. Eliminate sources of food and water, as well as breeding and hiding places, through good sanitation and housekeeping and food safety practices.
2. Physically prevent pests from entering the facility by closing off any openings in the building, practicing good grounds maintenance, and by inspecting incoming supplies.

If insect or rodent pests are detected, it may be necessary to resort to control measures. Control measures, however, must be considered as a supplement to, not as a substitute for, good sanitation.

Because most pesticides are toxic to humans, these poisons should be selected with great care and applied only after nonchemical methods have been tried. Precautions must be observed in the use, storage, and disposal of pesticides. Although the foodservice manager may choose to apply pesticides on occasion, a pest control operator (PCO) should always be employed for the more complex or hazardous extermination tasks.

Employing a PCO who practices IPM is the best way to overcome a pest infestation. The PCO selected must be licensed or certified, insured, and should belong to a pest control association. Although the technical work of pest control should be left to the PCO, the manager should know what methods and which chemicals will be used and where in or around the facility each will be used. The manager must establish two-way communication with the PCO and cooperate when it comes to correcting faulty sanitary procedures and structural faults pointed out by the PCO.

A CASE IN POINT

The "Now We're Cooking Restaurant" is located in the middle of town on the first floor of a landmark building, which has stood since World War I. The restaurant building is in a shopping area that includes several other restaurants. Fred, the manager, recently had the interior of the restaurant remodeled with marble floors and new ceilings. He chose materials that are modern and easy to clean. All of the new foodservice equipment is designed for cleanability. The cleaning procedures listed on Fred's master cleaning schedule in food preparation areas are written out in detail and completed by the employees as scheduled, with frequent self-inspections. Spills and grease are cleaned up immediately. Stock procedures include FIFO and metal racks that keep food six inches off the floor and away from the walls.

During a self-inspection two weeks after the remodeling, Fred finds cockroaches behind the sinks, in the vegetable storage area, in the public restrooms, and near the garbage storage area.

What factors might Fred have overlooked in his recent remodeling that could have led to this infestation?

What should Fred do to eliminate the roaches?

STUDY QUESTIONS

1. Define integrated pest management (IPM) and tell why it is the best approach to eliminate pests.
2. Describe six general preventive practices to keep pests out of your facility.
3. List at least four housekeeping practices that will reduce insect infestation.
4. Why does dining al fresco (out of doors) cause extra concern to foodservice managers?
5. List three signs that a foodservice facility is infested with cockroaches.
6. List three signs that a foodservice facility is infested with rodents.
7. Tell how access can be denied to rodents through grounds and building maintenance.
8. How should a manager deal with bird pests?
9. What are some of the problems and precautions involved with using, storing and disposing of pesticides in a foodservice facility?
10. What are the reasons for using the services of a pest control operator (PCO)?
11. List the criteria for selecting a PCO.
12. What are the steps involved in the teamwork approach to working with a PCO?

ANSWER TO A CASE IN POINT

Older buildings often have more nooks and crannies than newer buildings, which should be corrected with remodeling. Decaying exterior building materials and less than sanitary neighbors can make pest control in such buildings difficult. Fred needs to have his contractor seal crevices around the sinks and pipe openings with a proper sealant, such as caulking for the sink, and a steel plate or cement around the pipe openings. All cracks and crevices should also be sealed in the walls, floors, and door and window frames in the storage areas.

It is also important to increase the cleaning methods and their frequency in the restrooms. Patrons could be contributing to the pest problem there, by spilling water on the floor around the sinks, bringing food or drinks with them into the restrooms, or not reporting leaking sinks and toilets. Monitoring and more frequent cleaning of the restrooms must be added to the master cleaning schedule and included in self-inspections.

Garbage and trash needs to be taken out of the facility several times during the day, and the garbage and trash containers should be cleaned and sanitized after each load is emptied. Recyclables must be rinsed out if applicable, and stored in a sanitary way until they are picked up.

Fred should use glue traps, placing them in the problem areas and in other corners around the facility, to get an indication of the size of the infestation. If the infestation is minor, with only a small number of roaches present, Fred may choose to combine increased sanitation practices and use glue traps. He may consider using over-the-counter pesticides if the glue traps are not effective. Rather than resorting to do-it-yourself pesticides though, the best approach for Fred to take is to call a PCO who is licensed or certified. The PCO will be able to work with Fred to develop an IPM prevention and control program that includes advice to correct sanitation and structural flaws, nonchemical control methods, and if necessary, the proper and safe pesticides to use. As part of his IPM approach, Fred should contact his neighbors who share the building, so all of them can work together with the PCO and practice good sanitation habits.

MORE ON THE SUBJECT

NATIONAL RESTAURANT ASSOCIATION. *Pest Control.* Washington, DC: National Restaurant Association, 1986. #PB570. 16 pages. This informative, illustrated bulletin covers identification, control, and destruction of pests, including pesticides and safety, selecting a pest control company, and information sources.

NATIONAL PEST CONTROL ASSOCIATION. *How to Select and Use Pest Control Services.* Dunn Loring, VA: National Pest Control Association. This publication provides information on locating and working with pest control operators. Other publications relevant to foodservice operators are available by contacting The National Pest Control Association Resource Center, 8100 Oak Street, Dunn Loring, VA 22027.

LEXINGTON BOARD OF HEALTH. *Cockroaches: An Owner's Manual,* 1984; and *Housefly Control,* 1981. Available from the Lexington Board of Health, Lexington, MA, these pamphlets provide basic information on the life cycles of flies and cockroaches and methods for controlling these pests in foodservice environments.

PART IV

Accident Prevention and Crisis Management

13

Accident Prevention and Action for Emergencies

Many people feel that accidents happen because of chance or bad luck, but the truth is that accidents can be prevented. Preventing accidents and the ability to act wisely in emergency situations are as much a part of the business of operating a foodservice establishment as preventing foodborne illness.

Good sanitation practices are the first step to help eliminate safety hazards. A well-designed accident-prevention program combines sanitation and good management to eliminate physical hazards, to train staff in good safety habits, and to provide constant supervision to see that unsafe conditions and practices do not arise.

This chapter will answer the following questions and discuss how to set up an accident-prevention program:

- What are accidents?
- How are accidents caused?
- What kinds of accidents occur in a foodservice establishment?
- How can accidents be prevented?

No accident-prevention program is foolproof. Since accidents, injuries, and emergencies occur suddenly and require immediate action, managers and employees need to know specific procedures. This chapter will also cover how to act in first-aid emergency situations.

ACCIDENTS DEFINED

An *accident* is an unintended event, resulting in injury, loss, or damage, which may or may not result from unsafe practices. By this definition, if a dish slips from the hands of a dishwasher into the sink and breaks, resulting in the loss of the dish and the time required to clean up the fragments, an accident has occurred. If the dishwasher's hand is cut in the process of cleaning up the broken dish, then personal injury has resulted from the accident.

CAUSES OF ACCIDENTS

It is entirely possible to have a series of accidents in an environment as safe as human ingenuity can possibly make it. This is because the determining factor in most accidents is *human error.*

Human Hazards

Preventing accidents caused by human error involves recognizing, correcting, and avoiding hazardous conditions. The following are some common categories of human errors and examples.

- *Creating unsafe conditions:* Stacking boxes too high, climbing on boxes and tables instead of using stepladders, leaving spills on the floor, failing to clean greasy vent filters, and blocking passages with equipment.
- *Ignoring obvious hazards:* Picking up broken glass with bare hands, neglecting to use safety devices on grinders and slicers, lifting loads that are too heavy.
- *Not paying attention to the job at hand:* Bumping into people, dropping heavy items on their own feet, and closing doors and drawers on their fingers.
- *Unsafe practices:* Knives left lying around on counters or in sinks when not in use and pots and pans put on stoves so their handles stick out into aisles.

While the element of human error is the most significant cause of accidents, eliminating pre-existing hazards in the work environment is another factor involved in prevention.

Environmental Hazards

Environmental hazards are conditions that are unsafe in themselves or that encourage unsafe actions by employees. The following are common environmental hazards and examples.

- *Architectural and building features:* Steep and narrow stairways, dining rooms located a step or two up or down from the main floor, and floor tiles chosen for appearance only, which become slippery with small amounts of soil or water.
- *Furniture and equipment:* Unguarded meat slicer blades, narrow aisles, or insufficient traffic patterns due to the

arrangement of furniture and equipment.

If architectural and building hazards cannot be eliminated immediately, offsetting these hazards is necessary until economic or other considerations allow for renovation and remodeling. Such hazards should never be built into new structures or be allowed to survive renovation.

This includes the current trend in restaurant design to build into new establishments a raised dining area. The one or two steps a server must take with a heavy food- or dish-laden tray increases the likelihood of falling. It can also cause lower back stress from the repetitive strain of carrying trays up and down steps that interrupt the flow of walking.

THE SAFETY SURVEY

The human and environmental hazards discussed previously are typical examples for any foodservice operation. Before attempting to put together a comprehensive accident-prevention program, the foodservice manager should conduct a complete safety survey and an accident review.

In multi-location decentralized operations, you can learn from others' mistakes. Review the types of accidents that have occurred in the past at your establishment and others. This information can be found by reading existing accident reports and files. These reports can help you discover which types of accidents occur most often and if there are environmental hazards that still need to be corrected.

If reports are not available, interviewing employees, especially those who have worked in the operation for a long time, can be helpful. Employees can point out areas where near misses often take place, as well as indicating sources of actual accidents. Investigating the accidents that have occurred will help you correct environmental hazards.

Employees see things in and around the kitchen every day. Supervisors and managers may unintentionally reinforce unsafe acts by not correcting them at the time they are first witnessed. Through monitoring and proper supervision, unsafe practices can be corrected.

The manager should use a safety checklist to examine every aspect of the operation, including the design of the facility, the housekeeping standards maintained, the positioning and functioning of equipment, and the methods of work, because all of these can contribute to the frequency—and the prevention—of accidents. A sample list is shown in Exhibit 13.1; modifications should be made to fit individual establishments.

THE ACCIDENT PREVENTION PROGRAM

An analysis of the hazards discovered in the course of the safety survey should be done to determine whether they are human or environmental hazards. After determining the category, the manager can then follow the appropriate corrective action. The following changes are general precautions to promote safety.

Exhibit 13.1 Accident prevention checklist

Employee Practices

Are all employees informed of hazards existing in their work areas?

Are employees instructed on placement of hands to avoid injury when handling potentially dangerous devices such as slicers?

Do employees make use of all guards, hot pads, railings, and other protective devices available to them?

Do employees wear proper shoes that are nonskid and will protect feet from injury?

Do employees wear clothing that cannot get caught in mixers, cutters, grinders, fans, or other equipment?

Is at least one employee on each shift trained and certified in emergency first aid techniques?

Fire Prevention Equipment

Are fire extinguishers conveniently located where fires are most likely to occur?

Are extinguishers in plain sight?

Are extinguishers adequate in size and of the proper type to control a fire?

Have employees been instructed in the effective use of extinguishers?

Are extinguishers kept fully charged and inspected weekly for damage?

Are sprinklers or automatic alarms installed and inspected?

Has all fire prevention equipment been inspected to comply with local fire prevention agency requirements?

Floors

Are all floors in safe condition—free from broken tile and defective floorboards, worn areas, and items that may cause people to trip or fall?

Are spills and debris removed from the floor immediately?

Where floors are frequently wet, are heavy traffic areas provided with nonskid mats?

Are floors cleaned adequately and after each meal rush period to prevent slipping?

Are adequate floor drains provided and properly covered with gratings?

Are all carpets securely tacked or otherwise fastened in place to prevent people from tripping over raised edges?

Serving Area and Dining Room

Are serving counters and tables free of broken parts, wooden or metal slivers, and sharp edges or corners?

Is all tableware regularly inspected for chips, cracks, or flaws? Are defective pieces discarded in a safe manner?

Is the traffic flow coordinated to prevent collisions while people are carrying trays or obtaining food?

Are ceiling fixtures firmly attached and in good repair? Are tables and chairs in good repair?

Doors and Exits

Are sidewalks and entrance and exit steps kept clean and in good repair?

Will all exits open from the inside without keys to allow escape from the building?

Can an exit be reached from every point in the building without having to pass through an area of high potential hazard?

Are routes to exits, and the exits themselves, clearly marked?

Are passages to exits kept free of equipment and materials?

Are all exits outward opening?

Are doors hung so they do not open into passageways where they could cause accidents?

Stairs, Ramps, and Ladders

Are stairs and slopes clearly marked and lighted?

Do stairs have abrasive surfaces to prevent slipping and falling?

Are handrails on open sides of stairways provided?

Are center handrails provided for wide stairs?

Are stairways kept unobstructed?

(Continued)

Exhibit 13.1 *(Continued)*

Are the slopes of ramps set to provide maximum safety—not too steep?

Are ladders maintained in good condition and inspected frequently?

Do ladders have nonslip bases?

Ventilation

Is the ventilation adequate in receiving, storage, and dishwashing areas and in walk-in coolers and freezers?

Are all fans and their moving parts shielded or guarded?

Is gas-fired equipment properly vented?

Electrical Equipment

Does it meet the National Electrical Code specifications or local ordinances and bear the seal of the Underwriter's Laboratories?

Are regular inspections of equipment and wiring made by an electrician?

Are electrical switches readily accessible in emergencies?

Are switches located so that employees do not have to lean on or against metal when reaching for them?

Are cords maintained without splices, cracks, or worn areas?

Is electrical equipment protected against the entrance of water?

Are weatherproofed cords and plugs provided for outdoor equipment?

Are wet floors and areas subject to flooding avoided for placement of electrical equipment?

Are service cords long enough to eliminate the need for extension cords?

Are all switches, junction boxes, and outlets covered?

Does all equipment with cord and plug connections have grounded connections—either three-pronged plugs or pigtail adapters?

Lighting

Is lighting adequate in all areas?

Are light fixtures, bulbs, and tubes protected with screen guards?

Is proper heat-proof lighting provided over cooking areas, in vent hoods, and so on?

Hot Water Heating

Is hot water temperature properly controlled in lavatories and sinks and are mixing faucets provided to prevent scalding?

Are overhead pipes or fixtures high enough to prevent head-bumping?

Receiving Area

Are employees instructed in correct opening, lifting, and storing methods for each item that is received?

Are adequate tools available for opening crates, barrels, cartons, and other containers?

Storage Areas

Is there sufficient space for storage of everything, with nothing stored on floors, behind doors, in corridors or on stairways?

Are shelves adequate to bear the weight of the items stored?

Are heavy items stored on lower shelves and lighter materials above?

Is a safe ladder or step stool provided for reaching high shelves?

Are portable storage racks and stationary racks in safe condition—free from broken or bent shelves and standing solidly on legs?

Is there a safety device in the walk-in cooler to permit exit from the inside, and a light switch inside?

Hazardous Materials

Are toxic materials and hazardous substances properly stored and handled?

Are employees informed of hazards and precautions according to the HCS?

Are combustible and flammable materials stored and handled properly?

Are compressed gases, such as carbon dioxide tanks, stored properly in a cool, dry, well-ventilated fire-resistant area? Are they protected from falling over? Are pressure gauges on the tanks working?

Are MSDSS provided and available to all employees?

(Continued)

Exhibit 13.1 *(Continued)*

Waste Storage Area

Are garbage and waste containers constructed of leak-proof material?

Are containers adequate in number and size?

Are disposal area floors and surroundings kept clean and clear of refuse?

Are containers on dollies, or other wheeled units, to eliminate lifting by employees?

Food Preparation Area

Is adequate aisle space provided between equipment to allow reasonable work movement and traffic?

Is a nonsplintering, easily cleanable tamper provided for use with grinders?

Are hot pads, spatulas, or other equipment provided for use with stoves, ovens, and other hot equipment?

Are scabbards, sheaths, racks, or magnetic bars available for proper storage of knives and other sharp instruments?

Are machines properly guarded?

Do employees make use of tampers, hot pads, safe knife storage devices, and machine guards provided for their protection?

Are knives and other blades kept sharp?

Are employees properly instructed in the operation of machines, mixers, grinders, choppers, dishwashers, and so on?

Are cooking utensil handles always positioned so that they do not protrude over the edges of cooking units and into passageways?

If anything breaks near the food-preparation or service area, are employees trained to discard all food from the areas adjacent to the breakage to prevent contamination?

Are mixers in safe operating condition?

Are steam tables regularly maintained by competent employees?

Utensil-Washing Area

If conveyor units are used to move soiled items, are edges guarded to avoid catching people's fingers or clothing?

Are portable racks or bus trucks in safe operating condition—wheels and casters working, shelves firm?

Are dish racks kept off the floor to prevent people tripping and falling over them?

Are racks, hooks, and gloves provided so that dishwashing employees do not have to put their hands into sanitizing baths of hot water or chemicals?

Are drain plugs mechanically operated or provided with chains so that employees can drain sinks without placing hands in sanitizing solutions?

Transportation

Are vehicles used in transport of food supplies equipped with all recommended or required safety devices—lap and shoulder belts, neck restraints, and so on?

Are driver and occupants instructed in use of the safety features?

Is defensive driving instruction provided?

Are vehicles provided with a safety partition to prevent slipping or shifting of merchandise forward against the driver?

Are shelving, recessed storage racks, and straps used to secure cargo and prevent sliding, slipping, falling, or breakage?

Source: Adapted from *Safety Operations Manual*. National Restaurant Association. Copyright © 1988 National Restaurant Association. Reprinted by permission.

Built-In Safety

Changes in the physical environment that increase the safety of the facility include: (1) the correction of dangerous conditions; and (2) providing safety features.

Eliminating dangerous conditions can be as simple as rearranging the tables in the dining room to allow wider aisles for traffic, installing a railing on a stairway, or putting down abrasive strips or non-skid mats on slippery floors in areas of heavy traffic. Other changes may involve more work and expenditures of money and time, as well as the advice of experts.

If correction of an environmental hazard cannot be accomplished immediately, short-term corrective actions, such as training, warning signs, and guards should be furnished. Areas with low overhead clearance or with unexpected changes in floor level should have warning signs posted, and the step or low overhang should be painted a bright color. Low overhead pipes or corners where people could bump their heads should be padded.

Safety Training

The greatest challenge in the prevention of accidents is simply making people realize that accidents *are* preventable and that each individual has a responsibility to prevent accidents. It is necessary to implement a safety training program for all staff to prevent situations and behavior that lead to accidents. This training should be combined with instruction on sanitation since many sanitation techniques are also safety habits. New employees should learn safe methods of work when they go through their job orientation. An operation might elect to conduct drills to rehearse the manner in which it would handle specific emergency situations.

The National Safety Council, the American Red Cross, and state and local health, police, and fire departments are sources of posters, filmstrips, and other materials that can be used in the safety training process itself or as memory refreshers afterward. Some of these organizations may also provide training specifically for your employees, or as part of classes that you or your employees can attend on an individual basis.

Safety Supervision

One of the best memory refreshers is a word from the supervisor. A responsibility of management is to monitor all work practices and the environment to prevent accidents from occurring by either instructing and supervising employees on the job while they perform operational procedures, or correcting environmental hazards to make the facility safe.

OCCUPATIONAL SAFETY AND HEALTH ADMINISTRATION

Since 1970, employers have been required by an Act of the Occupational Safety and Health Administration (OSHA) to provide employees with safe working conditions. This law sets standards for a hazard-free working environment, safe equipment, and job procedures designed with safety in mind. The Hazard Communication Standard (HCS), discussed in Chapter 10, is also required and regulated by OSHA.

All employers must report an accident that results in death or a catastrophe. A *catastrophe* is defined as a sudden disaster that is large or significant. OSHA officers may inspect restaurant facilities to check for compliance with the 1970 Act. During these unannounced visits, the operator will be asked to accompany the inspector and point out safety steps taken in the operation.

The penalties for violations were raised by law in 1990. According to the new law, a minimum fine of $5,000 is mandatory for intentional violations, with a maximum fine of $70,000 for intentional or repeat violations. A maximum fine of $7,000 applies for all other violations, including failure by an employer to post the required OSHA notice. A thorough accident-prevention program that includes employee training is a positive approach to dealing with OSHA regulations.

FOODSERVICE ACCIDENTS AND PREVENTION

The nature of the foodservice industry requires that many items of equipment be sharp or hot. Floors can quickly become wet or greasy if a clean-as-you-go approach is not taken by servers and other kitchen staff. The most common kinds of foodservice injuries and accidents are directly related to the presence of these conditions, coupled with human error.

Lacerations

Lacerations are injuries that break the skin. They result from accidents involving knives, food slicers, choppers, and mixers; poorly designed equipment with sharp edges; and broken glass or other items.

Knives

Dull blades are implicated in more cutting accidents than sharp blades. Since dull blades are difficult to work with, more pressure must be applied to cut with them; the chance of a dull knife slipping or being dropped is increased. Knives should be sharpened frequently and at regular intervals.

A knife should be used only for the special purpose for which it is intended. Knives should be sorted by size and stored separately from other utensils with their handles all in the same direction. Never leave knives in sinks or on counters where they may be forgotten. An unsuspecting employee reaching and grabbing for utensils in a sink full of soapy water can be cut by knives left to soak.

Protective steel and fiber gloves are available for employees to wear when cleaning knives and slicer blades. They do not take the place of caution, however. These gloves also require special cleaning and sanitizing.

Injury can result if a foodhandler attempts to catch a knife that he or she has dropped. Instead, step out of the way while the knife drops to the floor.

Power-Driven Cutting Equipment

Power-driven slicers, grinders, choppers, blenders, and mixers can cause severe lacerations, some of which can result in the loss of fingers. The employees who use these pieces of equipment should read the manufacturer's operation manual in order to follow precautions and directions for use, disassembly, and cleaning of electrically powered equipment.

Meat slicers and other grinding, chopping, or cutting equipment have to be kept sharp. You must also properly operate each device. For example, do not use slicers or choppers to try to cut up a frozen product.

Guard devices that prevent machine operators from touching moving blades must be a feature on all slicing, grinding, and chopping equipment, as well as for all equipment with moving or electrically charged surfaces, including fans, conveyors, and mixers, which can also injure employees.

A very effective safety device for dangerous equipment is a *spring switch,* or "dead-man control," which allows the machine to operate only while the switch is depressed; if the operator releases the switch, the machine stops. Guards and switches should not be disconnected or removed from equipment. These guards should also not be difficult to install nor create other obstacles to regular use.

Nonsplintering tampers, pushers, plungers, or blocks should be provided for pushing food into grinders and slicers. Substitute tools for this purpose are not acceptable and present danger to the user.

Jewelry, loose sleeves, or ties should never be worn when operating power-driven equipment. These items can be drawn into the equipment.

Glass

Dishes and glasses that are washed, rinsed, and sanitized manually should be washed separately; that is, all dishes at one time, and then all glasses. This will help prevent glasses from being crushed by heavier objects and breaking in the dishwater. If glass does break in the dishwater, the sink should be drained and the pieces carefully removed by an employee wearing protective gloves.

A separate garbage container should be kept for broken glass and clearly marked so that employees who take out the garbage and trash will not cut themselves on protruding pieces.

Glasses and cups should not be stacked inside one another; it is neither a sanitary practice, nor a safe one. It is possible that the cups and glasses can get stuck and then break when an employee tries to separate them.

Glasses, bowls, cups, or plates should never be used to scoop ice. The risk that glasses or dishes will break or chip while scooping the ice is high. Besides creating the risk of injury, this practice creates a physical hazard to food safety.

Other Laceration Hazards

Nails, staples, and protruding sharp edges present further hazards to employees unpacking boxes and crates. Use special tools to remove these hazards and make the opening process easier.

Burns

Burns are another common foodservice injury. Burns and scalds of varying degrees of severity can result from contact with the hot surfaces of grills, ovens, stove burners, steam tables, fryers, and any other heating equipment that might be in use. Spattered, splashed, and spilled hot food and drink can also burn skin, as can steam and hot water.

Hot-Food Preparation Areas

Hot equipment that is not designed for easy accessibility, pots and pans with loose handles, and loose connections on steam equipment are examples of conditions that can lead to burns.

Pan handles should always be pointed away from traffic, but within reach for foodhandlers without the risk of knocking over other pans. Stoves should not be crowded with hot pans of food between preparation and serving. After preparation, the food items should be moved immediately to service holding units or they should be cooled for storage. This will maintain food safety, as well as reducing burn risks for foodhandlers.

Pot holders and thermal gloves should be used for handling hot pans and cookware, but they must be completely dry so as not to transfer heat to the hands of the wearer or introduce the possibility for steam burns. Plastic gloves should not be worn by kitchen employees when they are working with hot food preparation because the gloves can melt onto the skin and cause burns.

Layout and planning that allows ample space for hot equipment and takes traffic patterns into account promotes safety in this area. Hot pots and pans should be moved minimal distances, or not at all.

Steam is another potential burn hazard. To avoid risks when removing the lid of a steam kettle, an employee should stand with his or her face away from the pot. Employees should also stand to one side and turn their faces when removing pans from a steam table and when opening steam cabinets. The lid from the steam kettle and pans from a steam table should be removed gradually, in a back to front motion, in order to allow the steam to escape in small amounts at a time and so the lid can act as a shield.

Deep Fryers

Deep fryers are a source of many severe burns. The usual temperature of the hot oil in a fryer is at least 300°F (149°C), and this oil can splash on a foodhandler.

Just-washed food items should dry before they are placed in oil and large frozen food items should not be put directly in a deep fryer. Because oil and water do not mix, the oil can cause the water or ice crystals on the surface of the food item to boil explosively, sending droplets of oil in the air all around the fryer, or actually igniting the oil droplets creating airborne "fireballs." Food items should not be dropped into hot oil, but placed in with tongs or in frying baskets. Frying baskets must not be overloaded, which also causes splashing.

When cleaning fryers, follow manufacturer's instructions. The hot oil should be allowed to cool first to 100°F (37.8°C). After it has cooled, use extreme caution to drain the oil into the proper containers, marked for oil recycling if applicable, with tight fitting lids. Do not use plastic containers unless the oil has cooled completely. These containers must be removed from the food preparation area immediately, so other employees cannot trip over or spill pans of hot oil.

Daily filtering of hot oil should be done before the oil is heated up to cooking temperatures. Users of solid frying fat must use special care when melting fresh fat to add to the fryer. Do not pack it in the fryer and then turn on the heat. Rather, melt it in a large pot before adding it to the oil already in the fryer.

Traffic Patterns

Traffic in hot-food preparation areas should be kept to a minimum. Patterns through the kitchen should allow sufficient space for avoiding hot equipment. Aisles and passageways should be kept clear at all times. Servers and kitchen employees need to follow a clear traffic pattern among themselves as well to avoid collisions when carrying hot liquids and trays of food through the kitchen and through the dining room.

Burn Hazards to Patrons

Patrons can also be burned as a result of employee actions. Cups or dishes containing hot foods should never be filled all the way to the edge, where they can be easily spilled or splashed. Servers should warn patrons if foods and plates are very hot.

Hazards from flaming food also exist. Flaming dishes require scrupulous attention to safety (see Exhibit 13.2).

Slips and Falls

Anything that is not already on the floor can fall. People, pots, boxes, bottles, knives, and napkins can all fall to the floor, causing damage, injury, or inconvenience. The bruises, strains, sprained muscles and joints, fractures, and more

Exhibit 13.2 Los Angeles Fire Department rules for serving flaming foods

Los Angeles operators must have special fire department permits to serve flaming foods or drinks. In addition to the required permit, the following safety measures must be followed:

1. Fuels to heat food may be of jellied or semi-solid type. Liquid fuel in excess of one ounce per table will not be permitted.
2. The preparation of flaming foods and drinks is restricted to the table being served. They shall not be transported or carried through rooms or areas while burning.
3. The person who prepares the flaming food or drink shall have a wet towel immediatelyvailable for use in smothering the fire in event of emergency.
4. The serving of flaming drinks or desserts shall be done in a safe manner without flamboyancy or the type of showmanship which would create high flames. The pouring, ladling, or spooning or burning liquids is restricted to a maximum height of eight inches.
5. Crepes suzette shall not be served to more than six persons with one "set up." The pan used shall not be more than one inch in depth. The total amount of flammable beverage used in one "set up" shall not exceed one ounce.

Source: Los Angeles Fire Department

serious injuries that result when people fall account for a large proportion of foodservice injuries.

Flooring Hazards

Foodservice employees can lose their footing on floors that are either covered with slippery material or that have been made slippery by the presence of food, grease, or water. Carpeting needs to be tacked down well. Mats in high traffic areas, like the kitchen, entryway, or aisles in the dining room should also be secured and have tapered edges so they do not slide.

Wet spots on floors are inevitable in a busy establishment, so spills must be cleaned up immediately. Frequent and effective floor cleaning will reduce the number of slips and falls. Floor drains that are kept clean will work effectively to keep floors dry.

The danger from slippery floors is compounded because the employees frequently hurry through their tasks in response to the rush of business at mealtimes, often carrying loads of varying sizes as they move back and forth. These loads can obstruct their vision and limit their ability to move around each other or around equipment and furniture.

Managers should instruct their servers to take shorter strides, not to run, and to always look where they are going. If a server has a large order going out to the dining room, divide the order into manageable loads with a second server helping with another tray. One server alone should not have to carry an awkward load on a large tray that can obstruct his or her line of vision.

Employees should wear shoes that have rubber soles, or ones made from neoprene, like boat-type shoes, to prevent slipping. They should not wear shoes with leather soles, since these are absorbent. Employees should wear shoes that have closed heels and toes, which will help to prevent cuts or crushing injuries as well. The shoes should not have high heels, platform soles, or any exposed metal plating.

Slip and Fall Hazards to Patrons

Customers of foodservice operations can also be victims of slip and fall accidents, which can result in liability for the operator. Managers and employees need to watch out for hazards, such as spills and objects in pathways, and correct them as soon as possible. If wet floors cannot be avoided, such as when they are being cleaned, then signs should warn patrons and employees of the hazard.

Inanimate Objects and Storage of Supplies

Inanimate objects fall because they have been stacked too high, left on vibrating surfaces, perched precariously on shelves, or dropped. They can cause injury by falling on unwary people who dislodge them as they pass by. If fallen objects land in foot traffic areas, they can, in turn, cause people to trip.

Boxes, chairs, tables, other shelves, and cans are not constructed to withstand the weight put on them by climbing and should not be used for that purpose. Ladders and stepladders exist to facilitate climbing and reaching, but must be used correctly. The top rung of a ladder must not be used as a step or for balance. Ladders should be placed on an even and dry surface. The spreader should hold the front and back sections securely open. All the steps, rungs, and cleats of the ladder must be in good condition as well.

Other Common Accidents and Injuries

The nature of foodservice operations contributes to the frequency of several other types of accidents in addition to those just mentioned.

Door Hazards

Windows on swinging doors can help servers and table bussers avoid collisions. One door should be marked *out* and the other *in*. Also instruct employees not to stand near these doors if they are not coming in or out. Supplies should never be stored behind doors.

Lifting Hazards

Heavy supplies should be placed on lower shelves and lighter ones on top shelves to avoid injuries caused by overreaching and incorrect lifting. If a box is too heavy for one person, he or she should be instructed to ask another employee to help. Incorrect lifting can cause serious back injury.

Lifting should be done from a knee-bending position with an erect back, not by bending from the waist, and one foot placed slightly in front of the other. The knee-bending position will allow the leg muscles, and not the back, to lift the weight. Do not twist to change direction, instead move feet in the direction of travel. Hand trucks should be available if heavy boxes must be moved often.

Tableside Hazards

Another type of hazard involves the use of tableside, food-preparation, or service displays, such as chopping foods with knives that are thrown up in the air. When this is done with caution, it can be an entertaining experience for customers. However, careful training and use of safe techniques are necessary to avoid injuries and to reduce liability.

Bottle corks from champagne, sparkling wine, or any sparkling beverage should not be allowed to pop. The proper way to open a bottle at the tableside is for servers to wrap the bottle in a towel, then hold it against their leg or their side with the palm of one hand over the cork. The bottle should not be pointed toward anyone or anything. The server should twist the bottle while holding onto the cork until the cork releases. With this safe method, the excess carbon dioxide is released as a streaming mist, so none of the beverage is wasted.

Electrical Appliance Hazards

Foodservice operations use large numbers of electric appliances, which can be hazardous if in poor condition or mishandled. Improperly grounded equipment and equipment placed in damp or wet areas can seriously shock unwary operators. It is also important to unplug electrical equipment before cleaning and sanitizing it, to avoid the risk of electrical shock.

Microwave Oven Hazard

An invisible hazard is radiation emitted from microwave ovens. Although the safety standards and acceptable radiation emission levels are specified by the federal government, it is up to the foodservice manager to monitor proper maintenance procedures, such as keeping the oven clean and not tampering with interlocks.

If the instruction manual indicates that the oven should be inspected by a trained technician at regular intervals, the manager should make sure such inspections take place. Spring hinges, improper door closures, or inoperable shut-off switches are of special concern during each inspection.

Fires

More fires occur in foodservice establishments than in any other kind of business operation. Managers should regularly check their operations, including employee procedures for hazards that could result in fires. Special fire protection equipment should be provided in all areas where fires are likely to occur, particularly in the kitchen near the grills and deep fryers. Directions in the use of fire extinguishers and evacuation procedures for guests and employees should be a part of every new employee's orientation.

Electrical Fires

Over one-third of all accidental fires in foodservice establishments are caused by the electrical system, faulty wiring and equipment, or by improperly operated or positioned electric appliances. Electrical fires can be prevented by regular checks. Managers should look for frayed wiring and should have an electrician periodically check for overloaded circuits, inadequate ventilation around motors, and improper grounding. Main electrical shut offs need to be well marked by the electrician. Equipment discrepancies should be corrected immediately, by the manager or an electrician, if possible, or the supplier or manufacturer should be contacted.

Arson

Arson can be the result of vandalism or a criminal attempt to cover up a robbery. Foodservice operations have also been targets for arson by disgruntled employees or customers. Fires caused by arson are usually only preventable by good security. A written memo of any threats by former employees or customers should be kept and stored with other personnel records in a fire-rated safe.

Grease Fires

Hot oil in fryers is not only a source of fires, but it can increase the severity of fires begun in other ways. Oil in fryers, in ventilation systems, and on walls, equipment, and other surfaces is highly flammable. Kitchen range hoods, filters, and other surfaces must be cleaned regularly to avoid grease buildup.

Oil bursts into flame at its *flammable limit,* usually between 425° and 500°F (218° and 260°C). Flammable limits vary with different kinds of oil or fat, in addition to its age and condition. The flammable limit temperatures should be stated on the label of the container. Do not use deep-frying oil, shortening, or fat for longer than the manufacturer suggests.

Smoking Material Fires

Burning cigarettes tossed carelessly into trash containers or discarded near storage areas can smolder unnoticed for hours, and then become a roaring blaze after closing when no one is there to discover it. It is essential to control smoking areas to guard against fires. Smoking should be forbidden in restrooms, with warning signs posted. However, as a caution, it is wise to place metal receptacles in restrooms. Smoking for employees should be prohibited in dry storage areas and food preparation areas, both for sanitation, as discussed in Chapter 4, and for safety.

Other Fire Hazards

Special attention should be given to the appropriate storage of flammable items. For example, aerosol cans should not be stored near grills, ovens, heat lamps, toasters, or any heat-producing piece of equipment.

Poor housekeeping leaves combustible material available to feed a fire. Crowded service areas and blocked or poorly marked exits can cause an unfortunate incident to escalate into a full-scale tragedy. The severity of fires can be increased by the lack of proper fire protection equipment, such as extinguishers, sprinkler systems, and smoke or heat alarms.

Fire Protection Equipment

The fire department should always be called first, before using the fire extinguisher. Waiting to call may result in vastly increased damage and loss if the extinguisher proves to be inadequate. If the fire is severe, evacuate the building first, and then call the fire department from another building or outside location.

Extinguishers should be accessible in a general location, in plain sight, and not too close to any area where a fire is likely to occur, since this location could make it dangerous to approach the extinguisher in the event of fire. It should be remembered that fire extinguishers are for very small fires and cannot be relied upon to put out fires of any great size.

Fire extinguishers for different classes of fires are not the same, and the manager should know the difference and purchase the proper kinds of extinguishers (see Exhibit 13.3). Your local fire department or fire extinguisher supplier may be able to demonstrate how to use your extinguisher, or they can give you information on operating directions for your particular model. Place the extinguishers near the areas where those types of fires are likely to occur.

Class A fires involve the burning of paper, cloth, wood, plastic, and rubber. The extinguishers that can be used to fight this type of fire are the A, A/B, and A/B/C.

While a grease fire is classified as a Class B fire, it is also classified as a special hazard. The best choice is a B/C extinguisher that discharges the dry chemicals sodium bicarbonate or potassium bicarbonate. A multi-purpose

Exhibit 13.3 Descriptions of the classes of fires and fire extinguishers

CLASSES OF FIRES

Class A fires include: wood, paper, cloth, cardboard, plastics
Examples: a fire in a trash can; a fire that catches drapes or tablecloths in the dining room

Class B fires include: grease, liquid shortening, oil, flammable liquids
Examples: a fire in a deep fryer; spilled flammable chemicals in kitchen

Class C fires include: electrical equipment, motors, switches, and frayed cords
Examples: a fire in a toaster; a fire in the motor of a grinder

CLASSES OF FIRE EXTINGUISHERS

A	AB	BC	ABC
Use on Class A fires only	**Use on Class A and Class B fires only**	**Use on Class B and Class C fires only**	**Use on Class A, Class B, or Class C fires**
Water, stored pressure Water pump tank Multipurpose dry chemical, stored pressure Multipurpose dry chemical, cartridge operated Aqueous film-forming foam (AFFF), stored pressure Halon, stored pressure	Aqueous film-forming foam (AFFF), stored pressure*	Dry chemical, stored pressure Dry chemical, cartridge operated Carbon dioxide, self-expelling* Halon, stored pressure*	Multipurpose dry chemical, stored pressure* Multipurpose dry chemical, cartridge operated* Halon, stored pressure
	***Not recommended for deep fryer fires**	***Not recommended for deep fryer fires**	***Not recommended for deep fryer fires**

Source: Adapted from *Safety Operations Manual.* National Restaurant Association.

A/B/C dry chemical extinguisher is not recommended.

Electrical equipment causes Class C fires, which can be best controlled with B/C or A/B/C extinguishers.

Since fire extinguishers are mechanical devices, they require periodic maintenance and care. Each extinguisher should have an inspection tag on it to indicate the month and year of the last inspection. Inspections should be performed on a regular basis, at least every six months or according to local fire code requirements. Fire code requirements may also cover which types of extinguishers are to be used in a facility.

Other fire-protection equipment includes several kinds of heat-activated sprinkler systems to suppress the fire; alarm systems; and detection systems that alert the occupants of a building and the fire department to the presence of a fire. Any detection system should be carefully and properly engineered by professionals for maximum efficiency. Places for installing devices to achieve effective results can be identified by the supplier of your particular system.

Automatic sprinklers, which use water or dry chemicals, are the most effective way to control fires, especially fires that occur when the building is unoccupied. A professional needs to be consulted for assistance in the design, planning, and inspection of both sprinkler systems and detection units.

TAKING ACTION IN AN EMERGENCY

In spite of all of the manager's and employees' efforts, some accidents will probably occur, and employees or customers may be injured or become ill. It is important to take action in an emergency. Acting may mean calling for help if necessary, and it may include providing first aid.

Legally, a victim has to give consent before a person trained and certified in first aid performs any procedures. For conscious victims, ask if they need help, and then ask their permission to do what is necessary. For unconscious victims, the law assumes that they would give consent. As the foodservice manager, people will expect you to be the one to initiate action and take charge, even if you do not administer first aid yourself.

At least one employee trained *and certified* in first aid, including how to administer the Heimlich maneuver and how to give cardiopulmonary resuscitation (CPR), should be present in the facility at all times. Keep emergency phone numbers current and near or on all phones. Include the local central emergency number and numbers for police, fire, poison control, ambulance service, and the closest hospital or emergency medical facility.

When you reach your emergency system, be it via 9-1-1, or by calling the police or fire emergency numbers, you will be asked a number of questions relating to the emergency, which you should be prepared to answer. Give a complete address and phone number. Know as much as possible about the accident, injury, or illness, and how many people need help. Tell the dispatcher what first aid procedures are being given and write down all instructions. Allow the dispatcher to hang up first after all the necessary information is given.

A foodservice establishment should be equipped with a complete first-aid kit. OSHA requires either a kit equipped according to the advice of a company physician, or telephone and physical access to community emergency services. Some states also have laws specifying the supplies that can or must be included in this kit; in general, the kit may contain individually wrapped sterile dressings, adhesive tape, slings, inflatable splints, scissors, and ice packs, (see Exhibit 13.4). Operators have the option to subscribe to a first-aid service that inspects and restocks the kits, or they can construct and supply their own.

The location of the kit should be conspicuous and its contents should be clearly marked on the outside or near the kit. Items should either be replaced at regular intervals or as they are used up.

Emergency action procedures should be included in the employee manual and during training sessions. (More information on how to act in an emergency situation is given in Chapter 14, which covers

Exhibit 13.4 First aid kit

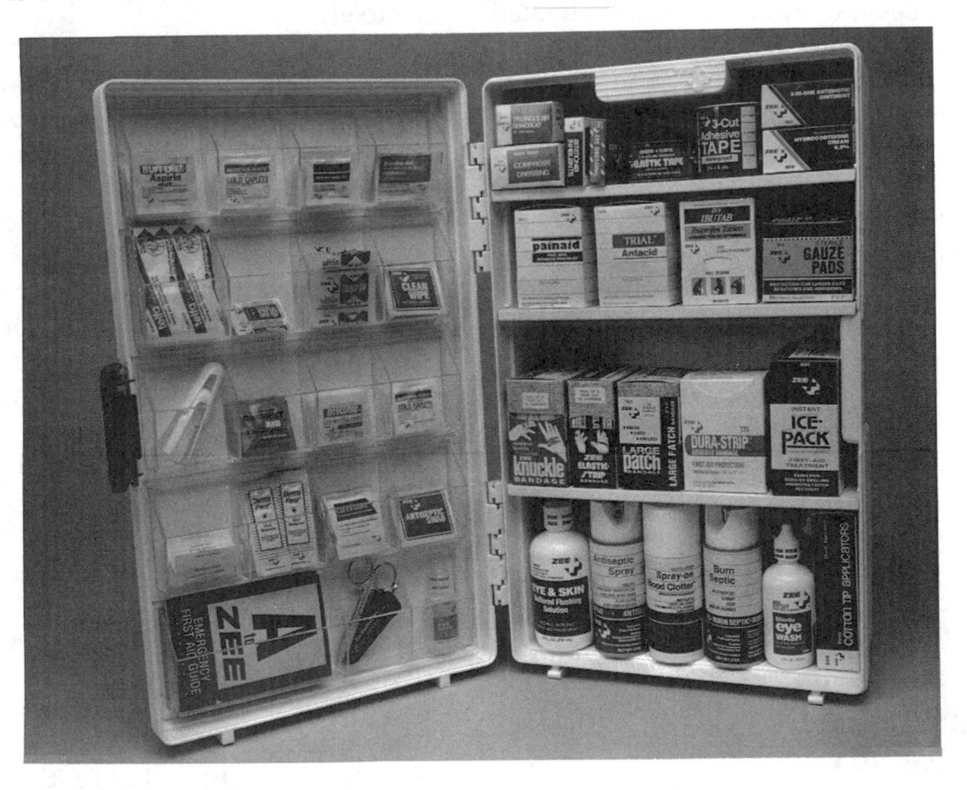

Photograph courtesy of Zee Medical Company, Inc., Irvine, California

crisis management.) The following procedures describe appropriate action for emergencies.

1. *Don't panic—remain calm.* This may seem simplistic, but it's worth noting that too often a common action is a *reaction* that delays help. Worse, panic often spreads quickly.
2. *Call for help when it is clearly necessary.* Post emergency numbers—including your own—throughout the facility and on every telephone.
3. *Be ready to give first aid.* If waiting will worsen the extent of the injury or illness, provide the first aid procedures for which you have been trained and certified, or have one of your employees who is currently certified do the necessary procedure, which will prevent further harm to the victim.
4. *Keep the person comfortable.*
5. *Keep people who are uninvolved away from the victim.*
6. *Keep a record of the incident.* This should include the date, the victim's name, the nature of the injury or illness, who else was involved, what action by employees or bystanders was taken, and how quickly help arrived.

First Aid in Emergencies

Ideally, someone trained and certified in first aid should be available in the facility at all times. Some injuries and illnesses do occur with some frequency in foodservice operations. It is a good idea to obtain first aid training through your local Red Cross chapter or a local hospital. You may be able to arrange in-house training sessions.

Certification for both CPR and the Heimlich maneuver, as well as basic first aid, are good for only one year, and must be renewed. Keep in touch with your local Red Cross chapter for the latest changes in techniques and procedures.

Posters illustrating first-aid procedures can be obtained from the Red Cross and the National Restaurant Association. The foodservice manager can review procedures with employees, and prominently display the posters in the operation. Among the types of emergencies that may occur in a foodservice establishment are choking, cardiac arrest, lacerations, burns, and poisoning.

Choking

Your dining room is crowded and noisy. Your cashier is ringing up totals. Your servers are going in and out of the kitchen with orders. Suddenly you notice a patron in the corner is choking on a piece of food and turning blue. What do you do? Brain damage can occur within four to six minutes of oxygen deprivation from choking if nothing is done to help the victim.

Choking occurs in foodservice operations because people are talking, drinking, and eating fast, without chewing properly. Young children are particularly likely to have this type of emergency. If the victim has made the universal sign of choking, both hands to the throat, and cannot speak or cough, the technique to use is called the *Heimlich maneuver* (see

Exhibit 13.5 First aid for choking victim

American Red Cross First Aid:

When an Adult Is Choking

1 Ask, "Are You Choking?"

2 Shout, "Help!"

Call for help if victim—
- Cannot cough, speak, or breathe.
- Is coughing weakly.
- Is making high-pitched noises.

3 Phone EMS for Help

- Send someone to call an ambulance.

4 Do Abdominal Thrusts

- Wrap your arms around victim's waist.
- Make a fist.
- Place thumbside of fist on middle of victim's abdomen just above navel and well below lower tip of breastbone.
- Grasp fist with your other hand.
- Press fist into abdomen with

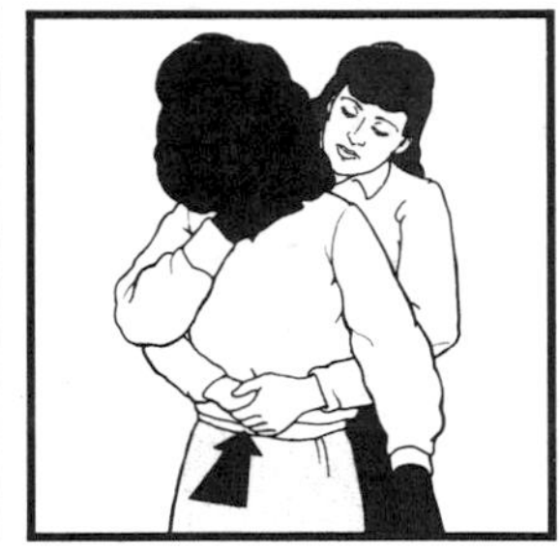

Repeat abdominal thrusts until object is coughed up, or victim starts to breathe or cough.

(Continued)

Exhibit 13.5). Do not let him or her go off alone to the restroom, as victims have been known to pass out and die before anyone knows what has happened to them.

If the victim can talk, cough, or breathe, *do not* pat them on the back or interfere in any way. If the victim cannot talk, cough, or breathe, call for help immediately. After first aid is given, the victim may need follow-up care, as the esophagus or airway may become inflamed and cause him or her further problems.

If the person is conscious, ask if he or she is choking. If the person cannot talk but points to his or her throat, a certified employee should administer the

Exhibit 13.5 *(Continued)*

If victim becomes unconscious, lower victim to floor.

5 Do a Finger Sweep

- Grasp tongue and lower jaw and lift jaw.
- Slide finger down inside of cheek to base of tongue.
- Sweep object out.

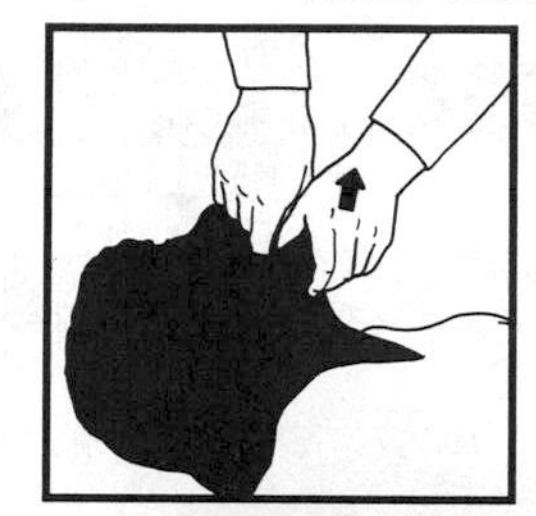

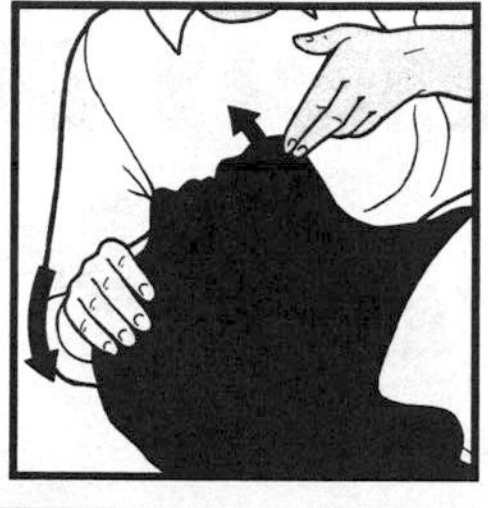

6 Open Airway

- Tilt head back and lift chin.

7 Give 2 Full Breaths

- Keep head tilted back.
- Pinch nose shut.
- Seal your lips tight around victim's mouth.
- Give 2 full breaths for 1 to 1½ seconds each.

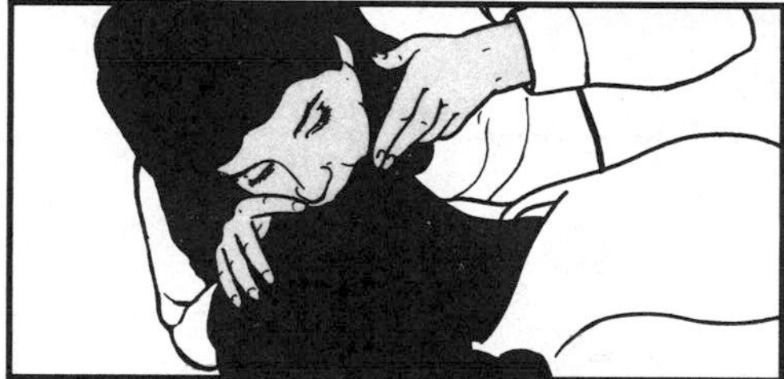

8 Give 6 to 10 Abdominal Thrusts

If air won't go in—

- Place heel of one hand against middle of victim's abdomen.
- Place other hand on top of first hand.
- Press into abdomen with quick upward thrusts.

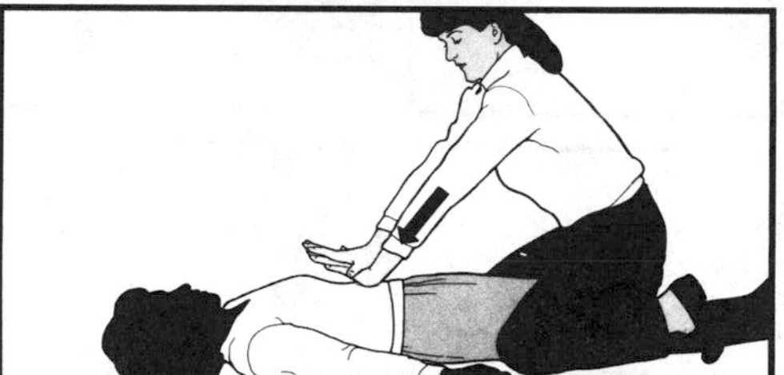

Repeat steps 5, 6, 7, and 8, until airway is cleared, or ambulance arrives.

Local Emergency (EMS) Telephone Number: ____________________

Everyone should learn how to perform the above steps and how to give rescue breathing and CPR. Call your local American Red Cross chapter ________________ (chapter telephone number) for information on these techniques and other first aid courses. Caution: Abdominal thrusts (the Heimlich maneuver) may cause injury. Do not practice on a person.

American Red Cross

Source: American Red Cross. Poster 1043. June 1989. Reprinted by permission

Exhibit 13.6 First aid for unconscious victim

American Red Cross First Aid:

When an Adult Stops Breathing

1 Does the Person Respond?

- Tap or gently shake victim.
- Shout, "Are you OK?"

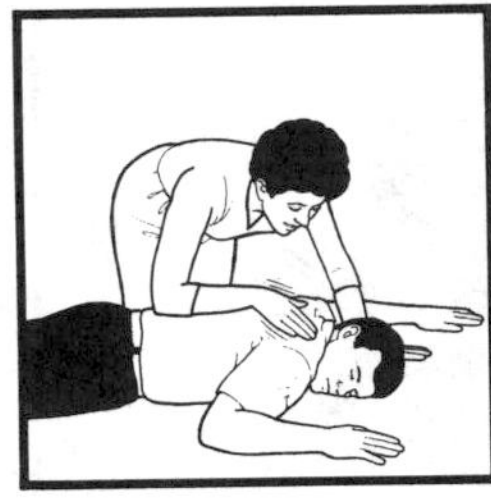

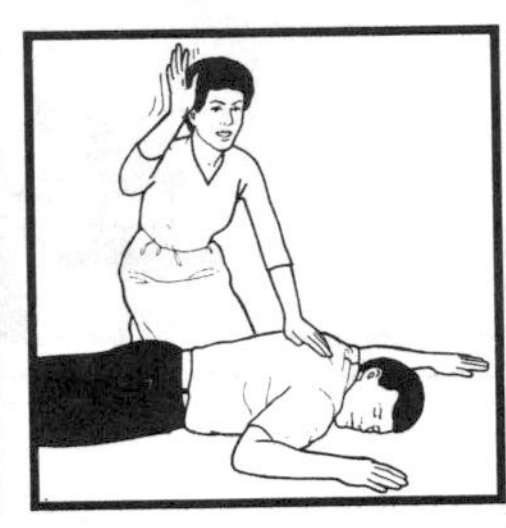

2 Shout, "Help!"

- Call people who can phone for help.

3 Roll Person Onto Back

- Roll victim toward you by pulling slowly.

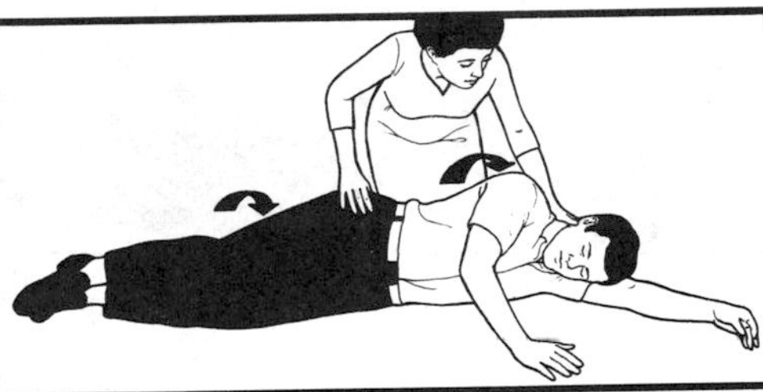

4 Open Airway

- Tilt head back and lift chin.

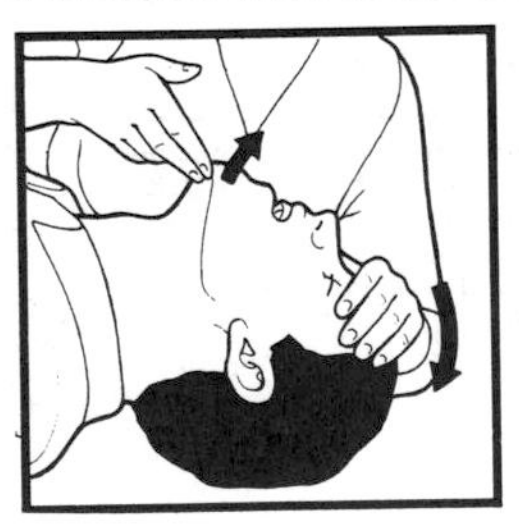

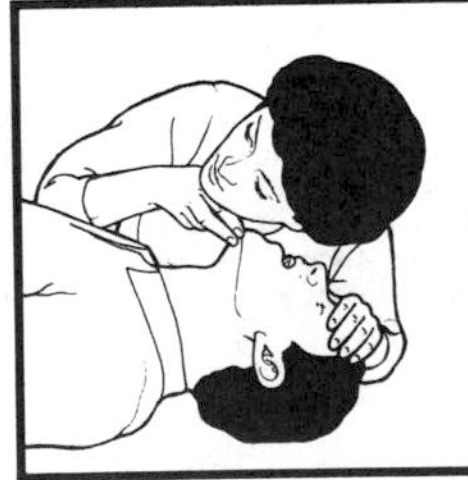

5 Check for Breathing

- Look, listen, and feel for breathing for 3 to 5 seconds.

(Continued)

Exhibit 13.6 *(Continued)*

6 Give 2 Full Breaths

- Keep head tilted back.
- Pinch nose shut.
- Seal your lips tight around victim's mouth.
- Give 2 full breaths for 1 to 1½ seconds each.

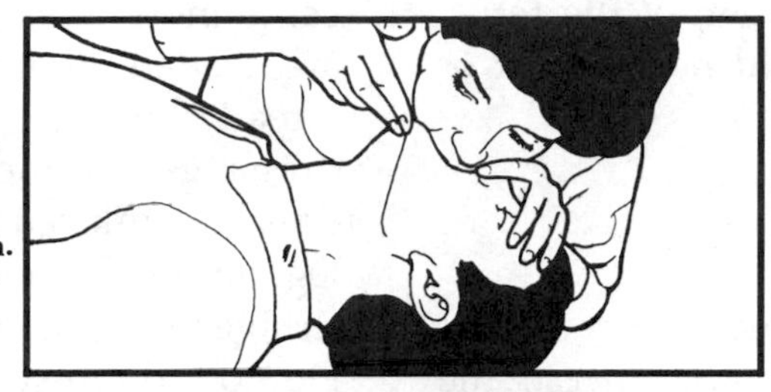

7 Check for Pulse at Side of Neck

- Feel for pulse for 5 to 10 seconds.

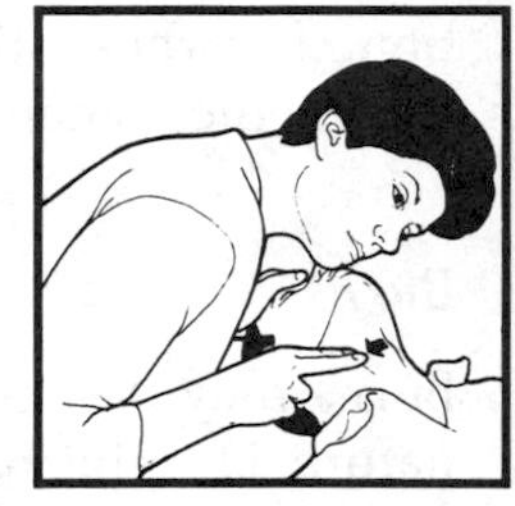

8 Phone EMS for Help

- Send someone to call an ambulance.

9 Begin Rescue Breathing

- Keep head tilted back.
- Lift chin.
- Pinch nose shut.
- Give 1 full breath every 5 seconds.
- Look, listen, and feel for breathing between breaths.

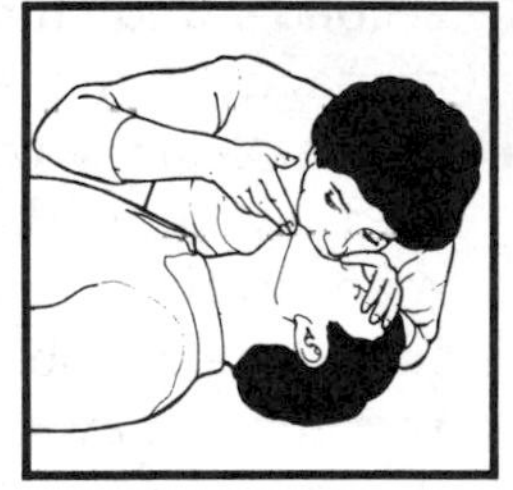

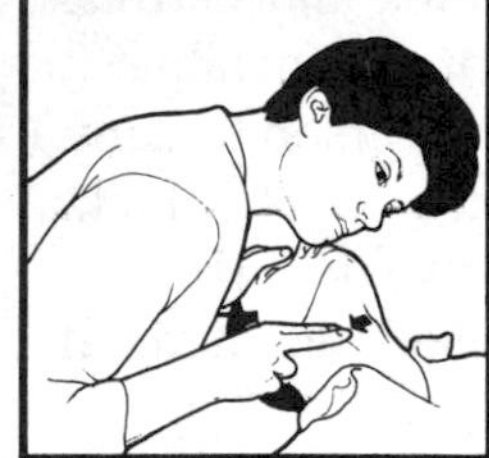

10 Recheck Pulse Every Minute

- Keep head tilted back.
- Feel for pulse for 5 to 10 seconds.
- If victim has pulse but is not breathing, continue rescue breathing. If no pulse, begin CPR.

American Red Cross

Local Emergency (EMS) Telephone Number: ________________

Everyone should learn how to perfom the steps above, how to give first aid for choking, and CPR. Call your local American Red Cross chapter ________________ (chapter telephone number) for information on these techniques and other first aid courses.

Source: American Red Cross. Poster 1042. June 1989. Reprinted by permission

Heimlich maneuver. If the victim is unconscious, he or she has probably stopped breathing. Since the steps for the Heimlich maneuver and mouth-to-mouth resuscitation are somewhat different for infants, children, and adults, it is important that only trained and certified persons perform these procedures (see Exhibit 13.6).

For pregnant women, the steps for the

Heimlich maneuver are the same as for adults. The thrusts associated with the Heimlich maneuver are upward, not downward in the direction of the fetus, so this procedure should not harm it or the mother.

Cardiac Arrest

If a customer or employee suffers a heart attack that results in *cardiac arrest,* the condition where the heart stops pumping, call for medical assistance immediately. You should have employees on the premises during all hours who are trained and currently certified in CPR, because CPR may be necessary to keep the victim alive until help arrives. As with victims of any emergency, heart attack and cardiac arrest victims must be kept comfortable. It is also not uncommon for a person to deny that he or she is having a heart attack. CPR should be applied *only* when there is no breathing and no pulse, and medical help has not yet arrived.

Lacerations

If an employee or patron has suffered a deep wound, it is important for someone to control the bleeding first while someone else calls for emergency assistance. Why control the bleeding first? Because bleeding from a deep cut can cause death in a matter of minutes.

To control bleeding, keep the victim as quiet and still as possible. Either you, if you are certified in basic first aid, or a certified employee will need to cover the wound with a sterile thick compress dressing from your first aid kit, and apply continuous pressure directly and firmly with your hands. *Do not* apply a tourniquet, or remove anything, including glass, from the wound as this may cause more damage. If further dressings are needed, do not remove the dressings already used, just apply additional dressings on top of the original one. Raise the injured area, such as an arm or leg, above the level of the heart to slow the bleeding, but do not move the victim. Do not administer treatment or medication. Let those things be done by the paramedics when they arrive.

Burns

Burns may be either heat or chemical in nature. Heat burns are classified as first-, second-, and third-degree based on the damage to the skin. *First-degree* burns are the least severe. They are characterized by pink discoloration or redness, mild swelling, and pain. *Second-degree* burns are red, blotchy, and have blisters present. They are deeper than first-degree burns. *Third-degree* burns are usually white or charred black, but may also look like second-degree burns. The pain may be either worse or nonexistent.

For first- and second-degree burns, flush the area in cool, running water, or use a cool, wet cloth. Cover the burned area loosely with a dry sterile dressing and raise the affected limb above the heart if this does not increase the pain. Call for medical help if the victim becomes unconscious, and suggest that the victim see a doctor to avoid infections from the burn.

For a third-degree burn, get medical help immediately. While the victim may not feel pain in the affected area, the damage is severe. Do not remove any clothing that is stuck in the wound and do not put anything on the burn. The only measures you can take to aid the victim are to try to keep the person calm to avoid shock, and to have a trained and certified person perform mouth-to-mouth resuscitation or CPR, if necessary.

For chemical burns, call for medical help immediately. Follow the product-specific advice listed on the material safety data sheet (MSDS) that is provided for all chemicals on-site. (See Exhibit 10.8 in Chapter 10.) Keep the victim calm until medical help arrives.

Falls

If an employee or customer falls in your establishment and cannot get up, *do not move him or her.* Do not do anything but make the victim comfortable until medical help arrives.

Poisoning

Some chemical hazards in foods can lead to poisoning. If you think that poisoning has occurred, do not wait for symptoms to appear. Call the Poison Control Center immediately and follow their instructions.

SUMMARY

The elimination of accidents and the pain and expense they cause are as much a part of the creation of a safe foodservice establishment as sanitation is. Accidents can be prevented. The implementation of a well-designed accident-prevention program can significantly reduce the number of accidents and injuries in a foodservice operation.

An *accident* is an unexpected event resulting in injury, loss, or damage. Accidents are caused by people who either behave in unsafe ways or fail to adjust to unsafe conditions.

Common employee accidents in foodservice establishments result in injuries such as lacerations, slips, falls, and burns. The nature of the equipment in a foodservice operation, and the type of work that is performed, may contribute to the frequency of these injuries.

One of the most serious hazards in a foodservice facility is the danger of fire. The cost of fire in lost lives, injuries, and financial loss makes fire prevention an absolute necessity.

The foodservice manager must develop an effective accident-prevention program, based on a thorough understanding of the needs of his or her own establishment. The program should include correction of existing physical hazards, the training of employees in safe practices, and constant supervision to see that unsafe conditions do not develop and that employees are working safely.

Training and certification in first aid should be a vital part of a foodservice safety program. An employee certified in first aid should be available at all times. First-aid training should include CPR, the Heimlich maneuver, artificial respiration, how to control bleeding, and how to treat burns. A well-stocked first aid kit should be available, and emergency telephone numbers should be near all phones.

A CASE IN POINT

A fire started in the kitchen of the Caravan County Mental Health Center. The fat in one of the fryers burst into flame when old grease and dirt dripped down from the hood. The kitchen supervisor told all the foodhandling employees to evacuate the kitchen. The assistant cook was sent to inform the manager of the situation, and the cook was sent to phone the fire department. The kitchen supervisor remained behind so he could put out the fire with the nearby extinguisher.

Do you think this emergency was handled properly?

Which type of extinguisher should the supervisor use to put out the fire?

STUDY QUESTIONS

1. What do sanitation and accident prevention have in common?
2. What is an *accident?*
3. What steps are included in a good accident-prevention program?
4. How do OSHA regulations affect employee safety?
5. What is the most important cause of accidents?
6. What should be done about physical hazards that cannot be immediately eliminated?
7. Are sharp knives more dangerous than dull ones? Why or why not?
8. What is the safety significance of a spring switch on dangerous equipment?
9. What are common causes of accidental fires in foodservice establishments?
10. What is meant by the *flammable limit* of fat or oil?
11. What are the steps to follow when calling 9-1-1?
12. What steps should be taken if a customer is choking on a piece of food?
13. How long does certification for CPR and the Heimlich maneuver remain valid?

ANSWER TO A CASE IN POINT

This case illustrates the connection between sanitation and safety. All equipment must be clean to help ensure food safety, but some of the sanitary practices that protect food are also employee safety practices.

Fat is a flammable substance, and safety is an essential consideration in its use. A clean fryer hood is a preventive measure against kitchen fires. Cleaning the hood over the fryer needs to be described on the master cleaning schedule, assigned to an individual, and then inspected by the manager or supervisor.

A grease fire is classified as a Class B fire, but it is also classified as a special hazard. This means a multi-purpose A/B/C dry chemical extinguisher is not recommended. The best choice is a B/C extinguisher that discharges the dry chemicals sodium bicarbonate or potassium bicarbonate.

The staff at Caravan County Mental Health Center handled the emergency well. The supervisor's decision to evacuate the kitchen and to call the fire department immediately were correct measures to take in this situation.

MORE ON THE SUBJECT

NATIONAL RESTAURANT ASSOCIATION. *Safety Operations Manual.* Washington, DC: National Restaurant Association, 1988. #MG850. 216 pages. This manual covers elements of safety programs, self-inspection, sample checklists, insurance, security systems and devices, OSHA, and sources of assistance, as well as first aid and other related topics. The three-ring format allows operators to include their own safety plan for reference.

NATIONAL RESTAURANT ASSOCIATION. *Safety Self-Inspection Program for Foodservice Operations.* Washington, DC: National Restaurant Association, 1983. #MG833. 32 pages. This self-inspection program will answer all your questions from employee practices to equipment usage. It will teach you how to evaluate your safety practices and where and how to improve them.

U.S. DEPARTMENT OF HEALTH AND HUMAN SERVICES. *Health and Safety Guide for Eating and Drinking Places.* Washington, DC: DHEW, Public Health Service, Centers for Disease Control, National Institute for Occupational Safety and Health, May 1976. DHEW Publication No. (NIOSH) 76-163. This guide is geared toward helping the foodservice operator comply with the Occupational Safety and Health Act of 1970. It provides some health and safety guidelines, warns of some frequently violated OSHA regulations, and includes a safety checklist.

AMERICAN RED CROSS. *American Red Cross Standard First Aid Workbook.* Washington, DC, rev. September 1989. #329380. 178 pages. This workbook contains 16 units that cover emergency action principles, CPR, the Heimlich maneuver, artificial respiration, and first aid for cuts, burns, shock, fractures and other injuries. It is part of the Red Cross certification course. Contact your local chapter for information on classes.

14

Crisis Management

No industry is free from crisis. Any belief that your operation will never experience a crisis should be put aside; crises are inevitable. Many experts contend that it is not a matter of *whether* an operation will experience a crisis, but rather a matter of *when.*

While a crisis can occur, swift and unannounced, at any time, the best time to deal with a crisis is *before* it occurs. An operation is better prepared for crises when it has existing, tangible plans than when a crisis occurs and plans must be devised suddenly and under extreme stress.

Although a wide range of crises can occur, all contain two common threads: an extreme sense of urgency and the need to make swift decisions. Events are likely to occur spontaneously, and quick responses become imperative. Different crises require different types of solutions. An operation must be able to categorize the different types of situations to devise appropriate crisis-management plans. When all possibilities, probabilities, and eventualities of an operation are examined, plans for crisis management can be made.

It is critical that the foodservice operator recognize and address the need for crisis management. In this chapter we will:

- Define a foodservice crisis
- Identify three stages of crisis management
- Discuss methods for establishing a crisis management plan

- Identify methods of dealing with outbreaks of foodborne illness.

CHARACTERISTICS OF A FOODSERVICE CRISIS

A foodservice *crisis* can be defined as an abrupt change in the course of operations, or a turning point or decisive moment in a situation. Comparatively, an *accident* was defined as an unintended event that results in injury, loss, or damage. An accident could be thought of as a crisis, but a crisis by definition is all-encompassing and involves series of events rather than a single event. One crisis commonly sets other crises in motion; a crisis seldom occurs in isolation. Planned responses to *unforeseen* circumstances are needed.

To establish a flexible definition of a foodservice crisis, you need to know these characteristics of a crisis:

- *Escalating intensity.* As a crisis progresses and the full range of its effects become apparent, it may grow in intensity. However, if you take quick, decisive, and well-planned action, you usually can minimize possible adverse reactions.
- *Media or regulatory scrutiny.* Your operation will come under close study by the media, on a local or national level, and by regulatory personnel. While this could signify a loss of control for you, it can also work to your advantage. By working closely and openly with the media and regulatory agencies, you could actually improve the image of your operation.
- *Interference with normal operations.* A crisis can result in many disruptions in operations, from laboratory testing of food and employees to the total closing of an establishment.
- *A reduced public image.* The publicity that your operation will receive, either through media coverage or simply through word-of-mouth, can be detrimental. Even when the crisis has subsided, it is possible that an operation will always be associated with the crisis in a negative light. By being in control of what seems like an insurmountable problem, you have the potential to turn around this negative image.
- *Damage to bottom line.* A crisis can cause great losses to your operation. Medical charges, lost wages, and lost business are just some of the costs you will likely incur.

ESTABLISHING A CRISIS-MANAGEMENT PLAN

Crisis management can be defined as the organized and systematic efforts of an operation to prevent, react to, and learn from crises. Crisis management seeks to:

- Decrease the likelihood of potential crises
- Contain and resolve existing crises
- Learn from the crisis experience.

The crisis plan should preserve human life, properties, and supplies. It should also serve to correct existing problems that led up to it. Effective crisis management includes three stages: preparation; management and resolution of the crisis; and evaluation during and after the crisis. Exhibit 14.1 lists actions to take at each of these stages.

Exhibit 14.1 Actions to take at each stage of crisis management to deal with a foodborne illness outbreak

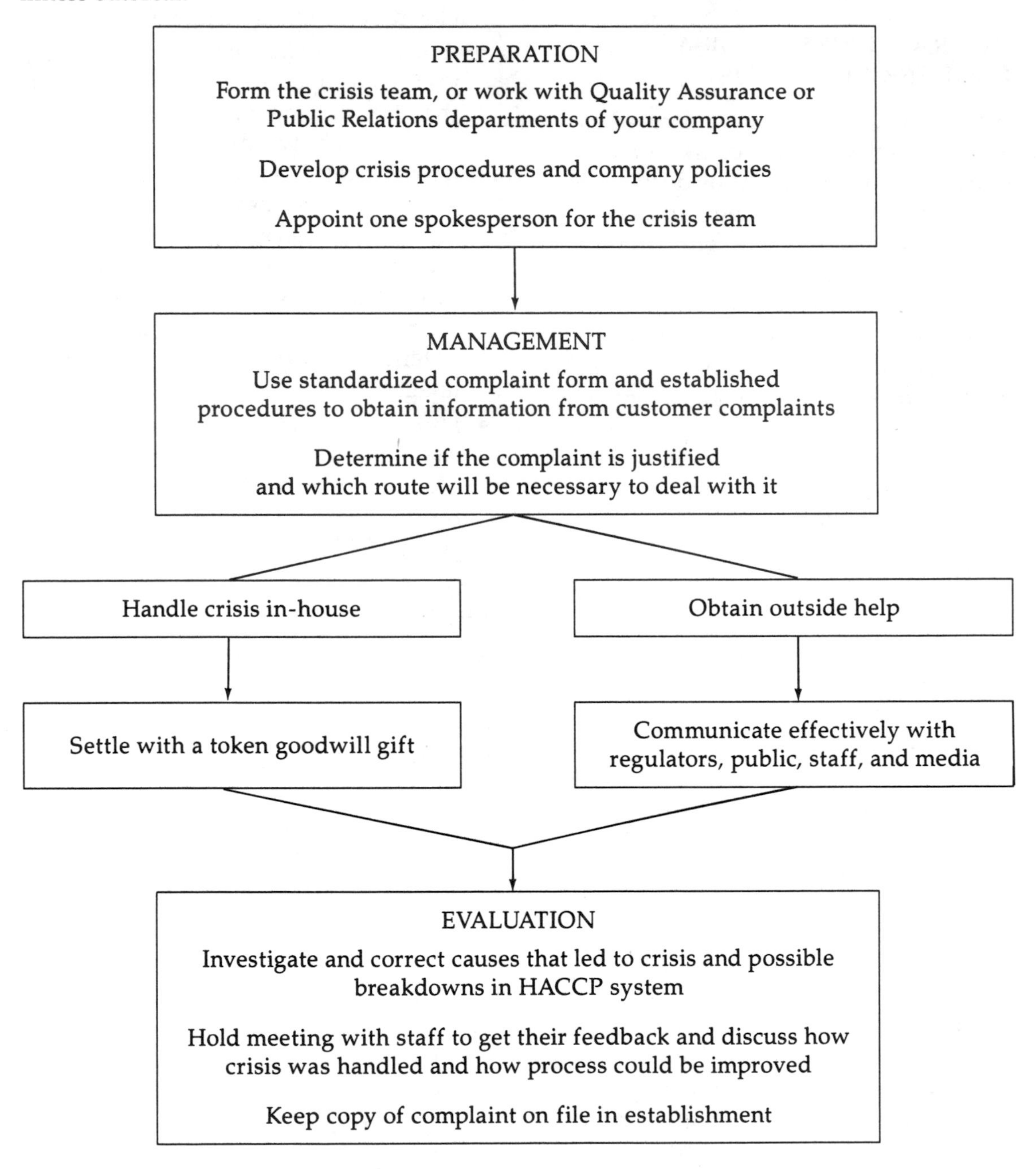

Source: Adapted from *Make a S.A.F.E. Choice: A New Approach to Restaurant Self-Inspection.* Copyright © 1987 National Restaurant Association. Reprinted by permission

When no crisis situation exists, you have the luxury of being able to carefully define a strategy for use when one *does* occur. Devise lists of crises that could occur in your establishment, such as a foodborne illness outbreak, fire, or burglary. Use worst-case scenarios to cover all the bases. Fortunately, these worst-case projections rarely occur; however, they are useful in illustrating and helping plan for the seemingly endless potential for damage that can occur.

Troubleshooting

Troubleshooting is locating and eliminating sources of trouble that could occur in procedures and operations. By establishing critical control points at each step in the flow of food, you have already done some troubleshooting. You have actively and systematically identified potentially hazardous foods in your operation and how they are handled at each step from receiving to serving. At this time, you can institute any corrective measures you deem necessary. With a HACCP-based food safety system in place, you can lessen the chance for potential crises.

Delegation

As you devise your crisis management plan, designate individuals from various aspects of your operation to serve as the *crisis team.* The manager can serve as the head of the team. In major foodservice chains, the crisis team will probably be comprised of the appropriate individuals from Quality Assurance or Public Affairs, who will designate their own leader and spokesperson. However, in either case *all* the employees in an operation should be trained in the policies and procedures to be followed in the event of a crisis. For potential foodborne illness cases, they should know how to obtain the basic information that makes up a complaint. Then employees should refer the complaint to the crisis team, which will handle the complaint.

If the manager is the head of the team, he or she should delegate responsibility and give the crisis team the necessary authority to act and to direct other employees. If the crisis team is formed from a Quality Assurance or Public Affairs department, the manager and employees involved will need to act as they are directed by the team.

Make certain that any statements to the complainants, to regulatory personnel, or to the media come *exclusively* from the head of the crisis team or its designated spokesperson. This will ensure that consistent information is given.

Make sure that *standardized* procedures and forms are used if it is necessary to take customer complaints. Such standardization ensures that important information is not forgotten or omitted. Exhibit 14.2 illustrates a sample complaint report. An attorney or an insurance agent might recommend including additional information on such a report.

Before a crisis occurs, you have the opportunity to test your crisis team and the procedures you have devised. Hold a mock crisis to see how your plan works. Evaluate the results of your trial run, and make any necessary changes to prevent problems encountered here from happening in a real crisis.

Exhibit 14.2 A standardized customer complaint form

Foodborne illness/complaint report

Complainant name: ______________________ Phone __________ Work

Address: ______________________ Phone __________ Home

____

Others in party? ______________________ (Get names and addresses) Use back of form if additional space is needed.

____

____

Onset of symptoms: Date __________ Time __________

Symptoms: ☐ Nausea ☐ Diarrhea ☐ Fever ☐ Blurred vision
☐ Vomiting ☐ Dizziness ☐ Headache ☐ Abdominal cramps

Other ______________________

Medical treatment: Doctor (Hospital) ______________________
Name Address Phone number

Suspect meal: ______________________

Location: ______________________

Time & date ______________________

Identification (brand name, lot number) ______________________

Description of meal: ______________________

Leftovers: ______________________ (Refrigerate, *do not freeze*)

Other foods or Beverages consumed, before or after suspect meal.

Date	Time	Location	Description

Other agencies notified?

Agency	Person to contact	Phone

Remarks:

Report received by: __________ Date: __________ Time: __________

Referred to: ______________________

Source: *Make A S.A.F.E. Choice: A New Approach to Restaurant Self-Inspection,* page 17.

Communication

A crisis requires that you use effective methods of communication. As with any facet of crisis management, it is essential to have an organized plan for communication in place before a crisis can occur. *Communication* is an exchange of information through speech or writing. Establish a policy for both internal (employees) and external (the public, regulatory agencies, and media) communications.

Share updated information with your employees as appropriate. By informing them about what has happened and what is currently happening, you keep rumors to a minimum. Your leadership can have a reassuring and calming effect on your employees. You may also wish to include parents in your messages if your employees are minors. If your operation is part of a chain, you will need to communicate with other operations within that chain.

To establish a plan for external communication, you will need to take into account the methods in which you will keep the regulatory officials, your suppliers (if appropriate), and the public informed of the crisis.

It is essential to appoint an official spokesperson who is comfortable communicating with the public. This individual should be the operator or manager; for chains or larger foodservice organizations a representative of Quality Assurance, Public Affairs, or another appropriate department should be designated. The spokesperson will be responsible for issuing communications as necessary. Designating a spokesperson will also help to protect you from well-meaning employees who might choose to share unverified information with the public.

In handling communications on any level, it is essential to be honest. Don't attempt to conceal what is occurring in your operation. Such concealment can only fuel rumors and cause you to lose your credibility.

OUTBREAKS OF FOODBORNE ILLNESS

Should an outbreak of foodborne illness occur, your operation will need to respond to and lessen the effects of the complaints it receives. Make sure that you respond at once with the necessary sense of urgency that the situation requires. Your plan of action should include the steps that follow.

1. Obtain complete and reliable information.
2. Evaluate the complaint.
3. Deal positively with regulatory personnel and with the media.

Dealing with Customers

A customer who believes that his or her health has been harmed by your operation has the right to take you to court. Your objectives are: (1) to prevent the problem from happening; (2) if the problem is real, to keep it from growing, continuing, or recurring; and (3) to regain that customer's goodwill and patronage. The first step is to realize that the customer might have a legitimate grievance. Make sure that your concern is evident and sincere to the customer. Do not assume that the complain is unjustified; doing so will only serve to further alienate a customer who is already upset, and may lead to more illnesses if steps are not taken to recognize and correct the situation.

Reassure the complainant that you and your employees have a food safety program in place. Offer to assist in any way possible. When you are sincere, professional, and listen to the complaint, you can possibly relieve some of the customer's resentment, and you can reduce some of your liability. *However, make sure that you do not accept or admit responsibility or liability.* While you might express concern that the customer is feeling ill, it would *not* be in your best interest to say, "I regret that our food made you ill." Say instead, "I am sorry that you are feeling ill. We will investigate your complaint and get back to you."

Obtaining Information

Your legal responsibilities and liability will vary from place to place and from incident to incident. When the manager or a member of the crisis team takes a customer's complaint, they need to use the standardized form to record as much pertinent information as possible. It is essential that the person taking the complaint not pressure the complainant or offer additional information.

When dealing with an ill customer, be polite. Do not argue with the customer. Allow the complainant to express his or her experience without interference. Do not introduce symptoms, since some people can be influenced toward illness. Simply record the information that the customer gives you.

Note the time that the symptoms began so the health department or laboratory can determine the onset time of the illness. This information is helpful for them to identify the probable disease through its incubation period. It might also provide a way to clear your establishment of involvement. Try to secure a food history; that is, ask the person what they ate before and after they dined in your establishment. Once again, this could help clear your operation of involvement.

Questions you can ask the customer are as follows:

1. Exactly which food items and beverages did the customer consume?
2. Did the customer eat somewhere else either before or after eating at your establishment?
3. At what time was the food eaten?
4. Where was the food eaten, at home or in the restaurant?
5. Does the customer have some of the food left over, and if so, can they return it to be analyzed by a laboratory? Was the food held in the refrigerator or freezer?
6. Did the customer seek medical attention? If so, at what time and where did the customer go? What is the name of the physician the customer saw?

The customer can provide you with valuable information about your facility and your food preparation.

When speaking to an especially irate customer, allow him or her to vent anger and frustration. Frequently, merely allowing a customer to do so enables you to then discuss the issue calmly.

Make sure that you do not play doctor. Do *not* diagnose, interpret symptoms, or suggest treatments. As you collect the information offered, assure the customer that you or an appropriate party will contact him or her when you have something to report; then make sure that you do so.

Evaluating Complaints

When you evaluate the complaint, you will have to make some judgment calls. You will need to determine if you have a legitimate illness on your hands, or merely a disgruntled customer. You will have to use your instincts and the available data to evaluate the complaint.

Evaluate the complainant's attitude and determine his or her level of sincerity. Resist any urge to argue the issue or to "pay off" the customer immediately. Instead, note the complainant's attitude on your complaint report.

Once you have all of the data, compare the implicated foods with other foods served during the same time period. Note any differences between the complainant's meal and the other meals served. If additional complaints are received, you most likely have a legitimate complaint on your hands.

At this point, also go back and reexamine your flow of foods in the HACCP food safety program. Check that critical control points are being monitored and that necessary corrective action is being taken. Identify any breakdowns in these procedures and implement changes to correct those that need it.

If you decide that the complaint is an isolated incident, you can conduct your own in-house investigation. Develop an in-house policy regarding token gifts for patrons in such cases. You might give meal coupons to the customer, for example, to promote goodwill.

For an outbreak, do not attempt to handle the problem alone. In addition to conducting an in-house investigation, seek outside help. For example, you might ask for assistance from your attorney or insurance agent, or quality assurance group, if one is available to you at your operation.

Dealing with Regulatory Personnel

If you believe that the complaint is legitimate and you have medical or laboratory evidence available to you that identifies a foodborne illness or communicable disease, then it may be in your best interest to notify the health department. If multiple complaints or hospitalizations are involved, you should definitely notify the health department.

While it might seem unappealing to invite regulatory scrutiny, some local ordinances *require* you to report all suspected cases of foodborne illness. Establishing a cooperative stance will be of great benefit to you.

Earn a reputation with the regulators as a responsible operator; then if a crisis should arise, you will already have established good relations and will have an external communication line open. You should develop a plan prior to a crisis to deal with the health department or other regulatory agencies. These agencies possess the potential to help you or to harm you. They can minimize the effects of an incident, lessening publicity, or they can overreact to a situation, making strong media statements.

After being notified of a suspected case of foodborne illness, or an outbreak, the health department will probably conduct an official investigation of your operation. To determine the specific laws in your jurisdiction regarding your rights and responsibilities, consult your attorney. Generally, the health department is authorized to:

- Take reasonable samples of suspect foods
- Prevent the sale of suspect foods
- Require medical and laboratory examinations of suspect employees
- Exclude suspect employees from foodhandling duties
- In extreme cases, order the closing of the facility.

An inspection by the health department may clear your operation and protect you against certain claims. When investigators from the health department speak with you, be cooperative. Make available for review any appropriate records, such as customer charge slips or dealer invoices. Give the investigators reasonable access to observe whatever they request. For example, when a *Salmonella* outbreak in 1990 occurred at a hotel in Illinois, the management actively participated in all phases of the investigation. Thereby, the management reassured the public of its concern about its customers and their welfare. This active stance helped to lessen some of the impact of the crisis.

As previously mentioned, you need to be honest. However, do not volunteer more information than is necessary. Consult your attorney about the legal limits on protecting disclosures relating to trade secrets, and so on.

Dealing with Media

Anything that is of public interest will involve the media; a report of foodborne illness can be newsworthy. You should be prepared to talk to reporters from newspapers, and from radio and television stations. The public views the media as a source of information, and in many cases, the public has a right to know about the events that have transpired.

When speaking to reporters, be cooperative. Answer questions honestly, stating the facts about what has occurred. Explain what is being done to contain the crisis and to limit its impact. Provide accurate and timely information.

It is critical that you make your presentation to the media as professional as possible. Have your facts in order *before* you speak; a crisis is not the time to make off-the-cuff remarks. You might choose to issue a prepared statement. In any case, make sure comments and written statements are delivered only by the official spokesperson.

When speaking to reporters, do not allow yourself to become provoked. Answer questions simply and without jargon. By remaining calm, you appear to be in control of the situation. Make sure

Exhibit 14.3 Guidelines for managers handling calls from reporters

1. Do not hesitate to accept or return calls from reporters. Being unavailable is a negative image to present to the public, who will view your establishment as denying or not responding to the crisis.
2. Find out all the details on the story the reporter is working on by asking the following questions:
   - What is the subject and scope of the story?
   - Has the reporter seen official evidence or documents?
   - Has the reporter spoken to anyone else?
   - What information does the reporter want from you?
3. After asking the questions do not respond immediately. Do not volunteer information. Set an exact time to either get back to the reporter, or to have the company or crisis team spokesperson return the call.
4. If you are the spokesperson, inform your crisis team. If you are part of a larger chain, make certain that you follow your organization's policy toward press contacts.
5. Gather available and appropriate data in a straightforward way and write it down so you will remember exactly what you said. This will ensure that you can repeat the same story and that you will have a permanent record for your files.
6. Try to anticipate additional questions that may come out of your information and write answers for these.
7. Give some key points to develop your perspective. For example, tell what actions are being taken by your establishment to deal with the crisis.
8. Prepare a few meaningful quotes to be used in the story.
9. If you do not know the answer to a question, admit it. Do not say "no comment." If you cannot answer due to legal or other considerations, explain this to the reporter.
10. Never lie to a reporter.
11. *Do not make remarks "off the record."* Expect reporters to print or mention everything you tell them.
12. Do not expect to review or edit the story before it becomes public.
13. If you like the story, call the reporter and tell him or her. This will help to develop a good relationship and to increase your opportunity to receive positive coverage in the future.
14. If there are discrepancies in the facts, first discuss it with the reporter and request that a public correction be made.
15. If this approach is not successful, you can call the editor, or you can write a letter-to-the-editor to set the facts straight. Also check with a lawyer or your legal department for other steps you can take to set the public record straight.

that your responses to questions are not defensive. Don't attempt to bluff; if you don't know the answer to something, admit it. Arrange to secure the information and to present it at a later time. Consider the times you have heard officials respond, "No comment" to a journalist's inquiries. This response is frequently perceived as an admission of guilt, so do not use it. Exhibit 14.3 gives further suggestions for how to deal with reporters.

Keep in mind that it is possible to use the media to your advantage by presenting yourself as sincere, as concerned, and as taking immediate action. The method in which your media relations are handled is often interpreted as the way in which the crisis is being handled.

OTHER CRISES

In addition to an outbreak of foodborne illness, an operation might experience other crises. Some common foodservice crises could include emergency situations, such as fires, robberies, injuries from accidents, lack of water or electricity due to power outages, and product recalls. In each case, it is critical that you have a plan in place *before* any crisis can occur.

Security

Written procedures concerning crime prevention should be given to all employees. Using a staff-team approach will create a series of checks and balances. For example, two or more employees should be present for closing, and instructed to lock all doors after the last customer has left.

Contact a security system supplier or a security service to find out all the available devices and which ones will be most appropriate for your foodservice. In general, a supplier will sell you a system that you are then responsible for operating, while a service will provide the system and staff to operate and maintain it. The supplier or service can inspect your facility to determine where vulnerable points of entry exist and where security devices need to be placed to be effective.

In addition, providing valet parking service for customers can be another safety measure where adjacent parking is unavailable. Asking advice from the local police concerning the type of crime in the area around the restaurant can provide information on the security systems needed.

In the event of a robbery, instruct your employees not to chase thieves or risk injury to stop a crime in progress. Tell your employees to do exactly what the robber tells them to do, and then to call the police when it is safe. They should tell the police where the robbery occurred, at what time, and as many other details about the physical appearance of the robber as possible.

If a bomb threat is received over the telephone, tell employees to listen to the caller without interrupting. Try to keep the caller on the line as long as possible to obtain information from them to aid the police in their investigation. For a bomb threat, be prepared to follow an evacuation plan.

Natural Disasters

For natural disasters like floods, earthquakes, and electrical storms, you can contact appropriate governmental agencies, such as weather bureaus and disaster relief, to find out precautions and measures to follow. These can include using dry ice to keep cold storage foods cold during a power outage, or bolting down large equipment to prevent damage from an earthquake. It is a good idea to keep a battery-operated flashlight and portable radio on hand for these kinds of emergencies.

In the case of product recalls, you might appoint one employee from the crisis team to serve as a recall officer. This employee would then carry out all the necessary tasks required should the FDA or another regulatory agency request a product recall.

Once you have established plans for crisis management, you can alter procedures to address specific circumstances. Your preparation for one type of crisis may help to prepare you for another type. Constantly examine your crisis management plan and note changes in personnel, equipment, inventory, the facility (such as an expansion or renovation), or operations (such as extended hours or patio service). By reviewing all these aspects, you will make sure that your plan is current at all times.

EVALUATING A CRISIS

Crisis management does not end when an immediate crisis is over. Once the crisis has ended, conduct a post-crisis evaluation. One of your goals should be to make sure that the crisis does not occur again.

In evaluating a foodborne-illness crisis, determine the breakdowns in operations at each step in the flow of foods that caused the crisis. A crisis usually occurs with a succession of events, rather than with one isolated incident. Look for changes to critical control point monitoring to determine if they were or were not met; determine if crisis resolutions were met; identify any unforeseen problems that occurred. Then make any necessary changes in operations that will prevent the recurrence of a problem, and inform your employees of these procedural changes.

Also evaluate the manner in which your operation handled the crisis. The following are questions you could ask yourself and your crisis team:

1. Were you satisfied with the way the situation was handled?
2. Was there something more that you could have done?
3. Were there errors made in the handling of the crisis?
4. Was anything overlooked?

If any changes should be made in your crisis procedures, make them immediately, and communicate the changes to all of your employees. Conduct tests of these new procedures.

You need to realize that your employees have been under stress for the duration of the crisis. Make efforts to address and reduce this stress; keep communication open so that everyone in the operation can recover as quickly as possible. You may wish to schedule a meeting with your employees to give them praise for their actions during the crisis or to gain their feedback on how things went and how they could be smoother.

A crisis is not always a completely negative experience. A crisis can be a learning tool. It has the potential, for example, to introduce a positive, long-lasting change in your operation. By concentrating on the learning experience presented by a crisis, you are facilitating the healing process that follows all crises.

SUMMARY

As with any industry, the foodservice industry can encounter crises. Effective crisis management includes three stages: preparation of a plan, management of the plan, and its evaluation. Before a crisis occurs, a foodservice operation should have plans and procedures in place for crisis situations. A specially designated *crisis team* should be identified, and all employees should be trained in the manner in which to handle a crisis.

When an outbreak of foodborne illness or communicable disease occurs, an operator needs to obtain information from customers and evaluate each complaint. Ill customers should be treated with concern and courtesy, but the employee taking the complaint should not admit responsibility. When the complaint is investigated, the manager should secure any necessary outside assistance and notify the health department or another regulatory agency, if appropriate. Such agencies and the media must be dealt with in a positive manner; special care must be given to maintain a responsible stance.

After an immediate crisis is over, management should conduct post-crisis evaluation. Management should review both the circumstances that led to the crisis and the operation's handling of the crisis. Any necessary changes at any step in the flow of food and in crisis procedures should be implemented at this time.

A CASE IN POINT

When an irate customer called Classy Cafe, Sandy, the hostess, answered the call. The customer claimed to have become ill after eating at the Cafe the previous evening. Sandy put the caller on hold and took a standardized complaint report form from the stand by the phone.

She spoke to the customer in a polite way and asked the questions exactly as they were on the form. She wrote down all the information and told the caller that her manager would be in touch.

After the caller hung up, Sandy immediately took the report to her manager, Calvin. During the next two hours of her shift, Sandy received two more phone calls from ill patrons. After the third report came in, Calvin called the health department.

What should Calvin's next step be?
What will Calvin need to do after the crisis passes?

STUDY QUESTIONS

1. Why does advance planning assist an operation in a crisis?
2. What is the major foodservice crisis that an operation is most likely to experience?
3. What are five characteristics of a crisis?
4. Define *crisis management.*
5. What are the three stages of crisis management?
6. List three goals of crisis management.
7. How might worst-case scenarios be used to devise crisis management plans?
8. Explain the function of a *crisis team.*
9. Why should standardized procedures and forms be used in taking complaints?
10. What two forms of communication should be addressed when devising plans for communication during a crisis?
11. A health department is generally authorized to take what five actions?
12. How should an operator approach dealings with the media and reporters?
13. How might the media be used for positive results?
14. What are some other crises that a foodservice operation might encounter?
15. Why is post-crisis evaluation an important stage in crisis management?

ANSWER TO A CASE IN POINT

Now that Calvin has notified the health department of the foodborne illness complaints, he will need to communicate effectively with the regulators, the public, newspaper and television reporters, and his employees. He will need to prepare a statement on the events leading up to and surrounding the illnesses, if they are indeed legitimate.

Besides the investigation by the health department, Calvin will need to conduct his own investigation to discover possible breakdowns in his HACCP program. These breakdowns will have to be corrected at every step in the flow of foods.

After the crisis is resolved, Calvin will need to hold a meeting with his employees to get their feedback on how everything was handled. He will also need to diligently monitor all critical control points, especially any that were changed or corrected, in order to make sure that each is effective.

MORE ON THE SUBJECT

NATIONAL RESTAURANT ASSOCIATION. *The Foodservice Operator's Crisis Management Manual.* Washington, DC: The National Restaurant Association, 1987. #MG600. 12 pages. This manual prepares the operator and manager to handle emergencies, such as foodborne illnesses, injuries, crimes, and fires, by covering pre-crisis preparation, response during the crisis, and how to communicate with the media.

AMERICAN MANAGEMENT ASSOCIATION. "Crisis Management: An Ounce of Prevention . . . ," *Trainer's Workshop,* Volume 3, Number 1, January-February, 1989. AMA, Periodicals Division, 135 West 50 Street, New York, NY 10020. 64 pages. This publication is a training module divided into four sessions on assessing crises, defining crises, defining crisis management, and creating a crisis management strategy. It includes 20 visuals and worksheets, as well as a bibliography. Single issues of the publication can be purchased from the publisher.

FINK, STEVEN. *Crisis Management: Planning for the Inevitable.* New York: American Management Association, 1986. 245 pages. Steven Fink outlines the anatomy of a crisis, as well as crisis forecasting, intervention, and management. He also includes two chapters on crisis communications, and gives concrete examples of catastrophes that have hit companies and how the companies handled ensuing crises.

PART V

Sanitation Management

15

Dealing with Sanitation Regulations and Standards

Foodservice operators and managers are not alone in the battle for food safety. There are people and organizations that can work with you to help and guide you.

In this chapter, we will briefly examine the two basic systems of guidance and control that operate in our society to protect the sanitary quality of food. These two systems are:

- The official system of government regulatory agencies that administer and enforce regulations and standards for the foodservice industry
- The unofficial system of controls and industry established standards observed by the foodservice industry, trade associations, and professional groups.

Almost every facet of a foodservice facility is inspected, evaluated, or controlled in one way or another by government, trade, or professional organizations. The perceptive manager will recognize the constructive guidance that regulatory and advisory organizations offer and use that guidance to operate their establishment in the safest and most efficient way possible.

Despite the existence of several national model codes, adoption and interpretation of the codes may vary widely from one jurisdiction to another. Therefore, the manager should consult local regulatory agencies to find out specific interpretations and to become knowledgeable of local regulations that apply to his or her operation.

GOVERNMENT REGULATION OF THE FOOD INDUSTRY

Government became interested in securing the sanitary quality of food a long time ago. The first health ordinances in the Western world were adopted more than 400 years ago, and the nation's first health officer was appointed when the United States was still a group of British colonies. Today, government control is exercised at three levels: federal, state, and county or municipal.

In order to better understand how these agencies conduct their enforcement and control activities, it is important to distinguish between standards, regulations, ordinances, and codes.

- An *ordinance, law,* or *statute* establishes the legal authority for a federal or local government agency to publicize and put regulations into effect.
- A *regulation* is a governmental control that has the force of law.
- A *code* is a systematic collection of regulations, or statutes, and procedures. A code often equals either a regulation or a statute.
- A *standard* is a measure used as a comparison for quality.

The Role of Regulation

Several governmental agencies have adopted standards or models that define minimum sanitation regulations. Their purpose is to protect the public from foodborne illness. Food processors, wholesalers, retailers, and restaurant operators are required to meet standards, specific definitions, and interpretations for critical areas of their operations that require specific attention and effective control measures.

To help meet these needs, many state and local regulatory agencies also provide training for foodservice managers. These programs educate managers concerning regulations, principles, and practices of food protection.

The Limitations of Regulation

The U.S. Food and Drug Administration (FDA) has developed a "Model Food Service Sanitation Ordinance" to assist health departments in developing regulations for a foodservice inspection program. Although the FDA recommends adoption by the states, they cannot require it. Some state and local agencies develop their own codes rather than adopt the FDA's model code. If there is no state code on a specific issue, the current FDA "Model Food Service Sanitation Ordinance" is enforceable by law. State laws can be tighter, but not looser, than the FDA code. The lack of uniformity that these variations create can complicate and frustrate industry and government efforts to establish uniform food safety standards. For example, some jurisdictions require refrigerated food temperatures to be 45°F (7.2°C) or lower, while others require 40°F (4.4°C) or lower.

How to View Regulations and Standards

Although the federal, state, and local governmental agencies set regulations and standards, it is the operators, managers, and employees who must meet the boundaries set forth in these regulations. Continued efforts by foodservice management are necessary in order to maintain and ensure compliance with food safety standards. By meeting or exceeding minimum requirements, you ensure safe food, improved quality, and allow for a greater margin of error.

Federal Regulatory Agencies

Federal government agencies work to safeguard the sanitary quality of many food products that are purchased by a foodservice establishment. The federal agencies of significance to the foodservice operator and manager are described next.

Food and Drug Administration

The FDA is a major unit of the U.S. Department of Health and Human Services. Within the FDA, the Center for Food Safety and Applied Nutrition (Retail Food Protection Branch) has the responsibility for developing model ordinances and codes for adoption by state and local health departments. The "Model Food Service Sanitation Ordinance" with the recommendations that were established in 1976, is contained in the *Food Service Sanitation Manual* published by the FDA. This document is periodically updated with interpretations issued by the FDA to further define or clarify the provisions of the existing code, and reflect developments in both the foodservice industry and the field of sanitation. The *Applied Foodservice Sanitation* program and coursebook is consistent with the regulations of the FDA sanitation manual.

The FDA directly enforces mandatory provisions of the laws and regulations relating to foodservice operations on interstate carriers. Planes, boats, trains, buses, and their commissaries are inspected by the FDA.

Another important job of the FDA, which is shared in part by the U.S. Department of Agriculture, is the inspection of food-processing plants to ensure adherence to standards of purity and wholesomeness and compliance with labeling requirements. The milk industry is cooperatively regulated by the FDA and state agencies. The National Shellfish Safety Program is a cooperative effort between the FDA, state agencies, and the shellfish industry in order to regulate the shellfish supply.

U.S. Department of Agriculture

The U.S. Department of Agriculture (USDA) is responsible for the inspection and grading of meats, meat products, poultry, dairy products, eggs and egg products, and fruits and vegetables shipped across state boundaries. The USDA provides these services through the Food Safety and Inspection Service agency (FSIS). The grading and inspecting functions of the USDA are described in detail in Chapter 6 of this book. The

USDA also develops a considerable amount of educational material about meat, poultry, produce, and processed foods.

U.S. Centers for Disease Control

The U.S. Centers for Disease Control (CDC), located in Atlanta, Georgia, are field agencies of the U.S. Public Health Service. The Centers investigate outbreaks of foodborne illness, study the causes and control of disease, and publish statistical data and case studies in the *Morbidity and Mortality Weekly Report (MMWR)*. CDC provides educational services in the field of sanitation.

Other Federal Programs

Several other agencies of the national government play lesser roles in the control of food safety.

- *The Environmental Protection Agency.* The EPA sets standards for air and water quality, and regulates the use of pesticides—including sanitizers—and the handling of wastes.
- *The National Marine Fisheries Service.* The NMFS of the U.S. Department of Commerce implements a voluntary inspection program that includes product standards and sanitary requirements for fish processing operations.

State and Local Regulations

The laws that most affect the foodservice operator and manager on a day-to-day basis are those enforced by state and local health authorities. These agencies create regulations and interpretations of the federal code, and conduct inspections to ensure compliance.

From state to state and from city to city, the regulations vary in coverage and manner of enforcement. In a large city, the enforcement agency will probably be a local one. In a small municipality or in rural areas, a county or state health department may have the authority to enforce health codes. In any case, the manager must be familiar with the applicable laws and with the enforcement system in effect. Some health departments issue guidelines on foodservice sanitation that explain the law in terms that are easy to understand. *Guidelines* are statements of policy or procedures and are not the same as standards, regulations, or laws.

As mentioned earlier, many state and local ordinances are patterned after the "Model Food Service Sanitation Ordinance" recommended by the FDA. Following is a list of the main provisions of that ordinance illustrating the range of federal, state, and local governmental interests.

1. Food care: supplies, protection, storage, preparation, display and service, transportation
2. Personnel: health, personal cleanliness, clothing, practices
3. Equipment and utensils: materials, design and fabrication, equipment installation and location
4. Equipment and utensil maintenance: cleaning, sanitizing, and storage
5. Sanitary facilities and controls: water supply, sewage, plumbing,

toilet facilities, lavatory facilities, garbage disposal and solid waste management, integrated pest management (IPM)

6. Construction and maintenance of physical facilities: floors, walls and ceilings, cleaning facilities, lighting, ventilation, dressing rooms and locker areas, poisonous materials
7. Mobile foodservice units
8. Temporary foodservice units
9. Compliance procedures: issuance and suspension of permits
10. Examination and condemnation of food

Foodservice Inspections

To the foodservice manager, the law is most directly represented by the sanitarian, sometimes referred to as the *health official* or *inspector*. The *sanitarian* is an individual trained in sanitation principles and methods, and public health. The sanitarian is a representative of the state or local health department.

Plan Review Inspection

Often jurisdictions require advance review of plans and specifications for new construction or extensive remodeling of foodservice facilities. This procedure makes sure that applicable codes will be met as Chapter 9 discussed.

Once the foodservice building is completed, an initial inspection visit can be expected prior to issuance of an operating permit. In some cases, an inspection immediately prior to opening is conducted to ensure that all necessary food safety systems are in place. Several agencies may be involved in the visit, for example, the fire marshal or a representative from OSHA.

Operational Inspections

After a foodservice establishment has opened for business, inspections may be conducted periodically, monthly, or annually at any time during the span of a calendar year. The FDA recommends that inspections be held *at least every six months.* The exact frequency can be influenced by factors such as the workload of the local agency. The frequency of inspections may also vary depending upon the size and complexity of the operation, its history in terms of violations and illness, the susceptibility to foodborne illness of its clientele, and the thoroughness of the operation's HACCP program.

Public health inspections are intended to *supplement* the manager's self-inspection program; public health inspections do *not* assume the responsibility for maintaining a safe and sanitary operation.

Your local health code serves as the sanitarian's guide. By obtaining copies of local sanitation ordinances, regulations, and standards, an operator or manager will be able to develop appropriate procedures to meet all guidelines and provisions. Periodically compare these documents to your establishment's operational procedures to keep your operation free of health violations.

In the traditional inspection method, the sanitarian completes an inspection report that must be signed by the operator, manager, or a designated representative. This signature is an acknowledgment of receipt. A copy of the report is given to the foodservice manager or representative to keep on file in the establishment. Exhibit 15.1 shows the form suggested by the FDA for use by local health departments. Note the highlighted column of numbers on the right of each item that indicates the *relative weight* given to each kind of violation. The four- and five-point items are similar to critical control points in a HACCP system. The operation receives a *score* after any weighted violations are subtracted from the perfect score of 100.

Another type of inspection, which is seeing increased usage among health agencies, addresses some of the shortcomings of traditional inspections by focusing on the HACCP system of food safety, rather than on the facility alone.

When a food service's sanitation practices are consistent with or exceed the local laws, a routine inspection will confirm the operation's food safety program. A good sanitation program and well-trained employees result in safe food, which will also be reflected in good inspection reports.

When a sanitarian arrives at your operation, always accompany him or her during the inspection. Take advantage of the sanitarian's experience and expertise; learn from the information that he or she shares with you as the inspection progresses. In addition, accompanying the sanitarian means that you are available to answer any questions that are raised. The following suggestions will enable managers and operators to get the most out of sanitation inspections:

1. *Ask for identification.* Many sanitarians will volunteer their credentials. By checking identification, you will help ensure that only authorized persons are allowed to check your facility. Do not let anyone enter the back of the facility without proper identification. Ask the purpose of the visit; make sure that you know whether it is a routine inspection, the result of a customer complaint, or some other purpose.

2. *Cooperate.* Most sanitarians have learned to expect some defensiveness, resentment, and evasiveness from operators and employees alike. This can be interpreted as you having something to hide, which is counterproductive. Answer all of the sanitarian's questions to the best of your ability, and instruct employees to do the same. Explain to the sanitarian that you wish to accompany him or her during the inspection. This will encourage open communication and a good working relationship, which is vital if problems do arise.

3. *Take notes.* As you accompany the sanitarian, make note of any problem that he or she points out. Make your willingness to correct problems obvious. If it can be corrected immediately, do so. Taking down your own notes will also protect your rights *without* a combative confrontation. Note all comments that the sanitarian offers. If you are convinced that the sanitarian is incorrect about something, specifically

Exhibit 15.1 FDA Food Service Establishment Inspection Report form

DEPARTMENT OF HEALTH, EDUCATION AND WELFARE
PUBLIC HEALTH SERVICE—FOOD AND DRUG ADMINISTRATION

PURPOSE	
Regular	29-1
Follow-up	2
Complaint	3
Investigation	4
Other	5

FOOD SERVICE ESTABLISHMENT INSPECTION REPORT

Based on an inspection this day, the items circled below identify the violations in operations or facilities which must be corrected by the next routine inspection or such shorter period of time as may be specified in writing by the regulatory authority. Failure to comply with any time limits for corrections specified in this notice may result in cessation of your Food Service operations.

OWNER NAME	ESTABLISHMENT NAME

ADDRESS	ZIP CODE

EST. I.D. (1-10)	COUNTY	DIST.	EST. NO.	CENSUS TRACT	SANIT. CODE		YR.	MO.	DAY	TRAVEL TIME	INSPEC. TIME
				11-13	14-16	17-22				23-25	26-28

ITEM NO.		WT.	COL.
FOOD			
*01	Source; sound condition, no spoilage	5	30
02	Original container; properly labeled	1	31
FOOD PROTECTION			
*03	Potentially hazardous food meets temperature requirements during storage, preparation, display service, transportation	5	32
*04	Facilities to maintain product temperature	4	33
05	Thermometers provided and conspicuous	1	34
06	Potentially hazardous food properly thawed	2	35
*07	Unwrapped and potentially hazardous food not re-served	4	36
08	Food protection during storage, preparation, display, service, transportation	2	37
09	Handling of food (ice) minimized	2	38
10	In use, food (ice) dispensing utensils properly stored	1	39
PERSONNEL			
*11	Personnel with infections restricted	5	40
*12	Hands washed and clean, good hygienic practices	5	41
13	Clean clothes, hair restraints	1	42
FOOD EQUIPMENT & UTENSILS			
14	Food (ice) contact surfaces: designed, constructed, maintained, installed, located	2	43
15	Non-food contact surfaces: designed, constructed, maintained, installed, located	1	44
16	Dishwashing facilities: designed, constructed, maintained, installed, located, operated	2	45
17	Accurate thermometers, chemical test kits provided, gauge cock (1/4" IPS valve)	1	46
18	Pre-flushed, scraped, soaked	1	47
19	Wash, rinse water: clean, proper temperature	2	48
*20	Sanitization rinse: clean, temperature, concentration, exposure time; equipment, utensils sanitized	4	49
21	Wiping cloths: clean, use restricted	1	50
22	Food-Contact surfaces of equipment and utensils clean, free of abrasives, detergents	2	51
23	Non-food contact surfaces of equipment and utensils clean	1	52
24	Storage, handling of clean equipment/utensils	1	53
25	Single-service articles, storage, dispensing	1	54
26	No re-use of single service articles	2	55
WATER			
*27	Water source, safe: hot & cold under pressure	5	56
SEWAGE			
*28	Sewage and waste water disposal	4	57
PLUMBING			
29	Installed, maintained	1	58
*30	Cross connection, back siphonage, backflow	5	59
TOILET & HANDWASHING FACILITIES			
*31	Number, convenient, accessible, designed, installed	4	60
32	Toilet rooms enclosed, self-closing doors; fixtures, good repair, clean: hand cleanser, sanitary towels/hand-drying devices provided, proper waste receptacles	2	61
GARBAGE & REFUSE DISPOSAL			
33	Containers or receptacles, covered: adequate number insect/rodent proof, frequency, clean	2	62
34	Outside storage area enclosures properly constructed, clean; controlled incineration	1	63
INSECT, RODENT, ANIMAL CONTROL			
*35	Presence of insects/rodents—outer openings protected, no birds, turtles, other animals	4	64
FLOORS, WALLS & CEILINGS			
36	Floors, constructed, drained, clean, good repair, covering installation, dustless cleaning methods	1	65
37	Walls, ceiling, attached equipment: constructed, good repair, clean, surfaces, dustless cleaning methods	1	66
LIGHTING			
38	Lighting provided as required, fixtures shielded	1	67
VENTILATION			
39	Rooms and Equipment—vented as required	1	68
DRESSING ROOMS			
40	Rooms, area, lockers provided, located, used	1	69
OTHER OPERATIONS			
*41	Toxic items properly stored, labeled, used	5	70
42	Premises maintained free of litter, unnecessary articles, cleaning maintenance equipment properly stored. Authorized personnel	1	71
43	Complete separation from living/sleeping quarters. Laundry	1	72
44	Clean, soiled linen properly stored	1	73

FOLLOW-UP	RATING SCORE 75-77	ACTION
Yes 74-1	100 less weight of items violated ⟶	Change . . 78-C
No. 2		Delete . . . . . . D

Received by: name ____________
title ____________
Inspected by: name ____________

* Critical Items Requiring Immediate Attention. Remarks on back (80-1)

Source: U.S. Food and Drug Administration.

note what was commented upon. Then refer the comments that you believe to be in error to an outside sanitation expert or another representative from your local health department for a second opinion.

4. *Keep the relationship professional.* Don't offer food or drink before, during, or after an inspection. This could be construed as bribery.
5. *Be ready to provide records.* The sanitarian may request records, including employee files. You may ask why they are needed. You can also check with your lawyer about the limits of disclosure on confidential or proprietary information. Be sure you keep records of all chemicals and integrated pest management treatments used in the facility.
6. *Discuss any violations with the sanitarian.* The inspection report should be studied closely and any violations noted should be discussed in detail with the sanitarian. In order to make the best and most complete correction, you will need to know the exact nature of the violation. The sanitarian sees many foodservice operations and can offer advice on correcting violations, as well as provide general information on improving foodservice sanitation.
7. *Follow-up.* Go over your copy of the report carefully, then take it with you through your facility. Correct the problems. Determine why each occurred by evaluating HACCP procedures, the master cleaning schedule, and employee training efforts. Establish new procedures and practices or revise existing ones to meet or exceed the minimum sanitation requirements that will correct the problem permanently.

It is essential that violations discovered by sanitarians be corrected as required by your local health department. According to the FDA model code, if the establishment receives a score of less than 60, corrective actions must be taken on all noted violations within 48 hours. The larger violations should have a set time frame for their correction, as approved by the sanitarian. Failure to correct violations may subject an operator or manager to fines and unfavorable publicity. The sanitarian may have the authority to close the establishment if violations are excessive and it is determined that there is a clear and present danger to the public health.

One limitation of inspections is that the extent of coverage and the areas of greatest and least emphasis will vary between localities and even among individual sanitarians. For example, some are more concerned with refrigeration temperatures of potentially hazardous foods and others may concentrate on the general physical appearance of the facility. One sanitarian may examine the flow of food in detail, and another may give primary attention to the personal hygiene of employees. Due to these differences, it is good for foodservice managers to know that all areas of their operations are safe and exceed local standards.

By working with the health department, you confirm your responsible role as foodservice manager. (Refer to Chapter 14 on Crisis Management for a discussion of more benefits from establishing a good working relationship with the health department.)

HACCP AND THE REGULATORY COMMUNITY

A voluntary food safety program of self-regulation will assure government regulators and the public that an operation is aware of and is protecting food safety at each step in food production. In some jurisdictions, traditional health department inspections that stress the appearance of facilities or even spot checking temperatures have been replaced by inspections that focus on HACCP principles and that examine the procedures related to the flow of food through an operation. Inspectors may observe the way an operator receives, stores, prepares, and serves food, and may help designate or verify critical control points for each step. Many state and local health agencies train their inspectors in HACCP principles. Some use the HACCP system as a basis to determine the frequency with which different types of foodservice operations are inspected and even how the inspections are conducted. Exhibits 15.2 and 15.3 show the types of forms used during a HACCP-based inspection.

CONTROLS WITHIN THE INDUSTRY

Probably few segments of U.S. industry have devoted as much effort to regulating themselves as have foodservice, processing, manufacturing, and associated food enterprises. Scientific and professional societies, manufacturing and marketing firms, and trade associations have energetically pursued programs designed to raise the standards of the industry through research, education, and cooperation with government. The overall results in the area of sanitation have included the following:

- An increased understanding of foodborne illness and its prevention within the foodservice industry
- Improvements in the design of equipment and facilities for greater cleanability, effectiveness, and reliability
- Industrywide assistance in maintaining the sanitary quality of foods during processing, shipment, storage, and service
- Efforts to make foodservice laws uniform throughout the nation, and to give them practical applicability.

While countless organizations have contributed to these endeavors, those having relevance for the foodservice operator and manager are discussed in the remainder of this chapter. (Addresses for some of the following organizations are in Appendix F.)

Trade and Professional Organizations

The National Restaurant Association

The National Restaurant Association is the national trade association for retail and institutional foodservice establishments. Through its Technical Services Department, Committees, and Study

Exhibit 15.2 HACCP inspection form using food flowcharting

NEW YORK STATE DEPARTMENT OF HEALTH
Bureau of Community Sanitation and Food Protection

Hazard Analysis Critical Control Point Worksheet

Establishment Name ______________________ Name of Contact Person ______________________

Address ______________________ County ______________ Zip Code ______________

Date: mo. ___ day ___ yr. ___ TIME: Start ___:___ A.M./P.M. End ___:___ A.M./P.M.

Product ____

Ingredients ____

____

Sources ____

____

Time	Temp.	Procedure/Observation	Comment/Interpretation

Time/Temperature (°F) Chart

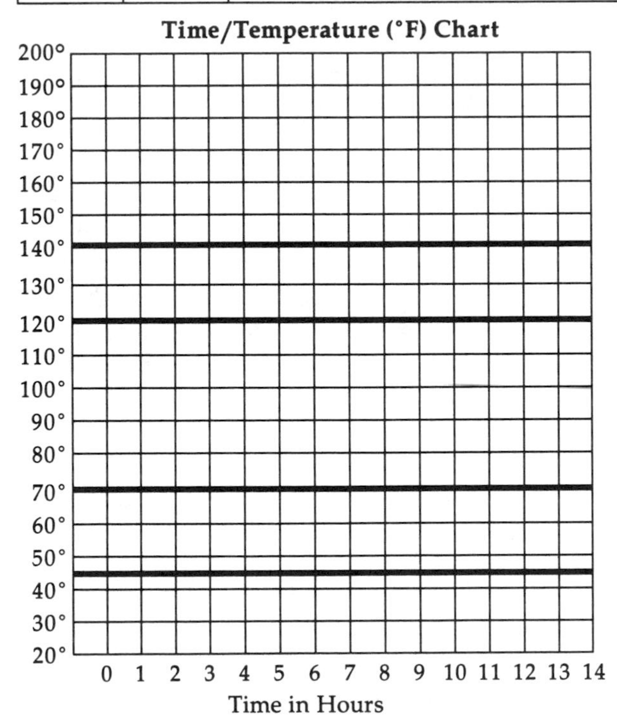

DOH-2615 (6/89)

Product Flow Chart

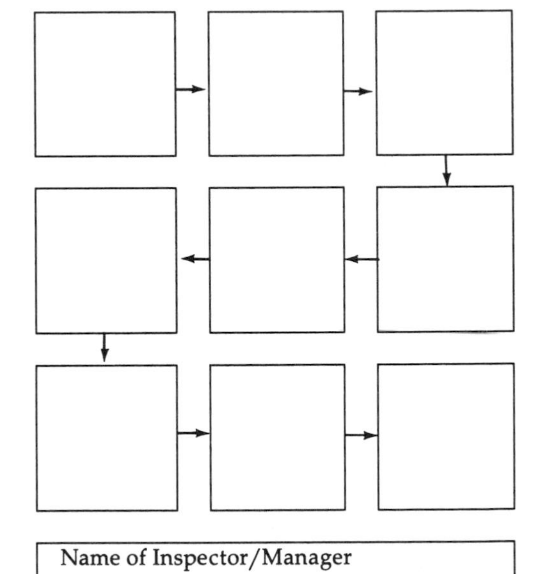

Name of Inspector/Manager

INSTRUCTIONS: Inspector may use this form to collect information for completion of the Hazard Analysis Critical Control Point Monitoring Procedure Report.

Source: New York State Department of Health

Exhibit 15.3 HAACP inspection form for monitoring critical control points

HAZARD ANALYSIS CRITICAL CONTROL POINT MONITORING PROCEDURE REPORT

COUNTY	DIST.	EST. NO.	MONTH	DAY	YEAR

Establishment Name ______________________ Operator's Name ______________________
Address ______________________ County ______________________
(T)(C)(V) ______________________ Zip code ______________________
Food ______________________

Critical Control Point	Priority	Process (Step)	Monitoring Procedure	Name or Title of Responsible Person
YES ☐ NO ☐		CONDITION AT DELIVERY	☐ Approved source (inspected) ☐ Temperature check less than 45°F ☐ Not spoiled ☐ Shellfish tag complete	
YES ☐ NO ☐		STORAGE	☐ Refrigeration: Product temperature less than 45°F ☐ Raw/Cooked/Separated	
YES ☐ NO ☐		DEFROST	Methods ☐ Under refrigeration ☐ Running water less than 70°F ☐ Microwave ☐ Less than 3 lbs., cooked frozen	
YES ☐ NO ☐		COOK	☐ Greater than or equal to 165°F ☐ Greater than or equal to 150°F ☐ Greater than or equal to 140°F ☐ Greater than or equal to 130°F How determined ______	
YES ☐ NO ☐		HOT HOLD	☐ Product greater than or equal to 140°F Temperature checks every ____ minutes	
YES ☐ NO ☐		COOL	120°F to 70° in 2 hours; 70° to 45° in 4 additional hours by the following methods: (Check all that apply.) ☐ Shallow pans ☐ Ice water bath and stirring ☐ Reduce volume by ______ ☐ Rapid chill refrigeration Temperature checks every ____ minutes	
YES ☐ NO ☐		PREPARE AND SERVE	Maximum total time between preparation and service ____ hours ____ minutes	
YES ☐ NO ☐		SLICE, DEBONE, MIX, ETC.	☐ Wash hands ☐ Use gloves, utensils ☐ Workers' health ☐ Wash and sanitize equipment and utensils ☐ Minimize quantity of food at room temperature ☐ Use pre-chilled ingredients	
YES ☐ NO ☐		REHEAT	☐ Rapidly heated to 165°F How determined ______	
YES ☐ NO ☐		HOLD FOR SERVICE	☐ Hot product greater than or equal to 140°F ☐ Cold product less than or equal to 45°F Temperature checks every ____ minutes	

Actions that will be taken when the monitoring procedures are not met: ______________________

I have read the above food preparation procedures and agree to follow and monitor the critical control points and to take appropriate corrective action when needed. If I want to change any monitoring procedure, I will notify the Health Department prior to such a change.

Signature of person in charge ______________________ Signature of inspector ______________________

Source: New York State Department of Health

Groups, the Association promotes food sanitation in cooperation with governmental, scientific, commercial, and educational institutions. The Association has a number of publications in the form of posters, pamphlets, and booklets available on the topics of sanitation, safety, pest control, the hazard communication standard, and employee training, among other significant topics for the foodservice industry. They also have videos available on back-of-the-house training, which covers sanitation and the flow of foods. The Association represents interests of the industry with its lobbying group in Washington, DC, and sponsors an annual conference and exhibition for the foodservice and hospitality industries.

The Educational Foundation of the National Restaurant Restaurant Association

The Educational Foundation of the National Restaurant Association is a not-for-profit foundation that provides educational services to the industry, including professional courses that comprise the "Management Development Diploma Program" for foodservice students, managers, and employees.

The Educational Foundation offers a comprehensive risk management series called SERVSAFE®. One part of the series promotes sanitation in the foodservice industry through the national food safety certification program, which is aimed at helping operators and managers train employees in food safety practices. SERVSAFE is a comprehensive program that provides training and testing, in addition to certification. The central component is this *Applied Foodservice Sanitation* coursebook. SERVSAFE includes employee manuals, a manager's kit, a video series, a color slide kit, posters and stickers, and a semi-annual newsletter, called the *INSTRUCTOR*, which gives relevant updates on topics covered in this coursebook.

Associations Active in Food Protection

National Environment Health Association (NEHA) is an organization of environmental specialists and professionals, including those responsible for food inspection services and environmental health programs. NEHA's Registered Sanitarian (RS) program has established a national standard based on education, experience, and testing, for sanitarian qualifications.

The International Association of Milk, Food, and Environmental Sanitarians (IAMFES) was a pioneer in the highly successful U.S. milk sanitation program. It also provides a publication on milk and food safety, as well as the generally accepted standard procedures on how to investigate a foodborne illness. The Conference for Food Protection works on issues submitted by members.

These three organizations, as well as the National Society of Professional Sanitarians (NSPS) and the American Academy of Sanitarians, are composed of food industry professionals and sanitarians concerned with environmental health protection policies at the national, state, and local levels. These organizations promote professional standards, recommend legislative policy, sponsor uniform enforcement procedures, and provide educational opportunities.

The Association of Food and Drug Officials

The Association of Food and Drug Officials (AFDO) develops and publishes food sanitation codes and encourages food protection through the adoption of uniform legislation and enforcement procedures.

The Council of Hotel and Restaurant Trainers

Foodservice trainers and human resource professionals have their own professional association, the Council of Hotel and Restaurant Trainers (CHART). CHART was formed in 1971 to provide members with a forum in which to grow professionally and to increase their effectiveness as trainers.

The Frozen Food Industry Coordinating Committee

The Frozen Food Industry Coordinating Committee has developed the *Code of Recommended Practices for the Handling of Frozen Food,* which describes procedures to be used from the processing stage to retail food service.

The National Pest Control Association

The National Pest Control Association (NPCA), consists of licensed and certified PCOs throughout the United States. NPCA provides guidelines and training materials for integrated pest management (IPM) treatment programs, for hazard communications, and for other topics relating to safety and sanitation.

NSF International

Equipment manufacturers who believe their equipment meets NSF *International* standards can request an evaluation from NSF *International,* formerly the National Sanitation Foundation, who will list the equipment if it meets their standards, and give it the NSF *International* mark. NSF *International* develops and publishes widely accepted standards for equipment design, construction, and installation, which are updated on an individual basis every five years. Listings of equipment that meets these standards are updated every six months. NSF's approach to determining the suitability of individual units of equipment for use in a food service exemplifies the kind of progress that can be achieved through voluntary programs on the part of the foodservice industry.

Underwriters Laboratories, Inc.

Underwriters Laboratories, Inc. (UL), performs a similar service, by listing equipment that meets NSF *International* standards. In addition, UL lists electrical equipment that passes their own safety requirements. (Chapter 9 discusses the role of NSF *International* and UL and details their standards for listing equipment.)

SUMMARY

Public and private organizations and agencies can offer valuable assistance to the foodservice manager or operator in meeting their commitment to food safety. It is up to the manager to use the help that is available, and to maintain a safe and sanitary food service.

Federal governmental agencies create and administer regulations and standards that affect the foodservice operator directly or indirectly. The model ordinance developed by the Food and Drug Administration (FDA) serves as a guideline for a large number of state and local regulations. In addition, federal authorities regulate the purity and safeness of foods in interstate commerce.

While state and local health codes vary throughout the nation, virtually all of them contain provisions governing food safety; personal hygiene; sanitary facilities, equipment, and utensils; safe operating practices; training; and enforcement procedures. The *sanitarian*, a representative from the federal, state, or local department of health, is a professional in sanitation and public health.

One highly effective way to produce safe food is to use the HACCP food safety system and to follow a daily program of self-inspection. The effectiveness of this system will be reflected in inspection reports. HACCP-based inspections, a departure from the traditional emphasis on facilities, are being adopted by industry and some regulatory agencies. Inspectors observe the way an operator receives, stores, prepares, and serves food, and may designate critical control points for each step. A positive approach to dealing with public health officials is always best.

Professional and trade organizations in the fields of food service and public health study business activities, investigate sanitation problems, and recommend guidelines for foodservice practices, equipment, and facilities. Professional associations also develop educational programs in sanitation and conduct research into the causes and prevention of foodborne illness.

A CASE IN POINT

Carolyn, the registered sanitarian from the city's public health department, was standing at the door of "Jerry's Place." Jerry greeted her and they both walked to the kitchen. Carolyn pulled out a HACCP worksheet. While Jerry explained that the cook was preparing a special of stir-fried chicken and vegetables, Carolyn noted the ingredients and their sources. Then she observed the preparation procedures and noted times and temperatures, which she plotted on a time/temperature graph. She also filled in a product flowchart and indicated the critical control points for the chicken special. Jerry told her what corrective actions his employees were trained to carry out if the standards of the critical control points were not met.

Next, Carolyn checked the concentration of the sanitizing solution in the three-compartment sink that Jerry's dishperson used for manual cleaning, rinsing, and sanitizing of equipment. She also checked the handwashing station for the food-handlers.

Jerry and Carolyn went to Jerry's office where they discussed the report. Jerry compared his flowchart for the stir-fried chicken and vegetables with the one Carolyn had done, and determined where changes could be made to improve the monitoring system.

Did Jerry handle the inspection correctly?

What does Jerry need to do following this inspection?

STUDY QUESTIONS

1. What are the two basic systems of guidance and control that work to guarantee the sanitary quality of food in this country?
2. Briefly describe the functions of the FDA with regard to food protection as described in this chapter.
3. Which governmental agency is responsible for inspecting and grading meat and poultry shipped across state lines?
4. Which level of government is more likely to *directly* affect foodservice operations—federal or local?
5. How do state and local foodservice regulations and standards differ from the federal ordinance?
6. What is the best way for a manager to prepare for an inspection from the local public health department?
7. What should a manager do during an inspection?
8. What are the two kinds of inspection reports?
9. What should a manager do following an inspection?
10. How would a health agency use the principles of HACCP to decide on the frequency and type of restaurant inspection used?
11. Name some of the significant industry organizations that help managers deal with sanitation regulations and standards, and employee training.

ANSWER TO A CASE IN POINT

An inspection serves as a confirmation of how well a food safety system is working. An ongoing sanitation program is the best way to receive a good inspection report and to verify the effectiveness of management's self-inspections. It is obvious in this case that Jerry and his employees have a food safety system in place.

Jerry knew that stir-fried chicken and vegetables are a potentially hazardous menu item and he had already devised a flowchart to establish control points for monitoring. He was cooperative with the sanitarian and worked with her to improve his existing system.

The next thing that Jerry needs to do is to continue conducting his own self-inspections and monitoring his operations procedures to test the effectiveness of the control measures. If corrective actions occur frequently, he may need to retrain some of his employees on how to follow the new procedures. He also needs to help them understand why the new way is better and to make a commitment to the new procedures. Jerry knows that the sanitation program is a continuous day-to-day priority and not something that is stepped up for inspections and then neglected.

MORE ON THE SUBJECT

NATIONAL RESTAURANT ASSOCIATION. *Current Issues Report: Non-uniformity of Regulations and the Foodservice Industry.* Washington, DC: The National Restaurant Association, 1986. #CI400. 8 pages. This pamphlet helps operators and managers understand the various regulations that affect their establishments.

NATIONAL RESTAURANT ASSOCIATION. *Sanitation Self-Inspection Program for Foodservice Operators.* Washington, DC: The National Restaurant Association, 1983. #MG869. 32 pages. This booklet contains 23 different inspection checklists that cover food safety and more.

U.S. DEPARTMENT OF HEALTH AND HUMAN SERVICES. *Food Service Sanitation Manual.* Washington, DC: Food and Drug Administration, 1978. 96 pages. This document contains a "Model Food Service Sanitation Ordinance, with 1976 recommendations from the FDA." It covers all aspects of sanitation, including plan reviews, inspections, and remedies for problems. It has useful appendices and an index.

SHERRY, JOHN E. H. *Legal Aspects of Foodservice Management.* Chicago, IL: The Educational Foundation, 1984. 334 pages. Chapter 2 contains information on the various regulatory agencies that affect foodservice operations and the related implications for managers.

16

Employee Sanitation Training

Training is teaching employees how to do a job properly. In your foodservice operation, it is up to you to train employees in a wide range of areas and to include sanitation as a key element in the overall training process.

The purpose of sanitation training and supervision is to provide employees with the knowledge, skills, and attitudes necessary to follow the policies and procedures of the food safety system you have established. The final responsibility for sanitation rests with the foodservice manager, but if employees are not familiar with sanitation and do not know how to use HACCP to maintain food safety, the manager's job of protecting the food is impossible. Sanitation training for all employees allows the manager to spend time on other matters important to the running of the foodservice establishment.

Employee training is also an important factor in every operation's financial statement. Necessarily, the foodservice manager must evaluate any facet of the business by asking, "How does this item contribute to my profits?" Training employees in safe and sanitary foodhandling practices may require time away from regular tasks for both employees and managers. It may also require the selection and use of professional trainers, videos, slides, films, books, and posters. Good sanitation training will have a positive return on investment in the long run. Benefits from good training include avoiding:

- The costs related to an outbreak of foodborne illness, such as doctors' and hospitals' bills, laboratory fees, and legal fees
- The loss of reputation and revenue sustained when an establishment is forced to close by the public health department after not meeting sanitation standards.

Employee morale is an important benefit of sanitation training. Most people want to do their jobs right. Once trained in sanitation procedures and then properly supervised, the payoff in good sanitation practices and employee pride is a *dual* benefit. Training can also help to reduce employee turnover, and increase customer appreciation. Both employees and customers who know that management has a commitment to sanitation will be more satisfied.

In order to derive these benefits from the training investment, the program must be carefully thought out, well-executed, and continually reinforced and evaluated. In this chapter, we will discuss:

- Assessing and analyzing training needs for sanitation
- Planning, executing, and reinforcing the training program
- Evaluating the success of the training effort.

ASSESSING TRAINING NEEDS

As the manager of a food service, it will be your responsibility to closely assess your operation's training needs and decide which areas require attention. In Chapter 5, you learned how to establish critical control points. Use those control points to target the steps that employees need to complete to ensure food safety. Establish standardized job descriptions and procedures covering each control point. If employees are *specifically directed in writing,* such as when and how to wash, rinse, and sanitize utensils, or to use separate utensils for raw and cooked products, they are instantly provided with training information that allows them to both perform their jobs effectively and to maintain food safety.

Determining the Subject

Exhibit 16.1 shows the steps to set up an employee sanitation training program.

Exhibit 16.1 Steps in planning a sanitation training program

1. Determine the specific subject to be taught that is appropriate to the audience.
2. Establish learning objectives.
3. Choose the method for training based on the subject and audience, and purchase necessary materials.
4. Select an instructor.
5. Schedule training sessions.
6. Select the training area that is appropriate to subject for each session.
7. Prepare the trainer for conducting each session.

The first step is to determine the specific subject matter to be covered. These subject areas should be classified into two groups: (1) those that all employees need to know; and (2) those that are unique to each job. For example, everyone needs to know the proper way to wash their hands, but the cooks, and not the buspersons, will need to know the entire list of steps used to clean the grill.

The *fundamentals of sanitation*—knowledge, skills, and attitudes—must be stressed for all new employees as part of each initial crew orientation on their first day on the job. Sanitation fundamentals include:

- The importance of sanitation in preventing foodborne disease
- Good personal hygiene practices that every employee follows before coming to work and during work
- Food protection, including proper foodhandling procedures
- Facilities and equipment, including special cleaning and sanitizing procedures required in particular areas and for specific equipment
- Habits and practices that promote on-the-job safety and help prevent accidents
- Proper methods of handling hazardous materials, in compliance with the OSHA Hazard Communication Standard (HCS).

After the general fundamentals are explained to all employees, specific sanitation procedures related to each job position should be analyzed by the manager and given to individual employees. *Task analysis* is breaking down a job position into its individual duties. For instance, task analysis for servers would identify all procedures necessary for the sanitary service of food (see Exhibit 16.2).

Establishing Learning Objectives

Defining expected outcomes and benefits is essential in order to determine and

Exhibit 16.2 Sample task analysis for table service employees

Sanitation-Related Tasks for Table Servers
(Teach individuals by demonstration)

- Remove all jewelry and change into a clean uniform in the employee locker room.
- Put on a hair restraint.
- Wash hands following the steps posted above the handwashing sink.
- When preparing beverages, use the stainless steel scoop to put ice in glasses. Only touch the bottom of glasses and the handles of cups or mugs. Put glasses on a tray to take out to the table.
- Wash hands before preparing salads or putting garnishes on plates. Pick up plates on the bottom or by the edges and carry out to table on a tray.
- If you remove soiled dishes from the dining room, wash hands immediately after putting the dishes on the soiled dish table.

establish the training program's objectives. These sanitation training objectives must be compatible with larger management goals, one of which is the serving of sanitary and wholesome food.

Set up objectives by analyzing the level of training that needs to be conducted and how it can be accomplished. In addition, you will need to evaluate the present skill levels and abilities of the people who are to receive the training. You will need to write up objectives that fit the needs of your particular foodservice operation and your particular group of trainees. For example, although each employee in a foodservice operation needs to receive training in sanitation, it would be useless to train at a highly technical level, or to provide specialized training to those who will never use it; however, if an employee is expected to work more than one job, he or she needs the objectives for both jobs.

Once objectives have been decided upon, they need to be clearly stated in writing. *Behavioral objectives* define specific actions the trainee will be able to do on completion of training. Objectives stated in behavioral terms are essential to successful sanitation training, since employee performance is the purpose and measure of training. When preparing objectives for a particular task, the words used should be behavioral words, for example, *operate* or *practice* rather than *be aware* or *notice*. Exhibit 16.3 shows sample sanitation objectives for a cook.

PLANNING THE TRAINING PROGRAM

Some large foodservice operations set up a separate training department and have a training director on staff. While a training department can be extremely effective in some operations, the manager alone can also provide an effective program through careful planning.

One of the most important roles of a manager is to set an example by his or

Exhibit 16.3 Sample sanitation objectives for a cook

After sanitation training the cook will be able to:

General

- Explain the fundamentals of sanitation and understand why it is necessary to learn sanitation principles and procedures.
- Understand how foodborne illness is caused.
- Practice good personal hygiene to avoid contamination of food and food-contact surfaces.

Specific

- Follow procedures to prevent foodborne illness during food preparation, storage, and serving using critical control points in the HACCP system of food safety.
- Use steps to avoid cross-contamination during foodhandling duties.
- Use a bi-metallic stemmed thermometer to measure internal temperatures of food during preparation.
- Follow all the necessary steps to clean and sanitize deep fryer, grill, and ovens.

her behavior and attitudes for all employees to follow. Although this might not be considered a method of formal training, it is a critical starting point. Setting a good example forms the backbone of an effective training program.

The ideal method of handling training needs for employees already on the job is to set up a program that is ongoing. Initial training of new employees followed by continual monitoring of progress is the best approach when it comes to teaching sanitation. The reason for this is simple: It is important to develop good habits early.

Training will not be effective unless it is organized. A *training program* is a structured sequence of information and activities that lead, step-by-step, to learning. The following sections present information on organizing available tools and techniques into your sanitation training program.

Understanding the Process

For knowledge to be applied on the job it must be *learned.* Training is really a two-step process. It requires that information is presented by one person, the trainer, and is then transferred by the learner into applied skills.

The trainer or manager must be knowledgeable about proper sanitation procedures and the reasons behind them. Once the basic knowledge is there, successful training is based on the interpersonal skills and the preparation of the trainer.

The learner must be motivated. The learner must also be encouraged and given feedback. *Feedback* is either constructive criticism given to an individual to correct a mistake, or praise given to reinforce proper performance of a skill or procedure. Finally, *follow-up* by the manager or supervisor is necessary to ensure compliance with sanitation procedures.

Training is a process that includes acknowledgment of each individual's learning skills and background, because different people learn at different rates. Training should be done in steps that include demonstration, practice, and feedback—particularly praise. Exhibit 16.4

Exhibit 16.4 Learning flow chart for training

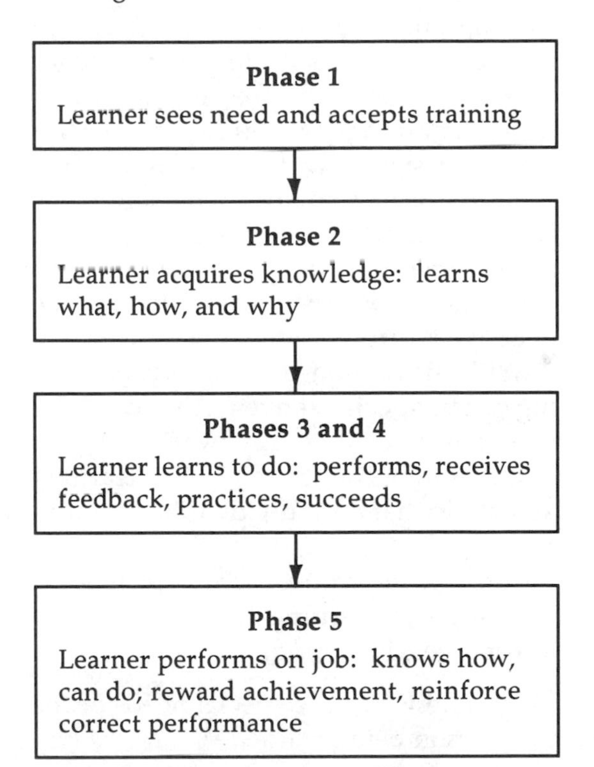

Source: *Supervision in the Hospitality Industry* by Jack E. Miller and Mary Porter. Copyright © 1985 by Jack E. Miller and Mary Porter. Reprinted with permission of John Wiley & Sons, Inc.

breaks this learning process down into four phases.

Choosing Methods

Methods of training can be divided into two techniques: (1) one-on-one; and (2) group training. Which technique to use for your training program will depend on the number of people that need to be taught, what the cost is, and more importantly, how the trainees are able to learn the subject quickly and thoroughly.

One-on-One Training

One-on-one training is efficient when a new employee needs to be trained or when one or two present employees need retraining. Sometimes a new recruit is assigned to the employee being replaced or to some other person in the same work area. This method is called *magic apron.* For example, a new assistant cook might be assigned to follow the head cook through the daily routine.

With the one-on-one technique, training can be specialized to fit the particular situation and the employee's job. By taking into account the level of learning and skills that an individual has prior to training, more or less time can be devoted to teaching each step or skill. This type of training has a number of advantages, including:

1. It allows for the special needs of the person being trained
2. It can take place on the job, so a special or separate location is not needed
3. It enables the trainer or manager to monitor employee progress
4. It provides the employee with immediate feedback
5. It directly involves the trainee, which makes the subject more interesting and the applications of the subject are readily demonstrated.

There are also some disadvantages. Although magic apron and one-on-one are widely used in foodservice establishments, it is possible for the new employee to pick up the experienced employee's bad habits along with the good. The new employee has no basis for differentiating sanitary practices from unsanitary ones, so this method is effective only if the employee doing the training has already received sanitation training that he or she can pass on to the new employee.

Group Training

If more than a few employees require sanitation training, holding group sessions may be a practical alternative to one-on-one training. The advantage is that a uniform program can be adapted and used for all training needs. Group training can include lectures, discussions, and role-playing sessions. This method also allows for the use of videos or other visual aids, that can be valuable tools to spark interest and relay information. Group interaction, question-and-answer sessions, and exchange of knowledge are all useful training tools.

One disadvantage of group training is its lack of personalized instruction. Trainees may have to learn more information than they actually need for their jobs, or those that learn at slower rates or that have fewer skills prior to training may not pick up all the material that is presented. This is because group sessions must usually meet the overall needs of the employees, not individual needs. Another disadvantage is that group sessions can be dull unless various techniques, such as videos, films, or role playing, are used.

Crash training, or an intensive effort to accomplish a lot in a short amount of time, is not an effective method because it is rarely well-planned and the results are short-lived. Most employees, regardless of their learning level, cannot absorb all the necessary sanitation knowledge they need in a single session.

Applied Training

The training method selected should be dictated by the behavior that is desired. If the manager wants an employee to wash, rinse, and sanitize the salad preparation counter, the best option is to demonstrate the process and then have the employee perform the task. Demonstration allows for immediate feedback. The manager can explain the key points of a task while performing it, then observe the progress as the employee practices the same task.

Job aids, such as posters that illustrate all the necessary steps in certain procedures can serve as ready references and reminders. For example, a poster that is hung up over the sink showing each step of manually washing, rinsing, and sanitizing pots and pans is an appropriate aid for an employee who is being trained in dishwashing procedures.

When the manager wants all the employees to recognize the connection between the time-and-temperature principle and foodborne illness, he or she can present to the group a slide show that includes diagrams of bacterial growth, and give a brief lecture on the everyday hazards to safe food.

For many operations, a mix of several training methods is useful. For example, a program that combines general sanitation training for servers might include:

1. Viewing a video on foodborne illnesses and foodhandling hazards that contribute to disease, which is followed by a group discussion with a question-and-answer period
2. Practicing sanitary procedures for serving foods and beverages after the trainer or manager has demonstrated the correct way to do it
3. Posting a written list on safe foodhandling techniques for serving and holding food before serving at a server station in the kitchen
4. Following-up by monitoring and supervising server trainees, and giving them reinforcement on the job with praise or other incentives, such as a certificate or pin they can wear.

Choosing Materials

It may be possible for an especially gifted instructor to use words alone to describe the fundamentals of foodservice sanitation. However, the fact that words have been spoken or written does not guarantee that anything has been learned. Additional instructional aids that incorporate written texts or workbooks, drawings, charts, posters, graphs, films, dramatizations, videotapes, slides, and transparencies raise the level of training effectiveness. If these aids are supported by active participation in written or oral exercises, and with on-the-job experience, the level of effective training is even higher.

Proper use of training materials saves time, adds interest, helps trainees to learn and remember what they have learned, and makes the manager's job easier. In order to use materials to their best advantage, the manager or trainer needs to read all accompanying instructor guidelines or suggestions that the distributor includes with the material.

The manager or trainer should be guided by the "three As" when making choices on which materials to use. To be useful, training materials must be:

1. *Accurate.* They must be factual, up-to-date, and complete. To ensure accuracy and authenticity, training materials should be purchased from professional or educational organizations, or an authority in foodservice training or the foodservice sanitation field.
2. *Appropriate.* They must be suitable for the purpose they are to serve and for the instructors and trainees who are to use them. Language and reading comprehension levels of the trainees need to be acknowledged and matched for written and presentation materials. Different materials may be necessary for a day shift of adults and a night shift of teenagers. If an English language barrier is a problem, training materials in other languages are available from different sources, such as regulatory agencies, The Educational Foundation, the National Restaurant Association, and other professional organizations. It is also necessary to determine if the materials are suited to the abilities of the instructor or manager who will conduct the program. If the instructor has a soft voice, a microphone might be used during a lecture or, if the instructor is not a public speaker, a videotape with narration could be an alternative to a lecture. The possible limitations of the training location will also determine training materials. If a videotape player is not available, then videotapes cannot be used.
3. *Attractive.* You cannot teach anything to people without first gaining their attention. In subjects that do not generally have a wide appeal (and sanitation can be one of these), it is often easier to make information exciting, interesting, and memorable with eye-catching audiovisual aids.

No single kind of training material can do the job effectively for all trainees at all levels. Many companies have developed written employee handbooks that are distributed to new employees and from which employees are expected to gain necessary information. These handbooks can be useful in aspects of orientation and as a guide to company policy.

Audiovisual programs and aids, ranging from videos to posters, are used widely for training in the industry, but do not constitute a *total* training program. The learner cannot be motivated solely by a film or poster. Rather, the manager should introduce each with a statement of objectives and expectations and discuss the presentation's content as it applies to the particular operation and to the employees involved. Practical demonstrations are vital for learning many sanitation procedures and should be used in addition to audiovisual aids.

Finally, a brief written, oral, or practical examination should be given to the trainees and scored by the manager. This exam can be one that has been prepared by the supplier of the training materials, or that comes from an educational/training organization, in order to make sure that it is validated and will accurately reflect the level of comprehension of the employees taking it.

Mistakes should be discussed and corrected immediately after the exam is scored. To follow-up, the manager can use signs, posters, bulletin boards, and pay-envelope stuffers as reminders or refreshers associated with sanitation training and proper sanitation practices.

Selecting an Instructor

A manager knows the operation best and is ultimately responsible for the sanitation in the establishment, making him or her the most likely employee trainer. If the manager chooses not to personally conduct the training, an instructor should be selected based on his or her knowledge of the subject of sanitation, familiarity with the organization's sanitation problems, skills in human relations, and proven ability to help others learn.

The selection of a trainer can be influenced by where the training takes place, either on-site in the foodservice establishment itself, or in a formal classroom situation at some separate location. A list of individuals other than the manager who can act as a trainer includes:

1. *The trainees' immediate supervisor.* Many times this person is a logical choice as trainer due to the working relationship between employees and supervisor and because of the possibility for immediate on-the-job feedback for the trainee.
2. *Staff trainer.* In large foodservice companies, a training professional is often available to provide sanitation instruction.
3. *Representative of the health department.* A local sanitarian and a public health educator are possible candidates as instructors in sanitation training. The district sanitarians of the state health department may assist in setting up training sessions if their schedule allows.
4. *Representative from a professional or educational organization.* These organizations and associations, like The Educational Foundation, often provide trainer preparation courses for operators and managers, and also have staff instructors that conduct courses on-site or at their own locations.

Very often, the most successful training is a result of the combination of these resource people, with one person, usually the manager, acting as the senior instructor and other outside instructors called in to present special concepts or to provide new perspectives. It is important that the trainer presents the principles of sanitation in a concrete form in situations that are sure to arise as the trainees perform their regular jobs.

Scheduling Sessions

Scheduling sanitation training for foodservice employees is essentially a matter of deciding priorities. Which employees should be trained first? Should it be a new employee, or an employee being given new responsibilities? Should it be a group of employees, such as servers who have never received formal training in sanitation? Or should it be all members of the staff, to call renewed attention to general health and safety rules?

A master training schedule with blocked-out time slots can be useful in determining priorities and in showing both the company's and the manager's commitment to the program. Both orientation and retraining should be included in the schedule. Special priority situations, such as a large group of new employees, an opening of a new restaurant, a reopening after remodeling, or preparations for a convention banquet will require adaptations to the master training schedule.

Training sessions should not be too long. The ideal length is probably about 30 minutes, but successful training sessions have lasted as long as an hour and a half, or ten minutes, depending on the types of activities conducted.

Written notices to announce each training session should indicate the date, time, and place; the subject; which individuals are to attend; and the name of the trainer who will conduct the session, if other than the manager. The manager needs to make sure that these announcements are posted where they will be seen easily and read by the employees. The trainees should also know that each session will begin on time and that absences must be approved by the manager prior to each session.

Training sessions can be scheduled at slow times during operation, before opening, after closing, or on a day that the restaurant is normally closed. If there is an area specifically for training purposes, scheduling can be more flexible.

Selecting the Training Area

Training should be conducted in an atmosphere in which employees feel at ease. On-the-job locations are typical for most sanitation training classes for employees below the managerial level. An on-site location easily allows for demonstrations of procedures, and provides opportunities to relate instruction to the specific needs of your particular foodservice operation. In most foodservice operations, an employee lounge, an executive office, or a separate section of the dining room not in use can serve adequately for group training sessions. The area chosen should be free of distractions, for example, other employees working or taking their breaks.

Part of the training should include practice, so employee work stations could be used as training areas, as well as the kitchen, particularly for one-on-one sessions.

The group training area should be an appropriate size so no one feels crowded. There should be a place for all participants to write or take notes. Seating should be adequate and comfortable, and arranged so as to encourage the exchange of information in open discussion. If a projector, sound equipment, or a videotape player is to be used, the room must have enough electrical outlets to accommodate the equipment and a screen to view the video, film, or slides. If visuals are to be used, it should be possible to darken the room so trainees can see better.

Implementing Training

In any training, it is important to take a straightforward approach. For the best results, keep the training simple. The following steps break the process down into manageable units:

1. Tell employees what they are going to be taught and why it is important to learn it.
2. Present the material to them.
3. Demonstrate steps and procedures.
4. Answer their questions.
5. Let them practice.
6. Give them feedback on their performance.
7. Review the material that you just taught and demonstrated.

For training to be successful, you must be comfortable with any questions employees might have. If you are uncertain of your subject matter, you will communicate this fact to your trainees, and they will have cause to doubt the accuracy of what you are saying.

Preparing the Trainer

Training employees usually requires some skills that are not developed overnight. The following are suggestions for developing communication skills and conducting successful training sessions:

1. Be sure you know the fundamentals of sanitation and the reasons why they are important. Also know all the procedures necessary in sanitary practices. Choose training techniques with which you feel comfortable. For example, if you will be lecturing, make an outline of your material that you can refer to, and practice your presentation until it feels natural to you. If you are using overhead slides, you might want to have them framed so that they will not stick together or get out of order. Make sure the room is ready and that all visual aids or tools are in working order and easy for you to use without interrupting the flow of the presentation.
2. During your presentation, maintain eye contact with the listeners.
3. Keep your delivery conversational and informal to achieve a comfortable atmosphere.
4. Vary the tone of your voice for emphasis and do not speak too quickly. A moderate rate will allow trainees to take notes and ask questions.
5. Use simple language. Unnecessary technical terms will cause listeners

to tune you out, so speak on their level.

6. Treat all questions and comments made by the trainees seriously by answering them quickly and in a straightforward way. You can give them feedback as well by saying, "That's a good question."
7. Ask questions often to allow trainees to put concepts into their own words. Give trainees enough time to fully answer questions.
8. Look for cues that may indicate the employees are not picking up the information or that they are bored and losing interest. These cues could include obvious fidgeting, looking at their watches, doodling on their notes, and not asking or answering questions. If slides do not hold your audience's attention in one session, you can use a videotape for the next one, and try other methods until you find the one that does hold their interest.
9. Keep the sessions short. You expect your trainees to be on time for scheduled sessions, so respect their free time by ending when you said you would.
10. Keep the training as related as possible to sanitation and the workers' specific jobs. If the range of subjects is too broad, people might not learn the procedures they need to know.

Motivating Trainees

In addition to establishing a relationship between the material you are teaching and the workers' actual jobs, you must also find a way to motivate the group to want to learn what you are teaching. It is the job of the trainer to capture and keep the attention of the group and to get them to remember the material. Here are some guidelines to follow in order to achieve this:

1. Use entertaining training materials. Introduce variety in the program by varying the materials you use. If you give a lecture with slides for one session, have the trainees role play or do demonstrations for the next, and then watch a video for the session after that.
2. Make each trainee feel that you are talking to him or her individually and that you have a real interest in whether the needs of each member of the group are being met. Tell trainees that you are interested in what they have to say. Letting employees know that training is a company priority will increase your program's chances of success.
3. Recognize employee achievement throughout the training program. Employee wall charts that track individual progress through the training course, and certificates of completion can serve as incentives.
4. Maximize employee participation during sessions.

Do not let teaching aids limit you in your approach to the material for each group session, be creative. You can put questions and answers in a game show format, creating teams and then having them compete against each other to give correct answers for points or small prizes. You can have one trainee draw sanitation terms or procedures on a chalkboard or flipchart while the rest of the group guesses what is drawn. If the group is not interested in drawing, they can act out the words and procedures. Or you can tell the trainees to write a rhyming riddle or song that describes foodborne illnesses or micro-organisms and have them read or perform them for the rest of the group to guess.

For more specific areas, such as preventing contamination of potentially hazardous foods, you can bring in a bulk recipe and divide the trainees into small sections to identify the procedures where the opportunity for contamination exists, the places where time-and-temperature violations could occur, and where instructions are unclear. Have the trainees identify critical control points and revise the recipe to include the food safety system.

EVALUATING TRAINING OUTCOMES

To *evaluate* the sanitation training program, the manager must carefully examine and judge the end results based on the pretraining objectives. The process of evaluating is important because it forces an examination of both the benefits of training to the employees and to the foodservice operation. Evaluation balances these benefits against the cost of the training in time and dollars.

Evaluation also requires that the manager or trainer answer two questions:

- "Did the training produce results on the job?"
- "If intended results were not produced, why not?"

Failure to answer these questions means that the manager will never know whether knowledge was learned but never applied, learned but applied incorrectly, or not learned at all.

One approach to evaluation is *objective,* which means facts are presented through the use of tests, both during and after training, and on the job. Another approach is observing performance through supervising and monitoring during the training, and as a continuing part of each manager's and supervisor's job. Evaluation works best when both approaches are used.

Objective Measurement

Objective measurement of individual progress can involve a variety of testing tools, but do not use tests as a threat to trainees. Tests and study questions are included in most professional training materials. If these are not available from your supplier, check with other organizations that provide professional education, or the instructor or trainer can create his or her own questions, geared to the material covered and the audience.

Written and oral quizzes and tests, used as either a formal classroom activity or an informal learning aid, will allow managers to determine the level of knowledge and material that employees achieved after individual training sessions. By evaluating the number scores, a manager can quickly assess not only the employees level of achievement, but if some areas of importance need to be reviewed.

Research by learning-theory psychologists stresses the importance of providing trainees with test results at the earliest practical opportunity after the test. Telling a trainee immediately after the test how well he or she responded can actually improve the trainee's comprehension of the material on the test, meaning that they will retain this information. The faster the trainee is notified that the answer is correct or what the correct answer is, the better the learning. While tests can reveal areas where comprehension is high and low, tests do not measure how trainees apply their knowledge into usable skills.

Measurement Based on Performance

The real measure of training in sanitation is the demonstrated ability of a foodservice employee to do his or her assigned tasks on the job. Evaluating the effectiveness of training is part of the responsibilities of each supervisor and manager. The supervisor in the kitchen, or the manager in the front of the house, can arrange for an employee's new skills to be observed during his or her work shift. The supervisor has a direct person-to-person contact with each employee, so a supervisor knows firsthand the training needs and successes and can report the employee's progress to the manager. Recently trained employees should be given ample opportunity as part of their day-to-day duties to perform what they have learned, so that both supervisors and managers can observe, monitor, and supervise their actions.

Praise is a part of all learning experiences and should follow the evaluation process. Praise indicates to employees that they are performing well and have been successful in the training program. Generous praise reinforces employees' positive attitudes about their job. When they see that management values their best effort, they will continue their good practices.

SUMMARY

One of the foodservice manager's most important responsibilities is to train employees in the principles and practices of foodservice sanitation. Effective training plays a significant role in protecting the public from foodborne illness.

The benefits from a properly designed and conducted training program are demonstrated by food safety, employee morale, and the bottom line. Sanitation training is necessary and needs to be a priority in all foodservice establishments. Assessing your sanitation training needs begins with examining your critical control points and writing job tasks that incorporate food safety procedures.

Year-round training programs on an individual or group basis for both new and experienced employees are essential for running a successful, quality foodservice establishment. Setting a good example is a primary role for the manager before, during, and after the training program is implemented.

The first step in establishing a sanitation training program for employees is to determine the specific subject matter to be covered. These subject areas should be classified as general for all employees to learn, and specific for a particular position. In order to determine the training program objectives, the manager needs to define the expected results. These objectives form the standards of achievement for the program, that can later be used to evaluate its effectiveness.

Training is a process that includes acknowledgment of each individual's learning skills and level. It should be comprised of demonstration, practice, and feedback to the trainee. *Feedback* is praise or constructive criticism based on the employee's performance. Follow-up by the manager through monitoring and supervising is a necessary step in the training process.

Training can be conducted using the one-on-one method or group training, depending on the number of trainees, the costs, and the priority of the needs. The length of time and location for training sessions will depend on the subject being taught and the audience. The trainer or manager needs to be prepared. The learner must be motivated. A combination of training techniques and materials is the best approach for sanitation training.

Two methods are generally used to evaluate the outcome of training. One is objective, using tests or quizzes to measure the level of trainee comprehension. The other is employee performance on the job observed by the supervisor or manager. Praise is an important aspect of learning and should follow evaluation of the program to reinforce the good sanitation practices of the trained employees. Wall charts that show daily or weekly progress, certificates, and pins to wear can all serve as incentives for employees to succeed.

A CASE IN POINT

Paul has two employees who must be trained in sanitation. One is a cook who used to be a busperson and needs to be retrained. The other is an inexperienced server. Paul maintains an ongoing sanitation training program, so he has already trained the rest of his employees and evaluated the effectiveness of his original program. Both of these employees require specialized training. The cook received general training in the fundamentals of sanitation and their importance about three months ago, but the server has no idea what sanitation is.

Paul decides to have the kitchen supervisor train the cook using the one-on-one method, so the supervisor can demonstrate procedures and the cook can practice them and receive immediate feedback. By relying on the supervisor to train, Paul can have both the cook and server trained at the same time. After using his standardized job description for the server to write out learning objectives, Paul sets up a sanitation videotape in his office for the server to watch. Then he goes to the back of the facility to talk about an expected shipment with a clerk. After 30 minutes, he returns to his office to give the server a test, and then goes to the kitchen to observe the supervisor and cook. Noting that everything looks good in the kitchen, he goes to the dining room.

Do you think Paul's training program for the cook and server will be effective? What else does Paul need to be doing?

STUDY QUESTIONS

1. What are the benefits for a foodservice operation that invests in a year-round program of sanitation training?
2. Why should training objectives be written in behavioral terms?
3. What classifications should be considered in determining the subject matter for an employee sanitation training program?
4. How are learning objectives established?
5. What are two training methods mentioned in the text and the advantages and disadvantages of each?
6. List three guidelines for choosing sanitation training materials.
7. What information should be included in a notice announcing a training session?
8. Describe the physical requirements for the location in which a training session will be held.
9. What is feedback and why is praise an important part of training?
10. Name four methods for motivating trainees.
11. How are training objectives related to training evaluation?
12. What two approaches are available to a manager in measuring the outcomes of sanitation training?

ANSWER TO A CASE IN POINT

Ongoing supervision and feedback throughout the training process corrects problems in performance and assures employees that they are doing a good job. While Paul got off to a good start, he didn't follow up on his efforts. The videotape on sanitation that the server watches may be packed with information, but by not stopping the tape to explain concepts or terms or to allow the server to ask questions, the server may not pick up or retain the knowledge presented. Paul needs to know that learning is better accomplished by interaction. The test he gave the server can be a good tool to measure the level of learning, but it will be more effective if he corrects it as soon as the server is done, and then discusses each question with the server.

Since the cook was given general training three months ago, a refresher session on sanitation would help reinforce its importance. Paul should also discuss the progress of the cook with the supervisor on a frequent basis to determine if more training is necessary.

Paul needs to present each trainee with some form of recognition at the completion of their training, such as a certificate or button to wear, which will let them know they are doing their jobs well and that management values the training program.

MORE ON THE SUBJECT

THE EDUCATIONAL FOUNDATION OF THE NATIONAL RESTAURANT ASSOCIATION. Chicago: The Educational Foundation of the National Restaurant Association, 1991. The Foundation offers the following employee sanitation training aids:

Serving Safe Food: A Guide for Foodservice Employees. Written for foodservice employees, this easy-to-read guide reinforces the key principles of foodservice sanitation, such as personal hygiene, receiving and storing food, safe foodhandling, and cleaning and sanitizing. Includes learning exercises and a certificate of completion.

Serving Safe Food Video Series. This four-part employee training program includes "Introduction to Food Safety: Employee Health and Hygiene," "Safe Foodhandling: Receiving and Storage," "Safe Foodhandling: Preparation and Service," and "Cleaning and Sanitizing." Each comes with a leader's guide containing learning objectives and questions for discussion.

ServSafe Manager Kit. This kit provides a comprehensive training package for foodservice managers and includes a step-by-step leader's guide, five *Serving Safe Food* guides, five certificates of completion, an employee wall chart for tracking progress through the training program, and nine posters for strategically placing in a foodservice operation.

LAGRECA, GENEVIEVE. *Training Foodservice Employees: A Guide to Profitable Training Techniques.* Chicago, IL: The Educational Foundation of the National Restaurant Association, 1988. 182 pages. This textbook helps managers train their employees for peak performance and how to maintain performance standards after training. This book is part of The Foundation's "Management Development Diploma Program."

MAGER, ROBERT F. *Preparing Instructional Objectives.* Belmont, CA: Fearon Publishers, 1975. 136 pages. This is a must for every trainer or manager and a cornerstone book in educational technology. In easily understood and interesting language, the book demonstrates how to define teaching objectives, how to state them clearly, and how to describe criteria by which to measure success.

RAE, LESLIE. *How to Measure Training Effectiveness.* New York: Nichols Publishing Company, 1986. 148 pages. This book suggests ways that training can be assessed and evaluated over the short- and long-term. The various approaches needed at each stage of training are summarized and can be applied to one-on-one or group training.

APPENDIX A

Signs of Acceptable and Unacceptable Quality in Fresh Fruits

	Signs of Good Quality	Signs of Bad Quality, Spoilage
Apples	Firmness; crispness; bright color	Softness; bruises. (Irregularly shaped brown or tan areas do not usually affect quality)
Apricots	Bright, uniform color; plumpness	Dull color; shriveled appearance
Bananas	Firmness; brightness of color	Grayish or dull appearance (indicates exposure to cold and inability to ripen properly)
Blueberries	Dark blue color with silvery bloom	Moist berries
Cantaloupes (Muskmelons)	Stem should be gone; netting or veining should be coarse; skin should be yellow-gray or pale yellow	Bright yellow color; mold; large bruises
Cherries	Very dark color; plumpness	Dry stems; soft flesh; gray mold
Cranberries	Plumpness; firmness. Ripe cranberries should bounce	Leaky berries
Grapefruit	Should be heavy for its size	Soft areas; dull color
Grapes	Should be firmly attached to stems. Bright color and plumpness are good signs	Drying stems; leaking berries
Honeydew melon	Soft skin; faint aroma; yellowish white to creamy rind color	White or greenish color; bruises or watersoaked areas; cuts or punctures in rind
Lemons	Firmness; heaviness. Should have rich yellow color	Dull color; shriveled skin
Limes	Glossy skin; heavy weight	Dry skin; molds
Oranges	Firmness; heaviness; bright color	Dry skin; spongy texture; blue mold
Peaches	Slightly soft flesh	A pale tan spot (indicates beginning of decay); very hard or very soft flesh

(Continued)

APPENDIX A (Continued)

	Signs of Good Quality	Signs of Bad Quality, Spoilage
Pears	Firmness	Dull skin; shriveling; spots on the sides
Pineapples	"Spike" at top should separate easily from flesh	Mold; large bruises; unpleasant odor; brown leaves
Plums	Fairly firm to slightly soft flesh	Leaking; brownish discoloration
Raspberries, Boysenberries	Stem caps should be absent; flesh should be plump and tender	Mushiness; wet spots on containers (sign of possible decay of berries)
Strawberries	Stem cap should be attached; berries should have rich red color	Gray mold; large uncolored areas
Tangerines	Bright orange or deep yellow color; loose skin	Punctured skin; mold
Watermelon	Smooth surface; creamy underside; bright red flesh	Stringy or mealy flesh (spoilage difficult to see on outside)

APPENDIX B

Signs of Acceptable and Unacceptable Quality in Fresh Vegetables

	Signs of Good Quality	Signs of Poor Quality, Spoilage
Artichokes	Plumpness; green scales; clinging leaves	Brown scales; grayish-black discoloration; mold
Asparagus	Closed tips; round spears	Spread-out tips; spears with ridges; spears that are not round
Beans (snap)	Firm, crisp pods	Extensive discoloration; tough pods
Beets	Firmness; roundness; deep red color	Gray mold; wilting; flabbiness
Brussels sprouts	Bright color; tight-fitting leaves	Loose, yellow-green outer leaves; ragged leaves (may indicate worm damage)
Cabbage	Firmness; heaviness for size	Wilted or decayed outer leaves (Leaves should not separate easily from base)
Carrots	Smoothness; firmness	Soft spots
Cauliflower	Clean, white curd; bright green leaves	Speckled curd; severe wilting; loose flower clusters
Celery	Firmness; crispness; smooth stems	Flabby leaves; brown-black interior discoloration
Cucumber	Green color; firmness	Yellowish color; softness
Eggplant	Uniform, dark purple color	Softness; irregular dark brown spots
Greens	Tender leaves free of blemishes	Yellow-green leaves; evidence of insect decay
Lettuce	Crisp leaves; bright color	Tip burn on edges of leaves (slight discoloration of outer leaves is not harmful)
Mushrooms	White, creamy, or tan color on tops of caps	Dark color on underside of cap; withering veil
Onions	Hardness; firmness; small necks; papery outer scales	Wet or soft necks

(Continued)

APPENDIX B (Continued)

	Signs of Good Quality	Signs of Poor Quality, Spoilage
Onions (green)	Crisp, green tops; white portion two to three inches in length	Yellowing; wilting
Peppers (green)	Glossy appearance; dark green color	Thin walls; cuts, punctures
Potatoes	Firmness; relative smoothness	Green rot or mold; large cuts; sprouts
Radishes	Plumpness; roundness; red color	Yellowing of tops (sign of aging); softness
Squash (summer)	Glossy skin	Dull appearance; tough surface
Squash (winter)	Hard rind	Mold; softness
Sweet potatoes	Bright skins	Wetness; shriveling; sunken and discolored areas on sides of potato (Sweet potatoes are extremely susceptible to decay.)
Tomatoes	Smoothness; redness. (Tomatoes that are pink or slightly green will ripen in a warm place.)	Bruises; deep cracks around the stem scar
Watercress	Crispness; bright green color	Yellowing, wilting, decaying of leaves

APPENDIX C

Refrigerated Storage of Foods

Food	Recommended Temperatures (°F/°C)	Maximum Storage Periods	Comments
Meat			
Roasts, steaks, chops	32–36/0–2.2	3 to 5 days	Wrap loosely
Ground and stewing	32–36/0–2.2	1 to 2 days	Wrap loosely
Variety meats	32–36/0–2.2	1 to 2 days	Wrap loosely
Whole ham	32–36/0–2.2	7 days	May wrap tightly
Half ham	32–36/0–2.2	3 to 5 days	May wrap tightly
Ham slices	32–36/0–2.2	3 to 5 days	May wrap tightly
Canned ham	32–36/0–2.2	1 year	Keep in can
Frankfurters	32–36/0–2.2	1 week	Original wrapping
Bacon	32–36/0–2.2	1 week	May wrap tightly
Luncheon meats	32–36/0–2.2	3 to 5 days	Wrap tightly when opened
Leftover Cooked Meats	32–36/0–2.2	1 to 2 days	Wrap or cover tightly
Gravy, Broth	32–36/0–2.2	1 to 2 days	Highly perishable
Poultry			
Whole chicken, turkey, duck, goose	32–36/0–2.2	1 to 2 days	Wrap loosely
Giblets	32–36/0–2.2	1 to 2 days	Wrap separate from bird
Stuffing	32–36/0–2.2	1 to 2 days	Covered container separate from bird
Cut-up cooked poultry	32–36/0–2.2	1 to 2 days	Cover
Fish			
Fatty fish	30–34/–1.1–1.1	1 to 2 days	Wrap loosely
Fish—not iced	30–34/–1.1–1.1	1 to 2 days	Wrap loosely
Fish—iced	32/0	3 days	Don't bruise with ice
Shellfish	30–34/–1.1–1.1	1 to 2 days	Covered container
Eggs			
Eggs in shell	40/4.4	1 week	Do not wash. Remove from container
Leftover yolks/whites	40–45/4.4–7.2	2 days	Cover yolks with water

(Continued)

APPENDIX C (Continued)

Food	Recommended Temperatures (°F/°C)	Maximum Storage Periods	Comments
Eggs (continued)			
Dried eggs	40–45/4.4–7.2	1 year	Cover tightly
Reconstituted eggs	40–45/4.4–7.2	1 week	Same treatment as eggs in shell
Cooked Dishes with Eggs, Meat, Milk, Fish, Poultry	32–36/0–2.2	Serve day prepared.	Highly perishable
Cream-Filled Pastries	32–36/0–2.2	Serve day prepared.	Highly perishable
Dairy Products			
Fluid milk	38–39/3.3–3.9	5 to 7 days after date on carton	Keep covered and in original container
Butter	38–40/3.3–4.4	2 weeks	Waxed cartons
Hard cheese (cheddar, parmesan, romano)	38–40/3.3–4.4	6 months	Cover tightly to preserve moisture
Soft cheese			
Cottage cheese	38–40/3.3–4.4	3 days	Cover tightly
Other soft cheeses	38–40/3.3–4.4	7 days	Cover tightly
Evaporated milk	50–70/10–21.1	1 year unopened	Refrigerate after opening
Dry milk (nonfat)	50–70/10–21.1	1 year unopened	Refrigerate after opening
Reconstituted dry milk	38–40/3.3–4.4	1 week	Treat as fluid milk
Fruit			
Apples	40–45/4.4–7.2	2 weeks	Room temperature till ripe
Avocados	40–45/4.4–7.2	3 to 5 days	Room temperature till ripe
Bananas	40–45/4.4–7.2	3 to 5 days	Room temperature till ripe
Berries, Cherries	40–45/4.4–7.2	2 to 5 days	Do not wash before refrigerating
Citrus	40–45/4.4–7.2	1 month	Original container
Cranberries	40–45/4.4–7.2	1 week	
Grapes	40–45/4.4–7.2	3 to 5 days	Room temperature till ripe
Pears	40–45/4.4–7.2	3 to 5 days	Room temperature till ripe

(Continued)

APPENDIX C (Continued)

Food	Recommended Temperatures (°F/°C)	Maximum Storage Periods	Comments
Fruits (continued)			
Pineapples	40–45/4.4–7.2	3 to 5 days	Regrigerate (lightly covered) after cutting
Plums	40–45/4.4–7.2	1 week	Do not wash before refrigerating
Vegetables			
Sweet potatoes, mature onions, hard-rind squashes, rutabagas	60/15.6	1 to 2 weeks at room temp. 3 months at 60°F	Ventilated containers for onions
Potatoes	45–50/7.2–10	30 days	Ventilated containers
All other vegetables	40–45/4.4–7.2	5 days maximum for most; 2 weeks for cabbage, root vegetables	Unwashed for storage

APPENDIX D

Storage of Frozen Foods

Food	Maximum Storage Period at –10° to 0°F (–23.3° to –17.7°C)
Meat	
Beef, roasts and steaks	6 months
Beef, ground and stewing	3 to 4 months
Pork, roasts and chops	4 to 8 months
Pork, ground	1 to 3 months
Lamb, roasts and chops	6 to 8 months
Lamb, ground	3 to 5 months
Veal	8 to 12 months
Variety meats (liver, tongue)	3 to 4 months
Ham, frankfurters, bacon, luncheon meats	2 weeks (freezing not generally recommended.)
Leftover cooked meats	2 to 3 months
Gravy, broth	2 to 3 months
Sandwiches with meat filling	1 to 2 months
Poultry	
Whole chicken, turkey, duck, goose	12 months
Giblets	3 months
Cut-up cooked poultry	4 months
Fish	
Fatty fish (mackerel, salmon)	3 months
Other fish	6 months
Shellfish	3 to 4 months
Ice Cream	3 months. Original container. Quality maintained better at 10°F (–12.2°C).
Fruit	8 to 12 months
Fruit Juice	8 to 12 months
Vegetables	8 months
French-Fried Potatoes	2 to 6 months

(Continued)

APPENDIX D (Continued)

Food	**Maximum Storage Period at –10° to 0°F (–23.3° to –17.7°C)**
Precooked Combination Dishes	2 to 6 months
Baked Goods	
Cakes, prebaked	4 to 9 months
Cake batters	3 to 4 months
Fruit pies, baked or unbaked	3 to 4 months
Pie shells, baked or unbaked	1 1/2 to 2 months
Cookies	6 to 12 months
Yeast breads and rolls, prebaked	3 to 9 months
Yeast breads and rolls, dough	1 to 1 1/2 months

APPENDIX E

Recommended Maximum Storage Periods for Goods in Dry Storage

Food	Recommended Maximum Storage Period if Unopened
Baking Materials	
Baking powder	8 to 12 months
Chocolate, baking	6 to 12 months
Chocolate, sweetened	2 years
Cornstarch	2 to 3 years
Tapioca	1 year
Yeast, dry	18 months
Baking soda	8 to 12 months
Beverages	
Coffee, ground, vacuum packed	7 to 12 months
Coffee, ground, not vacuum packed	2 weeks
Coffee, instant	8 to 12 months
Tea, leaves	12 to 18 months
Tea, instant	8 to 12 months
Carbonated beverages	Indefinitely
Canned Goods	
Fruits (in general)	1 year
Fruits, acidic (citrus, berries, sour cherries)	6 to 12 months
Fruit juices	6 to 9 months
Seafood (in general)	1 year
Pickled fish	4 months
Soups	1 year
Vegetables (in general)	1 year
Vegetables, acidic (tomatoes, sauerkraut)	7 to 12 months
Dairy Foods	
Cream, powdered	4 months
Milk, condensed	1 year
Milk, evaporated	1 year
Fats and Oils	
Mayonnaise	2 months

(Continued)

APPENDIX E (Continued)

Food	**Recommended Maximum Storage Period if Unopened**
Fats and Oils (continued)	
Salad dressings	2 months
Salad oil	6 to 9 months
Vegetable shortenings	2 to 4 months
Grains and Grain Products	
Cereal grains for cooked cereal	8 months
Cereals, ready-to-eat	6 months
Flour, bleached	9 to 12 months
Macaroni, spaghetti, and other noodles	3 months
Prepared mixes	6 months
Rice, parboiled	9 to 12 months
Rice, brown or wild	Should be refrigerated
Seasonings	
Flavoring extracts	Indefinite
Monosodium glutamate	Indefinite
Mustard, prepared	2 to 6 months
Salt	Indefinite
Sauces (steak, soy, etc.)	2 years
Spices and herbs (whole)	2 years to indefinite
Paprika, chili powder, cayenne	1 year
Seasoning salts	1 year
Vinegar	2 years
Sweeteners	
Sugar, granulated	Indefinite
Sugar, confectioners	Indefinite
Sugar, brown	Should be refrigerated
Syrups, corn, honey, molasses, sugar	1 year
Miscellaneous	
Dried beans	1 to 2 years
Cookies, crackers	1 to 6 months
Dried fruits	6 to 8 months
Gelatin	2 to 3 years
Dried prunes	Should be refrigerated
Jams, jellies	1 year
Nuts	1 year
Pickles, relishes	1 year
Potato chips	1 month

APPENDIX F

Professional and Trade Organizations

Association of Food and Drug Officials (AFDO)
Post Office Box 3425
York, Pennsylvania 17402

Centers for Disease Control
Public Inquiries Specialist
1600 Clifton Road N.E.
Atlanta, Georgia 30333

The Education Foundation of the National Restaurant Association
250 South Wacker Drive, Suite 1400
Chicago, Illinois 60606

International Association of Milk, Food and Environmental Sanitarians (IAMFES)
502 East Lincoln Way
Ames, Iowa 50010

International Council of Hotel and Restaurant Industry Educators (International CHRIE)
1200 17th Street, N.W., 7th Floor
Washington, DC 20036

International Food Manufacturers Association (IFMA)
321 North Clark Street, Suite 2900
Chicago, Illinois 60610

National Automatic Merchandisers Association (NAMA)
20 North Wacker Drive, 35th Floor
Chicago, Illinois 60606

National Environmental Health Association (NEHA)
720 South Colorado Boulevard, Suite 970
South Tower
Denver, Colorado 80222

National Pest Control Association (NPCA)
8100 Oak Street
Dunn Loring, Virginia 22027

The National Restaurant Association
Technical Services, Public Health and Safety
311 First Street, N.W.
Washington, DC 20001

NSF *International*
3475 Plymouth Road
Post Office Box 1468
Ann Arbor, Michigan 48106

Underwriters Laboratories, Inc.
333 Pfingsten Road
Northbrook, Illinois 60062

Glossary

Note: The bold number(s) in parentheses at the end of each entry refers to the chapter in which this term is used.

Abrasive A cleaning agent that contains a scouring agent used for cleaning some sinks, rusty metals, and badly soiled floors. Abrasives can mar some surfaces. **(10)**

Accident An unintended event that results in injury, loss, or damage, which may or may not result from negligence. Accidents involve environmental hazards and human error. **(13)**

Acid A medium, such as a food product or detergent, with a pH below 7.0. **(2, 10)**

Additives Any substance added to foods in processing or preparation that may become a chemical hazard, such as sulfites. **(3)**

Aerobe A bacterium that grows only in the presence of free oxygen. **(2)**

Air curtain A ventilation unit consisting of outgoing air that helps keep flying insects out of open doors or windows. **(12)**

Air gap An unobstructed, vertical distance through the air that separates an outlet of the potable water supply from a potentially contaminated source, such as a faucet in a sink. **(9)**

Alkali A medium, such as a food product or detergent, with a pH above 7.0. **(2, 10)**

Anaerobe A bacterium that grows only in the absence of free oxygen. **(2)**

Anisakiasis The disease caused by *Anisakis* parasitic roundworms. It causes vomiting, abdominal pain, coughing, and fever. It can also mimic the symptoms of appendicitis. **(3)**

Anisakis The parasitic roundworm that lives primarily in the vital organs of finfish. In humans, it causes the disease anisakiasis. **(3)**

Artificial respiration Mouth-to-mouth resuscitation used as a first aid method to help someone who has stopped breathing. **(13)**

Aseptic packaging A method of packing food so that it is free from pathogenic micro-organisms. **(6)**

Bacillary dysentery A disease caused by *Shigella*. Bacillary dysentery is also called shigellosis. **(3)**

Bacillus cereus Facultative, spore-forming bacterium found in soil, dust, and water, which produces two toxins that cause foodborne intoxications. The symptoms from one form of toxin are diarrhea and abdominal pain, which occur between 8 and 16 hours after ingestion. The symptom from the other toxin is vomiting, which occurs 1 to 5 hours after ingestion. **(3)**

Backflow The flow of contaminants from undrinkable sources into potable water distributing systems. **(9)**

Back siphonage One kind of backflow that occurs whenever the pressure in the potable water supply drops below that of the contaminating supply. **(9)**

Bacterium A living micro-organism made up of a single cell. **(2, 3)**

Behavioral objective A statement that defines what the trainee will be able to do on the completion of a training program. **(16)**

Bi-metallic stemmed thermometer A food thermometer that is used to measure internal temperatures of products or items during receiving, storage, preparation, serving, holding, chilling, and reheating by inserting a sensor area into the product. It is accurate to ±2°F (±1°C). **(6, 7, 8)**

Biological hazard The danger posed to food safety by the contamination of food with pathogenic micro-organisms or naturally occurring toxins. **(3)**

Blast chiller A special refrigerated unit that maintains temperatures to below −20°F (−28.8°C) and quickly freezes fresh and chilled food items for storage. **(7)**

Botulism The foodborne intoxication caused by *Clostridium botulinum* bacteria. The illness, which can be fatal, attacks the nervous system rather than the digestive tract, causing double vision, difficulty swallowing, and respiratory collapse, among its symptoms. **(3)**

Campylobacter jejuni Anaerobic, non-spore-forming bacterium that causes a foodborne infection. *C. jejuni* is an emerging pathogen. **(3)**

Campylobacteriosis The foodborne infection caused by *Campylobacter jejuni.* Typical symptoms are fever, headache, and fatigue, followed by abdominal pain and diarrhea. **(3)**

Cantilever mounting A sanitary way of installing immobile equipment by using a bracket that projects from the wall and suspends the piece of equipment at least six inches away from the wall and above the floor. **(9)**

Cardiac arrest The condition that results from a massive heart attack that makes the heart stop pumping. CPR is the first aid procedure used on someone who has cardiac arrest. **(13)**

Cardiopulmonary resuscitation (CPR) The first aid technique that is used on victims whose hearts have stopped pumping, and who may also have stopped breathing. Certification is required to perform this procedure. **(13)**

Carrier A person or animal who harbors disease-causing micro-organisms in the body without being noticeably affected, but can transmit the organisms to food or to other humans. **(3)**

Chemical hazard The danger posed to food safety by the contamination of food by chemical substances, such as pesticides, detergents, additives, and toxic metals. **(3)**

Chlorine Chemical used in the form of hypochlorites in sanitizing solutions. Chlorine compounds can tarnish and corrode metals like pewter, brass, and silverplate, if used in incorrect concentrations. **(10)**

Ciguatoxin A toxin that collects in predatory marine reef fish that have eaten smaller reef fish that have eaten toxic algae. In humans, it causes the disease ciguatera, characterized by vomiting, itching, dizziness, hot and cold flashes, and temporary blindness. **(3)**

Clean Free of visible soil including food particles and dirt. **(9)**

Cleanability Requirement for sanitary equipment and facilities in which an item or surface can be exposed without difficulty for cleaning and inspection. Construction and design allow removal of soil by normal cleaning methods. **(9)**

Clean-in-place equipment Equipment of such design that food-contact surfaces are not readily removable so that cleaning, rinsing, and sanitizing is done in place by having the solutions and rinse water circulated under pressure throughout the fixed system. The clean-in-place system is self-contained and self-draining. **(9)**

Cleaning The removal of soil and food matter from a surface. **(10)**

Cleaning agent A chemical compound specifically formulated for use on floors, walls, equipment, food-contact surfaces, or in dishwashing machines. **(10)**

Cleaning program The system that the operator or manager devises in order to organize all the cleaning tasks in the foodservice establishment. **(11)**

Clostridium botulinum Anaerobic, spore-forming bacterium that is most often found in canned goods and other sealed packages of food. It causes botulism. **(3)**

Clostridium perfringens Anaerobic, spore-forming bacterium found in the soil, dust, and in animal feces that causes a foodborne toxin-mediated infection of the same name. The symptoms for the illness are diarrhea and abdominal pain. **(3)**

Code A systematic collection of regulations, or statutes, and procedures. A code often equals either a regulation or a statute. **(15)**

Communicable disease An illness that is transmitted directly or indirectly from one human to another. **(3)**

Compressor In a refrigeration unit, the component that compresses air to provide energy to run the refrigerator. **(7, 9)**

Contamination The unintended presence of harmful substances or micro-organisms in food or water. **(3)**

Coving A curved, sealed edge between the floor and wall that makes cleaning easier and inhibits insect harborage. **(9)**

Crisis A turning point, decisive moment, or an abrupt change in operations. Characteristics of a crisis are escalating intensity, scrutiny, interference with normal operations, reduced public image, and damage to the bottom line. **(14)**

Crisis management The organized and systematic efforts of an operation to prevent, react to, and learn from crises. **(14)**

Critical control point An operation (practice, preparation-step, procedure), by which a preventive or control measure can be applied that would eliminate, prevent, or minimize (a) hazard(s). **(5)**

Cross-connection Any physical link through which contaminants from drains, sewers, or waste pipes can enter a potable water supply. **(9)**

Cross-contamination The transfer of harmful micro-organisms from one item of food to another by means of a nonfood-contact surface (human hands, utensils, equipment), or directly from a raw food to a cooked one. **(2, 3)**

Decline phase The period when the number of bacteria dying from a lack of nutrients and because of their own waste products exceeds the reproduction rate. **(2)**

Deep chilling Storage method that extends shelf life by holding fresh perishable and potentially hazardous food items at 26° to 32°F (−4° to 0°C) in special units. **(7)**

Delegation The second step of crisis management planning, where the head of the crisis team authorizes or appoints another team member to act and to direct other employees during a crisis. **(14)**

Digital thermometer A battery-powered food thermometer that reveals the internal temperatures of food products with a digital display that is accurate to 0.1 degrees. **(6)**

Dry storage Holding nonperishable food items, such as rice, flour, crackers, and canned goods, at 50 percent humidity and between 50° and 70°F (15.6° and 21.1°C). **(7, 9)**

Emerging pathogen A disease-causing agent increasingly identified as causing foodborne illness. These pathogens include *C. jejuni*, *E. coli*, and the Norwalk and Norwalk-like viral agents. **(3)**

Employee rules Policies written by the management covering personal hygiene, safe foodhandling practices, and prohibited habits for all employees during working hours. **(4)**

Environmental hazard Conditions that are unsafe in themselves or that encourage unsafe actions by employees, for example, slippery floors, boxes stacked too high in a storage room, or swinging doors in the kitchen. **(13)**

Escherichia coli Facultative, nonspore-forming bacterium that can cause gastroenteritis in humans. **(3)**

External communication An exchange of information between the head of a crisis team (or its designated spokesperson) and the media, the public, and regulatory agencies during and following crises. **(14)**

Facultative Bacteria that can grow either with or without free oxygen present. **(2)**

Feedback The response given by a trainer to a learner, either in the form of constructive criticism to correct a mistake, or as praise to reinforce proper performance of a skill or procedure. **(16)**

FIFO A stock rotation and storage principle that states First In, First Out, that is, use items in storage in the order in which they were delivered. **(7)**

Flammable limit The temperature at which hot oil will burst into flames. This varies with the type of oil or fat used and its condition and age, but it is around 450°F (111.1°C). **(13)**

Flowchart A chart that shows the flow of food through a food service. It is designed for potentially hazardous foods on a menu. **(5, 6, 7, 8)**

Flow of food The path that food travels in a foodservice operation. It begins with the decision to include a food item on the menu. **(5)**

Follow-up Actions taken after the implementation of a new program, such as HACCP or the master cleaning schedule, and after public health department inspections to observe how well procedures are being followed, to measure the results, to supervise the employees, and to correct any errors or breakdowns. After integrated pest management treatments, follow-up is the corrective action implemented to prevent infestations and to observe problem areas that may develop into infestations. Also, the manager's reinforcement of what has been learned after training has been given to an individual employee or an entire staff, by using monitoring, supervising, feedback, or retraining. **(5, 12, 15, 16)**

Food Any substance intended for use or for sale in whole or in part for human consumption, including ice and water. **(1)**

Foodborne infection An illness that results from eating food that contains live pathogenic bacteria or other micro-organisms. **(3)**

Foodborne intoxication An illness that results from eating food in which toxins produced by bacteria or molds are present. The bacteria themselves may be dead. **(3)**

Foodborne toxin-mediated infection A disease that results from eating a food containing a large number of disease-causing micro-organisms. Once ingested, the human intestine provides the perfect conditions for the micro-organisms to produce toxins. *Clostridium perfringens* cause this type of illness. **(3)**

Food-contact surface Any surface of equipment or utensils with which food normally comes into contact, or from which food can drip, drain, or splash back onto surfaces normally in contact with food. **(3)**

Food-splash surface Any surface of equipment or in the facility on which food can splash or spill. **(9)**

Footcandle A unit of lighting equal to the illumination one foot from a uniform light source. **(9)**

Freezer burn The loss of water from the surface of a frozen food, resulting in an undesirable appearance and texture. **(6)**

Freezer storage Holding of frozen perishable food items at temperatures of 0°F (−17.8°C). **(7)**

Fungi A group of micro-organisms that includes molds and yeast. Fungi are considered to be plants. **(2)**

Garbage Wet waste matter, usually food products, that cannot be recycled. **(9)**

Gastroenteritis An inflammation of the linings of the stomach and intestines, which can result from ingesting bacteria or certain viruses. **(3)**

Grade The rating of quality and palatability of meats, poultry, and eggs. The rating is a voluntary service paid for by the packers. Grades do not indicate wholesomeness of food. **(6)**

Guidelines A statement of policy or procedure that uses simple language to convey government standards and regulations concerning food safety. **(15)**

Harborage Shelter for pests. **(12)**

Hazard Analysis Critical Control Point (HACCP) A food safety and self-inspection system that highlights potentially hazardous foods and how they are handled in the foodservice environment. **(5)**

Hazard Communication Standard (HCS) The standard established by OSHA by which foodservice operators and managers must inform their employees of the potential hazards of all chemicals used in the facility through training programs and with labels and signage on or near the containers. Material Safety Data Sheets (MSDS) are part of this standard. **(10, 11, 12, 13)**

Hazard Unacceptable contamination (of a biological, chemical, or physical nature), unacceptable microbial growth, or unacceptable survival of micro-organisms of a concern to food safety, or persistence of toxins. **(5)**

Heimlich maneuver The first aid procedure used when someone is choking on a piece of food and cannot breathe. The technique is the same for adults and pregnant women, but is different for young children and infants. Certification is required to perform this procedure. **(13)**

Hepatitis A A contagious viral disease that is transmitted to food by poor personal hygiene or contact with contaminated water. It causes inflammation of the liver. **(3)**

Host A human, animal, or plant in which another organism lives and nourishes itself. **(3)**

Hygrometer An instrument that measures humidity of the atmosphere in a particular area, such as dry storage, of a foodservice establishment. **(7, 9)**

Immobile equipment Equipment that cannot be easily moved by one person and is cleaned and sanitized by disassembly. This includes countertop, floor mounted, and free-standing equipment, such as a reach-in refrigerator, a walk-in freezer, a deep fryer unit, or a meat grinder. **(9)**

Immuno-compromised An individual who is susceptible to becoming ill from a foodborne illness due to an existing disease or weakened physical condition. **(3)**

Incubation period The length of time it takes before the symptoms of a foodborne illness appear. **(3)**

Infestation Occupation of a facility by pests, particularly cockroaches and other insects, rats, and mice. **(12)**

Integrated pest management (IPM) A system of preventive and control tactics and methods used to control or eliminate pest infestations in foodservice establishments. **(12)**

Internal communication An exchange of information between the manager or operator and employees. This is an important part of crisis management. **(14)**

Iodine The chemical used in sanitizing solutions. It can only be used in solutions that have a pH of 5.0 or less. **(10)**

Irradiation The use of radiation in food processing to lengthen shelf life by eliminating pathogenic micro-organisms. It is considered a food additive and is regulated by the FDA. **(3)**

Laceration A cut that breaks the skin. This foodservice injury may result from knives, food slicers, choppers, and broken glass. **(13)**

Lag phase The period of adaptation of bacteria to a new environment, which is marked by slow growth in the total number of cells. **(2)**

Larva The immature stage of development of parasites and insects, including cockroaches and *Trichinella spiralis.* **(2, 12)**

Listeria monocytogenes Facultative, nonspore-forming bacterium found in soil. It grows well in damp places and at low temperatures. **(3)**

Listeriosis The foodborne infection caused by *L. monocytogenes.* It is characterized by severe symptoms and a high mortality rate for immuno-compromised individuals. **(3)**

Log phase The accelerated growth period of bacterial reproduction. **(2)**

Makeup air Clean air that replaces air taken out of a foodservice by exhaust hoods. **(9)**

Master cleaning schedule A summary of all the cleaning operations in a foodservice establishment, including what is to be cleaned, who is to clean it, when it is to be cleaned, and how it is to be cleaned with specific detergents and cleaning tools described. **(11)**

Mechanical plan A drawing that follows a design layout for construction or renovation of an establishment and includes exact measurement specifications. **(9)**

Media Newspapers, magazines, television, and radio stations. **(14)**

Medium An environment, such as food, in which bacteria live. **(2)**

Meningitis Swelling of the brain and spinal cord that is sometimes associated with listeriosis. **(3)**

Micro-organism A form of life that can only be seen with the aid of a microscope, such as bacteria, fungi, molds, parasites, viruses, and yeasts. **(2)**

Modified air packaging (MAP) Method of storing food products that includes sous vide, aseptic, and vacuum-packing. In all of these, air is removed from the package to prevent contamination and spoilage. **(6, 7)**

Mold Any of various fungi that spoil foods and have a fuzzy appearance. They reproduce by forming spores. **(2)**

Monosodium glutamate (MSG) A food additive used as a flavor enhancer. **(3)**

Neutral A medium, such as water, that has a pH of 7.0. **(2)**

Nitrites Preservatives used by the meat industry as a curing agent to prevent bacterial growth. They can pose a chemical hazard if meat is burned or overbrowned. **(3)**

Palatable Food that is acceptable in taste and flavor to be eaten. **(5)**

Parasite A micro-organism that lives on or inside a host and depends on the host for nourishment, such as roundworms. **(2)**

Parts per million (ppm) A unit of measure for water hardness and chemical sanitizing solution concentrations. **(10)**

Pasteurization A process that kills most pathogenic bacteria in food and beverages and slows down the growth of others with minimal chemical change by heating the food to a specific temperature for a specified length of time. **(3)**

Pathogen Any disease-causing agent, usually a living micro-organism. **(2)**

Perishable Subject to quick decay or spoilage unless stored properly. **(7)**

Personal hygiene Safe and healthy habits that include bathing, washing hair, wearing clean clothing, and washing hands often when handling and serving food and beverages. **(4)**

Pest control operator (PCO) The licensed or certified technician that implements and monitors prevention and control of pests in foodservice establishments. **(12)**

pH A measure of acidity or alkalinity of a medium, such as food products and cleaning agents, based on a scale from 1.0 to 14.0. **(2)**

Physical hazard The danger posed to food safety from particles or fragments of items that are not supposed to be in foods, such as chips of glass, metal shavings, and toothpicks. **(3)**

Porosity The extent to which a floor covering absorbs liquids. **(9)**

Potable Safe to drink, as in describing a water supply. **(9)**

Potentially hazardous foods Foods that are implicated most often in foodborne illness outbreaks. **(1, 2, 3, 5)**

Pounds per square inch (psi) A unit of measure for water pressure. For effective dishwashing machine spray action in the wash and rinse cycles, 15–25 psi are needed. **(10)**

Prostration Total exhaustion associated as a symptom of staph intoxication. **(3)**

Quaternary ammonium compounds (quat) Chemicals in noncorrosive chemical sanitizing solutions. Quats can be effective in both acid and alkaline solutions, but may also be more selective than other sanitizers. **(10)**

Reconstituted Dehydrated food products recombined with water, milk, or other liquids. **(7)**

Recovery rate The speed with which a hot water heater produces hot water, particularly for dishwashing machines. **(9)**

Recycling An activity of solid waste management in which items that can be reused are removed from trash dumpsters and set aside for pick-up by a recycler. These items include paper and cardboard packaging, polystyrene packaging, glass bottles, tin and aluminum cans, and plastic containers. **(9)**

Refrigerated storage Short-term holding of fresh, perishable, and potentially hazardous food items at internal temperatures of 45°F (7.2°C) or lower. **(7)**

Regulation A general governmental control having the force of law, that provides technical specifications for enforcement of statutes. **(15)**

Reservoir An alternate host or passive carrier of a pathogenic micro-organism. This may be soil, animals, or humans. **(3)**

Resiliency The ability of a floor covering to withstand shock. **(9)**

Risk The chance that a condition or set of conditions will lead to a foodservice hazard. **(5)**

Salmonella Facultative, nonspore-forming bacterium that is found in poultry, shell eggs, and humans, among other sources. It causes salmonellosis foodborne infection. **(3)**

Salmonellosis The infection caused by pathogenic *Salmonella* present in food. Symptoms include diarrhea, fever, chills, abdominal pain, and possibly a headache or vomiting. **(3)**

Sanitarian A representative of the public health department who is a professional in sanitation and public health. This individual conducts inspections, fills out reports, and may offer advice to foodservice operators on solving sanitation problems. **(15)**

Sanitary Free of harmful levels of disease-causing micro-organisms and other harmful contaminants. **(1, 10)**

Sanitary Assessment of the Food Environment (S.A.F.E.) A food safety and self-inspection system developed by the National Restaurant Association. It is based on the HACCP system. **(5)**

Sanitation The creation and maintenance of conditions favorable to good health. **(1)**

Sanitization The reduction of the number of disease-causing micro-organisms to safe levels on clean food-contact surfaces. **(10)**

Scale Buildup of lime deposits on water pipes and equipment from water hardness. An acid cleaner is recommended to remove the deposits. **(9)**

Scombroid poisoning A foodborne illness caused from histamine, which is produced when fatty finfish begin decomposing. This illness is marked by flushing, sweating, dizziness, nausea, and a burning, peppery taste in the mouth. **(3)**

Sealed A surface that is free of cracks or other openings that permit entry of moisture or air that can cause contamination or pest infestations. **(9, 12)**

Severity The seriousness of the consequences of the results of a hazard to food safety. **(5)**

Shelf life The length of time that a food product can be stored without losing quality or compromising food safety. **(7)**

Shellfish Soft-bodied seafood enclosed in hard shells including bi-value mollusks, such as mussels, clams, and oysters and crustacea, such as lobsters, shrimp, and crabs. **(3, 6, 7, 8)**

Shigella Facultative bacteria found primarily in humans that causes shigellosis foodborne infection. It does not form spores. **(3)**

Shigellosis The infection caused by *Shigella,* with primary symptoms of diarrhea, cramps, and fever. **(3)**

Sneeze guard A food shield that is placed in a direct line between the mouth of the average height of a person and the food being displayed in a food bar or buffet line, at a maximum of 48 inches off the floor. **(8)**

Solid waste Dry, bulky trash that includes glass, plastic, paper, and cardboard that can be recycled. **(9)**

Solvent cleaners Alkaline detergents that contain an agent to dissolve grease. **(10)**

Source reduction Cutting down on the amount of trash that a food service generates by using evaluation and elimination of unnecessary packaging and separating recyclable types of trash from those that go to landfills. This is an activity in solid waste management. **(9)**

Sous vide A method of packaging raw or partially cooked food, where the product is placed in a sealed pouch with the air removed. The pouch is cooked and refrigerated or frozen until needed, then it is reheated and served. **(6)**

Spoilage The breakdown of the edible quality of a food product. **(6)**

Spore In a bacterium, a thick-walled formation that is resistant to heat, cold, and chemicals, and that is capable of becoming a vegetative cell under favorable conditions. In mold, it is formed for the purpose of reproduction. **(2)**

Spring switch A switch that operates a slicer, grinder, chopper, or other sharp equipment only when it is depressed. Releasing the switch shuts the machine off. **(13)**

Standard A measure used as a comparison for quality to determine the degree or level that a requirement is met. **(15)**

Staphylococcal foodborne intoxication (staph)
The illness caused by the heat-stable toxin produced by *Staphylococcus aureus* in food. Symptoms include nausea, vomiting, diarrhea, dehydration, and cramps. **(3)**

Staphylococcus aureus Facultative bacterium that excretes heat-stable toxins to cause a foodborne intoxication. Humans are the main reservoir. **(3)**

Stationary phase The period of competition in bacterial development where the bacteria reproduces equal number to those that are dying. **(2)**

Sterile The absence of all living micro-organisms. **(10)**

Sulfiting agents Preservatives used to maintain the freshness and color of certain foods and some wines. They can pose a chemical hazard if consumed by people who are allergic to them, particularly people with asthma. **(3)**

Surfactants (surface active agents)
Components in detergent that reduce surface tension at the points where the detergent meets the soiled surface, which then allows the detergent to quickly penetrate and disperse the soil. **(10)**

Task analysis Dividing a job description into individual duties. **(16)**

Temperature danger zone The temperature range between 45° and 140°F (7.2° and 60°C) within which most bacteria grow and reproduce. **(2)**

Temporary foodservice establishment
One that operates in one location for a period of not more than 14 days consecutively and in conjunction with a single event or celebration. **(8)**

Time-and-temperature principle
Requires that all potentially hazardous food be kept at an internal temperature below 45°F (7.2°C) or above 140°F (60°C) during transport, storage, handling, preparation, display, and serving. Also, potentially hazardous food cannot remain at temperatures in the danger zone for more than a total of four hours. **(8)**

Toxin A poison. Specifically, a poison produced by a living micro-organism. **(2, 3)**

Training Teaching employees through a variety of methods how to do a specific job properly. **(16)**

Training program A structured sequence of information and activities that lead, step-by-step, to learning. **(16)**

Trichinella spiralis The parasitic roundworm that causes the foodborne illness trichinosis. These worms live primarily in pigs and wild game animals. **(3)**

Trichinosis The foodborne illness caused by *Trichinella spiralis* parasitic roundworms. Symptoms include vomiting, nausea and abdominal pain, followed by fever, muscle stiffness, and swelling around the eyes. **(3)**

Troubleshooting Locating and eliminating sources of trouble in procedures and operations as part of devising a crisis management plan. **(14)**

Utilities The water, electrical, gas, garbage, sewage, and trash disposal services that foodservice operators depend on to run their establishments. Also includes lighting and ventilating a facility. **(9)**

Vacuum packaging A process of packaging foods for storage that preserves food quality over an extended shelf life. **(7)**

Vegetative cell Capable of growing and reproducing. **(2)**

Vegetative stage The nonspore stage of some bacteria where growth and reproduction occur. **(2)**

Ventilation The removal of smoke, steam, grease, and heat from equipment and food-preparation areas and replacement with clean make up air. **(9)**

Vermin Pests that infest foodservice facilities, including insects, birds, and rodents. **(12)**

Viable Capable of living and reproducing. **(3)**

Virus The smallest and most simple life form known. Like bacteria, viruses cause illness in humans and other animals. Unlike bacteria, viruses need a host in order to live and reproduce. **(2)**

Water activity The availability of moisture, or water content, in a medium. It is expressed as A_w. **(2)**

Water hardness The amount of dissolved minerals in a water supply, which can cause problems for hot water sanitizing and the ability of certain chemical sanitizers to work, and can also cause scaling in pipes and dishwashing and ice machines. **(9, 10)**

Yeast Micro-organism that spoils food and requires sugar and moisture for survival. It is classified as a fungus. **(2)**

Index

J

K

L

M

N

O

P

T